PRECALCULUS

J. Douglas Faires
Youngstown State University

James DeFranza
St. Lawrence University

Brooks/Cole Publishing Company

I(T)P® *An International Thomson Publishing Company*

Pacific Grove • Albany • Belmont • Bonn • Boston • Cincinnati • Detroit
Johannesburg • London • Madrid • Melbourne • Mexico City • New York
Paris • Singapore • Tokyo • Toronto • Washington

GWO A GARY W. OSTEDT BOOK

Publisher: *Gary Ostedt*
Marketing Team: *Maureen Riopelle, Deborah Petit*
Marketing Representative: *Rich Pirozzi*
Editorial Associate: *Carol Ann Benedict*
Production Coordinator: *Kirk Bomont*
Manuscript Editor: *Linda Thompson*
Interior and Cover Design: *Sharon Kinghan*

Interior Illustration: *Scientific Illustrators*
Cover Photo: *David Bishop/Phototake, NYC*
Project Management and Typesetting:
    Integre Technical Publishing Co., Inc.
Cover Printing: *Phoenix Color Corporation, Inc.*
Printing and Binding: *R. R. Donnelley/Crawfordsville*

For more information, contact:

BROOKS/COLE PUBLISHING COMPANY
511 Forest Lodge Road
Pacific Grove, CA 93950
USA

International Thomson Publishing Europe
Berkshire House 168–173
High Holborn
London WC1V 7AA
England

Thomas Nelson Australia
102 Dodds Street
South Melbourne, 3205
Victoria, Australia

Nelson Canada
1120 Birchmount Road
Scarborough, Ontario
Canada M1K 5G4

International Thomson Editores
Seneca 53
Col. Polanco
11560 México, D.F., México

International Thomson Publishing GmbH
Königswinterer Strasse 418
53227 Bonn
Germany

International Thomson Publishing Asia
221 Henderson Road
#05–10 Henderson Building
Singapore 0315

International Thomson Publishing Japan
Hirakawacho Kyowa Building, 3F
2-2-1 Hirakawacho
Chiyoda-ku, Tokyo 102
Japan

Printed in the United States of America

10 9 8 7 6 5 4

Library of Congress Cataloging in Publication Data
Faires, J. Douglas
    Precalculus / J. Douglas Faires, James DeFranza
        p.   cm.
    Includes index.
    ISBN 0-534-25236-2 (alk. paper)
    1. Functions. I. DeFranza, James, [date]. II. Title.
QA331.3.F34   1997
512'.1—dc21   96-29911
                                          CIP

CONTENTS

CHAPTER 1   FUNCTIONS   1

1.1   Introduction   2
1.2   The Real Line   4
1.3   The Coordinate Plane   12
1.4   Equations and Graphs   17
1.5   Using Technology to Graph Equations   22
1.6   Functions   31
1.7   Linear Functions   46
1.8   Quadratic Functions   55
1.9   Other Common Functions   67
1.10 Arithmetic Combinations of Functions   75
1.11 Composition of Functions   82
1.12 Inverse Functions   90
Review Exercises   101
Calculus Preview Exercises   105

CHAPTER 2   ALGEBRAIC FUNCTIONS   108

2.1   Introduction   109
2.2   Polynomial Functions   110
2.3   Finding Factors and Zeros of Polynomials   122
2.4   Rational Functions   133
2.5   Other Algebraic Functions   148
2.6   Complex Roots of Polynomials   155
Review Exercises   162
Calculus Preview Exercises   164

CHAPTER 3   TRIGONOMETRIC FUNCTIONS   168

3.1   Introduction   169
3.2   The Sine and Cosine Functions   173
3.3   Graphs of the Sine and Cosine Functions   183
3.4   Other Trigonometric Functions   193
3.5   Trigonometric Identities   199
3.6   Right-Triangle Trigonometry   209
3.7   Inverse Trigonometric Functions   218
3.8   Applications of Trigonometric Functions   228
Review Exercises   241
Calculus Preview Exercises   243

CHAPTER 4   EXPONENTIAL AND LOGARITHM FUNCTIONS   246

4.1   Introduction   247
4.2   The Natural Exponential Function   248
4.3   Logarithm Functions   259
4.4   Exponential Growth and Decay   267
Review Exercises   272
Calculus Preview Exercises   273

CHAPTER 5   CONIC SECTIONS, POLAR COORDINATES, AND PARAMETRIC EQUATIONS   274

5.1   Introduction   276
5.2   Parabolas   278
5.3   Ellipses   284
5.4   Hyperbolas   292
5.5   Polar Coordinates   300
5.6   Conic Sections in Polar Coordinates   311
5.7   Parametric Equations   316
5.8   Rotation of Axes   322
Review Exercises   328
Calculus Preview Exercises   329

ANSWERS TO SELECTED EXERCISES   333

INDEX   387

PREFACE

The mathematical preparation of the students entering our colleges and universities is much broader than in the past. We presently have many more students arriving with a university-level calculus background, as reflected by the fact that the number of students taking the Advanced Placement Examination in Calculus has steadily increased to well over 120,000 annually. On the other end of the spectrum, there is an increasing number of students who, although they have taken the college preparatory courses in high school, are not quite prepared to do the type of analysis that is required to successfully complete a university calculus sequence. These students often have some knowledge of the elementary computational techniques of calculus, but have not been given a comprehensive analysis and elementary functions course while in high school. They need the calculus-context review that a good precalculus course should provide.

In addition to under-prepared recent high school graduates, universities, particularly the predominantly commuter campuses, have seen a substantial increase in serious non-traditional students in recent years. Many of these students had not considered science-oriented careers when in high school and need to take remedial mathematics courses before they can enter their intended major subject. Most universities now have a wide range of remedial courses to serve their needs, but precalculus should not be considered one of these. Precalculus is the course that puts this review material in the perspective needed for a student to succeed in calculus.

There can be no gap between the end of a precalculus course and the beginning of calculus, since the student in precalculus this term will be in calculus next term. To make the transition as smooth as possible, the terminology and level of exposition in precalculus should closely match that commonly used in calculus. There is a sufficient number of new and important terms, concepts, and applications for the student to master in calculus without the added difficulty of learning new ways to express old ideas.

A quick review of the multitude of university precalculus books on the market reveals that nearly all are well over 500 pages, which is far more material than can reasonably be covered in one term. There is so much algebra and trigonometry review material in most precalculus books that the student who is weak in remedial mathematics is justified in thinking that the precalculus course covers these topics in detail. When it does, it is at the expense of the graphing and function analysis that is needed for calculus. When the review material is not covered in detail in the course, the under-prepared student is unlikely to succeed, and will discover this too late to move to a remedial course.

This book leaves no gap between the end of precalculus and the beginning of calculus, and is of length that will reasonably permit it to be covered within one term. It includes material from algebra, geometry, and trigonometry that is sufficient to fill holes in a student's pre-precalculus background, but this review material is interwoven into the book rather than presented as a block in the beginning. This permits the student to review these topics in the manner they will be used in calculus, and does not mislead those who need more review than a true precalculus course can provide.

The terminology used in the book parallels that used in calculus books, and the examples and exercises are presented in the way the student will see them in calculus. There are exercises at the end of each chapter that point directly to calculus, so students can see the notions they will encounter in their next mathematics course.

Included in the book is a significant amount of material that is appropriate for use with graphing calculators and computer algebra systems. Although these devices are very helpful for developing mathematical intuition and visualizing important concepts, we have not assumed that they are essential for the understanding of the basic concepts of precalculus. We treat graphing devices as an important tool, but not as a substitute for the analysis that is so necessary for a complete understanding of calculus and its applications. Students who expect to successfully complete the calculus sequence need to feel comfortable with the graphs and behavior of all the basic functions and equations that are commonly seen in calculus.

Additional review material is presented in the Student Study Guide, which provides more algebra and trigonometry details and background material, as well as supplemental examples and many worked out exercises. The disk that accompanies this Guide contains this material in Maple worksheet format. This gives the students a choice of using the Guide in the traditional manner, or working through the material interactively.

Also included in the Student Study Guide are two copies of an examination that students can take to test their readiness for precalculus and for calculus. We expect students will be successful in a precalculus course based on this book if they can score 16 or higher on the 40 question test before taking the precalculus course. Students will be well-prepared for a university calculus sequence if they score 28 or higher on the examination and carefully review the problems they missed. We suggest that students try one of the examinations before taking their precalculus course and the other after completing it, and predict that there will be a significant improvement in their scores.

For those not having access to the Student Study Guide, the examinations are also available over the internet at

http://www.cis.ysu.edu/a_s/mathematics/faculty/JDFaires1/precalculus

They can be viewed or printed from this site.

To the Student

In a short time you will be taking a course in calculus, an exciting subject that has application in every quantitative discipline. It provides a systematic way to describe how a change in one quantity effects another.

Engineers use calculus to determine how the force of wind and water effects the stability of a structure, how heat and fluids flow, how current effects various electronic devices, and so many other applications that they are too numerous to list. Physical scientists use calculus to study population problems, the effect of toxic substances on the environment, and the rates of chemical reactions. In business, the growth of the economy is determined with methods based on calculus, and future trends predicted. In fact, most statistics that are such a common part of our every day life have their basis in calculus techniques. It is truly the keystone to quantitative study.

Calculus itself is not difficult to master, but it requires a sound background in, and conceptual feel for, the precalculus topics of algebra, geometry, trigonometry, and, most importantly, graphing. Many students seeing calculus for the first time at the university level have difficulty because they are not prepared for the way precalculus topics are used in calculus. The material in this book will give you a sound foundation for the study of calculus. You will see many topics that are familiar, but perhaps using new notation and from a different perspective that you have seen in the past. The topics in this book are directly related to the study of calculus, and the perspective we give is the same as the one that will be used in your calculus courses. Read the material carefully and work the exercises and you will master these concepts, and have a great head start in your study of calculus.

Acknowledgments

One of the interesting features of creating a new book is convincing your peers that you truly have something novel and worthwhile to offer. Good reviewers are extremely valuable in keeping authors on track, and pointing out oversights in their presentation. We received excellent advice on this project from people at a wide variety of institutions. A great deal of thanks for this is due to Gary Ostedt, who like all good editors chooses reviewers who are true professionals. The reviewers might not all totally agree with our philosophy for precalculus, but they made many valuable contributions to make the book better meet the goals we set. We sincerely thank the following.

June Bjercke	San Jacinto College
John Gosselin	University of Georgia
Glenda Haynie	North Carolina State University
Harvey Keynes	University of Minnesota
Estela Llinas	University of Pittsburgh at Greensburg
Lee McCormick	Pasadena City College
Nancy Ressler	Oakton Community College
Phillip Schmidt	University of Akron

We would also like to thank a number of the students at Youngstown State University for their help in preparing the manuscript for the book and the art for the supplements. Our primary assistant was Christy Conn, who created the TEX version of the manuscript, managed the student help, and did excellent work on all phases of the project. We also had the able assistance of Bob Komara and John Slanina in generating the art manuscript for the Solutions Manual, and number of others, including Jason Martin, Nakia Rimmer, and Erika Faires, did proofreading and error-checking at various stages of the project. Thank you all for your careful and cheerful work. We hope you gained as much from the experience as we did from you.

Doug Faires
Jim DeFranza

FUNCTIONS

CALCULUS CONNECTIONS

Contrary to some nasty rumors you may have heard, calculus is not a difficult subject. Most of the problems that students have in calculus can be traced directly to new applications of topics from precalculus. To illustrate this, let us consider a typical applied problem from calculus.

Suppose that you are working for a company that makes containers and are told to design an open plastic box with a square base that holds a specified volume V. Let's see how we can approach the solution to this problem.

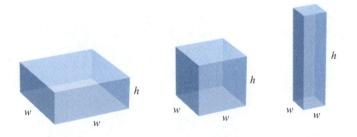

The figure shows some possible boxes that might meet the requirements. Should we choose one of these, and if so, which one? Although it may not have been stated explicitly that the box chosen should use the minimal amount of plastic, you can be quite sure that this is what your boss had in mind. We could produce a few boxes and choose the one using the least amount of material, but this would be time-consuming and a waste of material. Also, you might have a difficult time explaining why the box you chose is the best possible for the job.

The calculus solution to this problem requires that we first determine equations for the volume, $V = w^2 h$, and for the surface area, $S = w^2 + 4wh$. By solving for h in the volume equation we find that

$$h = \frac{V}{w^2}.$$

This can be substituted into the equation for surface area to produce

$$S(w) = w^2 + 4w\left(\frac{V}{w^2}\right) = w^2 + \frac{4V}{w}.$$

This describes the surface area of the box as a *function* of the width w. Each positive value of w gives a specific amount of required material $S(w)$. To solve the problem we need to choose w so that $S(w)$ is a minimum, which is easy to do if we have calculus. When we have this value of w, we can use the equation $V = w^2h$ to determine the appropriate value of h. The figure shows a typical graph of $y = S(w)$, which illustrates that a number w exists so that $S(w)$ is a minimum.

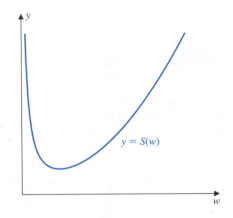

The difficulty in the problem is not in the calculus solution. It is the construction of the function $S(w)$ and convincing ourselves, and our boss, that we have done this correctly. This construction is a precalculus problem of the type we will consider throughout the book. In this chapter we see what functions are and what they look like and examine in detail some of those that are commonly used in calculus.

1.1   INTRODUCTION

The functions we will need in calculus—and, consequently, in precalculus—are transformations on the set of real numbers. To be certain that we are all working with the same concepts and notation, this section contains a short introduction and development of the set of numbers we call the *real numbers*. We begin with the most basic set of numbers, the *natural* numbers.

The set of **natural numbers** consists of the counting numbers, 1, 2, 3, ..., and is denoted by the symbol $\mathbb{N}$. Any pair of natural numbers can be added or multiplied and the result is another natural number, which is sometimes expressed by saying that set $\mathbb{N}$ is *closed* under the operations of addition and multiplication

The operation of subtraction is needed for many purposes, but the set of natural numbers is not closed under subtraction. For example, if we subtract the natural number

3 from 7, we get the natural number 4, but subtracting 7 from 3 is impossible if we must stay within the set of natural numbers. However, we can expand the set of natural numbers to the set of *integers*, a set that remains closed under addition and multiplication and is closed under subtraction as well.

The set of **integers** consists of the natural numbers, the negative of each natural number, and the number zero. The set of integers is denoted using the symbol $\mathbb{Z}$, which stands for the German word *zahlen*, meaning number. Addition, multiplication, and subtraction of two integers results in an integer, so the integers are closed under all these operations.

The set of rational numbers is introduced to ensure that division by nonzero numbers is well defined. A **rational number** has the form p/q, where p and q are integers and $q \neq 0$. The set of rational numbers is denoted $\mathbb{Q}$ (for quotient) and is closed under addition, multiplication, subtraction, and, with the exception of 0, division.

The rational numbers satisfy all the common arithmetic properties but fail to have a property called *completeness* because there are some essential numbers missing from the set. At least as early as 400 B.C. Greek mathematicians of the Pythagorean school recognized that $\sqrt{2}$, the length of a diagonal of a square with sides of length 1, is not a rational number. (See Figure 1.1.)

A precise definition of completeness requires the concept of limits, which you will see when you study calculus.

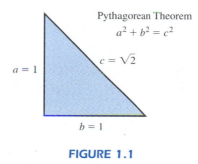

Pythagorean Theorem
$$a^2 + b^2 = c^2$$

$c = \sqrt{2}$

$a = 1$

$b = 1$

FIGURE 1.1

The nonrational, or **irrational**, numbers were originally said to be *incommensurable* because they could not be directly compared to the familiar rational numbers. There are many irrational numbers, including $\sqrt{5}$, π, and $-\sqrt[3]{3} + 1$. The discovery of irrational numbers resulted in a profound change in ancient mathematical thinking. Before that time it was assumed that all quantities could be expressed as proportions of integers, and integers in some cultures had a mystical or religious property.

The set $\mathbb{R}$ of **real** numbers consists of the rational numbers together with the irrational numbers. This set is described most easily by considering the set of all numbers that are expressed as infinite decimals. The rational numbers are those with expansions that eventually repeat in sequence, such as

$$\frac{1}{2} = 0.5\bar{0}, \qquad \frac{1}{3} = 0.\bar{3}, \qquad \frac{16}{11} = 1.\overline{45}, \quad \text{or} \quad -\frac{123}{130} = -0.9\overline{461538},$$

where the bar indicates that the digit or collection of digits is repeated indefinitely. Numbers with decimal expansions that are not repeating blocks are irrational.

1.2   THE REAL LINE

You will likely recall much of the material in the next few sections from your previous mathematics courses. If so, read it quickly, but don't become complacent. This material is included so that we will all be working with the same notation and definitions and so that you will have a convenient reference.

The fact that each real number can be written uniquely as a decimal provides a means of associating each real number with a distinct point on a *coordinate line*. We first choose a point on a horizontal line as the origin and associate with this origin the real number 0. Then we associate some point to the right of 0 with the real number 1. The positive integers are then marked with equal spacing consecutively to the right of 0. The negative integers are marked with this same spacing to the left of 0. Nonintegral real numbers are placed on the line according to their decimal expansions. Figure 1.2 shows a coordinate line and the points associated with certain real numbers.

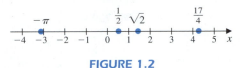

FIGURE 1.2

The coordinate-line representation of the real numbers is so convenient that we frequently do not explicitly distinguish between the points on the line and the real numbers that these points represent. Both are called the set of real numbers and are denoted $\mathbb{R}$.

Inequalities

The relative position of two points on a coordinate line can be used to define an inequality relationship on the set of real numbers. We say that a is less than b, written $a < b$, when the real number a lies to the left of the real number b on the coordinate line. This is also expressed by stating b is greater than a, written $b > a$, as shown in Figure 1.3.

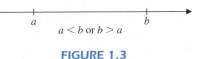

FIGURE 1.3

The notation $a \leq b$, or $b \geq a$, is used to express that a is either less than or is equal to b. The following properties of inequalities can be verified by referring to the coordinate line representation of the real numbers a, b, and c.

Inequality Properties

(i) Precisely one of $a < b$, $b < a$, or $a = b$ holds.

(ii) If $a > b$, then $a + c > b + c$.

(iii) If $a > b$ and $c > 0$, then $ac > bc$.

(iv) If $a > b$ and $c < 0$, then $ac < bc$.

Property (iv) implies that the inequality sign must be reversed when both sides of the inequality are multiplied by a negative number. For example,

$$4 > 3, \quad \text{but} \quad -8 = (-2)4 < (-2)3 = -6.$$

These rules are used frequently to solve problems involving inequality relations.

EXAMPLE 1   Find all real numbers x satisfying $2x - 1 < 4x + 3$.

Solution   To find the numbers x that satisfy the inequality, we isolate x on one side using the properties of inequalities. First add -3 to both sides of

$$2x - 1 < 4x + 3 \quad \text{to produce} \quad 2x - 4 < 4x.$$

Now subtract $2x$ from each side to give

$$-4 < 2x.$$

Multiplying both sides of the last inequality by $\frac{1}{2}$ gives the solution, $-2 < x$. This can also be written as $x > -2$, as shown in Figure 1.4.   ∎

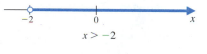

$$x > -2$$

FIGURE 1.4

EXAMPLE 2   Find all real numbers x satisfying $-1 < 2x + 3 \le 5$.

Solution   This inequality relation is a compact way of expressing that we have both

$$-1 < 2x + 3 \quad \text{and} \quad 2x + 3 \le 5.$$

Proceeding with the Inequality Properties, as in Example 1, we have

$$-4 < 2x \quad \text{and} \quad 2x \le 2,$$

so

$$-2 < x \quad \text{and} \quad x \le 1.$$

This last set of inequalities can be expressed compactly as $-2 < x \le 1$, as shown in Figure 1.5.   ∎

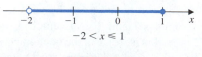

$$-2 < x \le 1$$

FIGURE 1.5

Intervals

Interval notation is preferred in calculus when we need to describe sets of real numbers lying between two given numbers.

Interval notation is a convenient way to represent certain important sets of real numbers. For real numbers a and b with $a < b$, the **open interval** (a, b) is defined as

$$(a, b) = \{x \mid a < x < b\},$$

and read "the set of real numbers x such that x is greater than a and less than b."

When the endpoints of the interval are included in the set, it is called a **closed interval** and denoted

$$[a, b] = \{x \mid a \leq x \leq b\}.$$

An interval that contains one endpoint but not the other is said to be *half-open* (although it could just as well be called half-closed). So the intervals

$$(a, b] = \{x \mid a < x \leq b\}$$

and

$$[a, b) = \{x \mid a \leq x < b\}$$

are half-open intervals.

The *interior* of an interval consists of all the numbers in the interval that are not endpoints. The intervals (a, b), $[a, b]$, $(a, b]$, and $[a, b)$ all have the same interior, which is the open interval (a, b).

In addition to the intervals with finite endpoints, the infinity symbol, ∞, is used when we wish to indicate that an interval extends indefinitely. Such an interval is said to be **unbounded**. The intervals

$$[a, \infty) = \{x \mid x \geq a\}$$

and

$$(a, \infty) = \{x \mid x > a\}$$

are *unbounded above* since they contain no largest real number, whereas the intervals

$$(-\infty, a] = \{x \mid x \leq a\}$$

and

$$(-\infty, a) = \{x \mid x < a\}$$

are *unbounded below*. The interval $(-\infty, \infty)$, which represents the set $\mathbb{R}$ of all real numbers, is unbounded both above and below.

In general, a square bracket indicates that the number next to it is in the interval, and a parenthesis indicates that the number next to it does not belong to the interval. The symbols $-\infty$ and ∞ are *never* next to a square bracket, since they are only symbols and do not represent real numbers.

Interval notation can be used to give an alternative expression for the answers to Examples 1 and 2. In Example 1 we found that the inequality was satisfied when $-2 < x$, that is, for x in the open interval $(-2, \infty)$. The answer to Example 2 was found to be $-2 < x \leq 1$. Using interval notation we can write this as $(-2, 1]$.

Table 1.1 summarizes the interval notation.

TABLE 1.1

Interval Notation	Set Notation	Graphic Representation
(a, b)	$\{x \mid a < x < b\}$	
$[a, b]$	$\{x \mid a \leq x \leq b\}$	
$[a, b)$	$\{x \mid a \leq x < b\}$	
$(a, b]$	$\{x \mid a < x \leq b\}$	
(a, ∞)	$\{x \mid a < x\}$	
$[a, \infty)$	$\{x \mid a \leq x\}$	
$(-\infty, b)$	$\{x \mid x < b\}$	
$(-\infty, b]$	$\{x \mid x \leq b\}$	
$(-\infty, \infty)$	$\mathbb{R} = \{x \mid -\infty < x < \infty\}$	

The next example involves a quadratic inequality that can be solved algebraically in a manner similar to the method used in Examples 1 and 2. It is easier, however, to use a graphical technique involving the coordinate line to solve this inequality. This technique is the method of choice for solving the numerous calculus problems that involve inequalities.

EXAMPLE 3   Find all values of x satisfying the inequality $x^2 - 4x + 5 > 2$.

Solution   This problem is solved by changing the inequality into one that has 0 on the right side and then factoring the term on the left side. The product of the factors is positive precisely when both of the factors are positive or both are negative.

Subtracting 2 from both sides of the inequality, we see that

$$x^2 - 4x + 5 > 2 \quad \text{implies that} \quad x^2 - 4x + 3 > 0.$$

Factoring $x^2 - 4x + 3$ as $(x - 3)(x - 1)$ gives

$$(x - 3)(x - 1) > 0.$$

Figure 1.6 is a *sign graph* for the inequality $(x - 3)(x - 1) > 0$, and it is used to determine where the inequality is positive and where it is negative.

To construct the sign graph, first determine where one of the factors is 0, namely, at $x = 3$ and $x = 1$. The linear factor $x - 3$ is positive to the right of 3 and negative to

Solving inequalities is an algebraic tool that is frequently needed in calculus. Problems of this type arise, for example, when determining the domains of functions or where graphs are increasing and decreasing.

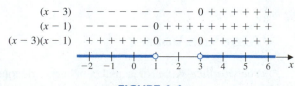

FIGURE 1.6

the left of 3. Similarly, the linear factor $x - 1$ is positive to the right of 1 and negative to the left of 1.

The product $(x - 3)(x - 1)$ is positive when both factors have the same sign and negative when they have opposite signs. So $x^2 - 4x + 3 = (x - 3)(x - 1) > 0$ precisely when $x < 1$ or $x > 3$. Consequently, the original inequality $x^2 - 4x + 5 > 2$ is satisfied precisely when $x < 1$ or $x > 3$.   ∎

The sign graph in Figure 1.6 can also be used to solve the inequality

$$x^2 - 4x + 3 = (x - 3)(x - 1) < 0$$

with no additional work. The solution, read from Figure 1.6, consists of those x where the factors have opposite sign, namely, $1 < x < 3$.

The answer to Example 3 can be expressed using interval notation, but it requires the introduction of the **union** symbol. The union of two sets A and B, written $A \cup B$, is the set of all elements that are in either A or in B (or in both). So, the answer in Example 3 is $(-\infty, 1) \cup (3, \infty)$.

The **intersection** of A and B, written $A \cap B$, is the set of elements that are in both A and B.

EXAMPLE 4   Find all values of x for which

$$\frac{x^2 - 1}{x(x - 2)} < 0.$$

Solution   This quotient is zero when $x = 1$ and $x = -1$ and is undefined when $x = 0$ and $x = 2$. On the top portion of the sign chart in Figure 1.7 we have the signs of the individual factors, $x - 1$, $x + 1$, x, and $x - 2$.

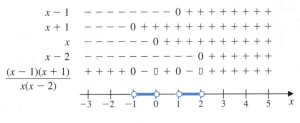

FIGURE 1.7

At the bottom of the chart we have the values of the quotient

$$\frac{(x - 1)(x + 1)}{x(x - 2)}.$$

The symbol ▯ is used to indicate where the quotient is undefined. The chart indicates that the solution to the inequality is $(-1, 0) \cup (1, 2)$.   ∎

Absolute Values

The *absolute value* of a real number x, denoted $|x|$, describes the distance on the coordinate line from the number x to the number 0. It is defined as follows.

The Absolute Value

$$|x| = \begin{cases} x, & \text{if } x \geq 0, \\ -x, & \text{if } x < 0. \end{cases}$$

For all real numbers x we have $|x| = |-x|$, so, for example, we have both $|2| = 2$ and $|-2| = 2$. This means, as shown in Figure 1.8(a), that the absolute value of a number gives a measure of its distance from 0. Using this as a guide, we can define the *distance* $d(x_1, x_2)$ from the real number x_1 to the real number x_2, as shown in Figure 1.8(b).

The absolute value of x can also be written as $|x| = \sqrt{x^2}$, so $d(x_1, x_2) = \sqrt{(x_1 - x_2)^2}$. You will see a similar formula for the distance between points in the plane in the next section. In calculus you will see a similar formula for the distance between points in space.

The Distance from x_1 to x_2

$$d(x_1, x_2) = |x_1 - x_2|.$$

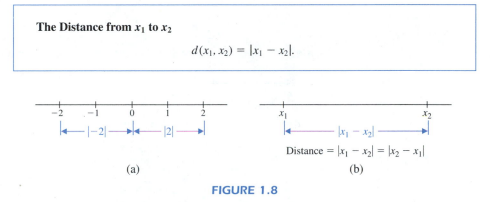

(a)   (b)

FIGURE 1.8

Since

$$|x_1 - x_2| = |-(x_1 - x_2)| = |x_2 - x_1|, \quad \text{we have} \quad d(x_1, x_2) = d(x_2, x_1)$$

for all real numbers x_1 and x_2. This implies, for example, that

$$d(5, 1) = d(1, 5) = 4 \quad \text{and} \quad d(-3.2, 4.5) = d(4.5, -3.2) = 7.7.$$

The distance formula can also be used to determine a formula for the midpoint of two numbers. (See Figure 1.9.)

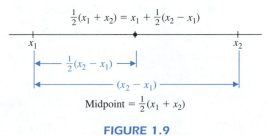

FIGURE 1.9

Suppose that we have the two numbers x_1 and x_2. If $x_2 \geq x_1$, the midpoint is also

$$x_1 + \frac{1}{2}|x_2 - x_1| = x_1 + \frac{1}{2}(x_2 - x_1) = \frac{1}{2}(x_1 + x_2).$$

Similarly, if $x_2 < x_1$ then the midpoint is

$$x_2 + \frac{1}{2}|x_2 - x_1| = x_2 + \frac{1}{2}(x_1 - x_2) = \frac{1}{2}(x_1 + x_2).$$

The Midpoint Formula

The midpoint of the line segment joining the two numbers x_1 and x_2 is $\frac{1}{2}(x_1 + x_2)$.

For example, the midpoint of the interval $[-1.2, 5.6]$ is

$$\frac{1}{2}(5.6 + (-1.2)) = \frac{1}{2}(4.4) = 2.2.$$

Listed next are some other useful properties of the absolute value.

Absolute Value Properties

(i) $|x| \geq 0$ for all values of x, and $|x| = 0$ if and only if $x = 0$.

(ii) $|x_1 x_2| = |x_1| \, |x_2|$ and, if $x_2 \neq 0$, $\left| \dfrac{x_1}{x_2} \right| = \dfrac{|x_1|}{|x_2|}$.

(iii) $|x_1 + x_2| \leq |x_1| + |x_2|$.

(iv) $|x| < x_1$ if and only if $-x_1 < x < x_1$.

(v) $|x| > x_1 \geq 0$ if and only if $x < -x_1$ or $x > x_1$.

EXAMPLE 5   Find all values of x satisfying $|2x - 1| < 3$.

Solution   By the relationship given in (iv) we have

$$|2x - 1| < 3 \quad \text{if and only if} \quad -3 < 2x - 1 < 3,$$

that is, if and only if both

$$-3 < 2x - 1 \quad \text{and} \quad 2x - 1 < 3.$$

This pair of inequalities holds precisely when

$$-2 < 2x \quad \text{and} \quad 2x < 4.$$

Dividing the inequalities by 2 gives the solution $-1 < x$ and $x < 2$, as shown in Figure 1.10.   ■

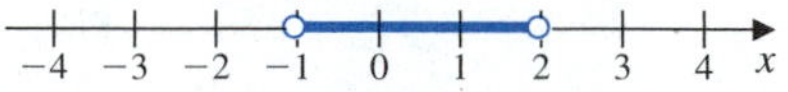

$$-1 < x < 2$$

FIGURE 1.10

The problem in Example 4 can also be solved using the coordinate line and the distance property of the absolute value of the difference of two real numbers. Since

$$|2x - 1| = \left| 2\left(x - \frac{1}{2} \right) \right| = 2\left| x - \frac{1}{2} \right|,$$

we have

$$|2x - 1| < 3, \qquad \text{equivalent to} \qquad \left| x - \tfrac{1}{2} \right| < \tfrac{3}{2}.$$

So, x satisfies the condition $|2x - 1| < 3$ precisely when $d(x, \frac{1}{2})$ is less than $\frac{3}{2}$. This is true precisely when $-1 < x < 2$, as shown in Figure 1.11.

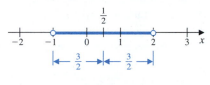

FIGURE 1.11

EXERCISE SET 1.2

In Exercises 1–6, express the interval using inequalities and give a sketch of the numbers in the interval.

1. $[-2, 4]$ **2.** $[0, \pi)$

3. $(-\sqrt{3}, \sqrt{2}]$ **4.** $(-\sqrt{3}, -\sqrt{2}]$

5. $(-\infty, 3)$ **6.** $[-1, \infty)$

In Exercises 7–10, express the inequalities using interval notation and give a sketch of the numbers in the interval.

7. $x \le 2$ **8.** $-1 < x \le 3$

9. $-2 \le x < 10$ **10.** $x \ge 3$

In Exercises 11–24, determine the values of x that satisfy the inequality.

11. $x + 4 < 7$ **12.** $-3x + 4 < 5$

13. $-2 < 3x - 2 < 5$ **14.** $-2 < 2x + 9 < 5 + x$

15. $(x + 1)(x - 2) \ge 0$ **16.** $x^2 - 5x - 20 < 4$

17. $(x - 2)(x + 4)(x - 3) \ge 0$

18. $x^3 - 6x^2 + 8x < 0$

19. $\dfrac{1}{x} \le 5$ **20.** $-2 \le \dfrac{1}{x}$

21. $\dfrac{3x + 1}{x - 1} \ge 0$ **22.** $\dfrac{(x - 1)(x + 2)}{x(x + 1)} > 0$

23. $\dfrac{2}{x - 1} \ge \dfrac{3}{x + 2}$ **24.** $\dfrac{2}{x - 1} - \dfrac{x}{x + 1} \le -1$

In Exercises 25–28, solve the equation.

25. $|5x| = 2$ **26.** $|2x + 3| = 1$

27. $\left| \dfrac{x - 1}{2x + 3} \right| = 2$ **28.** $|x - 1| = |2x + 1|$

In Exercises 29–34, solve the inequality and write the solution using interval notation.

29. $|x - 4| \le 1$ **30.** $|4x - 1| < 0.01$

31. $\dfrac{1}{|x + 5|} > 2$ **32.** $\left| \dfrac{3}{2x + 1} \right| < 1$

33. $|x^2 - 4| > 0$ **34.** $|x^2 - 4| \le 1$

35. Show that if $0 < a < b$, then $a^2 < b^2$.

36. Degrees Celsius (C) and degrees Fahrenheit (F) are related by the formula $C = \frac{5}{9}(F - 32)$.

   a. What is the temperature range in the Celsius scale corresponding to a temperature range in Fahrenheit of $20 \le F \le 50$?

   b. What is the temperature range in the Fahrenheit scale corresponding to a temperature range in Celsius of $20 \le C \le 50$?

37. Calculus can be used to show that a ball thrown straight upward (neglecting air resistance) from the top of a building 128 ft high with an initial velocity of 48 ft/s, has a height, in feet, of

$$h(t) = -16t^2 + 48t + 128$$

t seconds later. On what time interval will the ball be at least 64 ft above the ground?

1.3   THE COORDINATE PLANE

Calculus is the study of change, and graphs help us visualize the change between variables. To sketch graphs we first need a coordinate system in which to place the graphs.

In Section 1.2 we saw how the set of real numbers is related to points on a coordinate line and how this relationship permits us to solve a number of problems more easily. To sketch graphs we need a coordinate plane. In this section we consider ordered pairs of real numbers and their relation to the coordinate plane.

Each point in the plane is associated with an ordered pair (a, b) of real numbers. First an arbitrary point in the plane is associated with $(0, 0)$ and designated the origin. Then horizontal and vertical lines are drawn intersecting at the origin. The horizontal line is called the first-coordinate axis, or, more generally, the x-axis. The vertical line is called the second-coordinate axis, or the y-axis (see Figure 1.12).

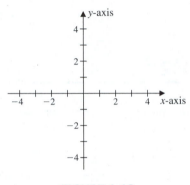

FIGURE 1.12

A scale is placed on both axes. The x-axis is the same as the coordinate line introduced in the previous section. It is labeled with positive numbers to the right of the origin and negative numbers to the left. The y-axis is similar, but the numbers above the origin are labeled as positive and the ones below the origin are negative. The ordered pair (a, b) is associated with the point of intersection of the vertical line drawn through the point a on the x-axis and the horizontal line drawn through b on the y-axis (see Figure 1.13). In general, we will not distinguish between an ordered pair and the point it represents in a coordinate plane.

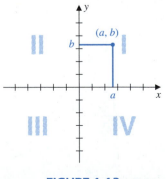

FIGURE 1.13

The x- and y-axes divide the plane into four regions, or quadrants. These quadrants are labeled as in Figure 1.13. The set of all ordered pairs of real numbers is denoted $\mathbb{R} \times \mathbb{R}$, or $\mathbb{R}^2$, which is sometimes called the cross product of $\mathbb{R}$ with itself. The plane determined by the x- and y-axes is called the *xy-plane*.

The coordinate plane is also called the Cartesian plane, and a rectangular coordinate system is called a Cartesian coordinate system. These names honor the versatile mathematician, philosopher, and physicist René Descartes (1596–1650), whose name in Latin was Renatus Cartesius.

EXAMPLE 1   Sketch the points in the coordinate plane associated with the ordered pairs $(1, 2), (-1, 3)$, $(-2, -\pi)$, and $(\sqrt{2}, -\sqrt{3})$.

Solution   These points are shown in Figure 1.14.   ■

In *La géométrie*, an appendix to his treatise on universal science *Discours de la méthode pour bien conduire sa raison et chercher la verité dans les sciences*, Descartes introduced the mathematical world to analytic geometry.

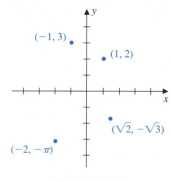

FIGURE 1.14

EXAMPLE 2   Find the points in the xy-plane that satisfy the inequalities

$$1 \le |x| \le 2 \qquad \text{and} \qquad -1 < y < 3.$$

Solution   The inequality $1 \le |x| \le 2$ implies that the distance from the x-coordinate of the point to the y-axis is between 1 and 2 units. These are the points that lie on or between the vertical lines $x = -2$ and $x = -1$, together with those that lie on or between the vertical lines $x = 1$ and $x = 2$. This region is shown in Figure 1.15(a).

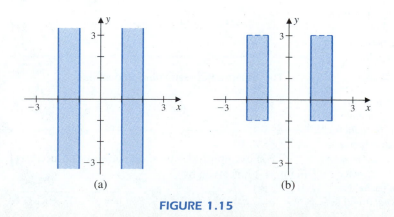

(a)   (b)

FIGURE 1.15

The points whose y-coordinates satisfy $-1 < y < 3$ lie strictly between the horizontal lines $y = -1$ and $y = 3$. So the points satisfying both these conditions are in the shaded regions shown in Figure 1.15(b). ∎

Distance Between Points in the Plane

The distance between two ordered pairs of real numbers (x_1, y_1) and (x_2, y_2) is found by considering the point (x_2, y_1) and applying the Pythagorean Theorem, as shown in Figure 1.16.

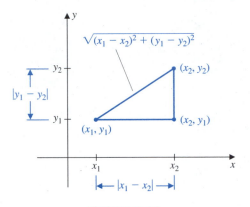

FIGURE 1.16

On the x-axis we have $d(x_1, x_2) = |x_1 - x_2|$, so the distance between (x_1, y_1) and (x_2, y_1) is

$$d((x_1, y_1), (x_2, y_1)) = |x_1 - x_2|.$$

Similarly, $d((x_2, y_1), (x_2, y_2)) = |y_1 - y_2|$. Now we apply the Pythagorean Theorem to obtain

$$d((x_1, y_1), (x_2, y_2)) = \sqrt{[d((x_1, y_1), (x_2, y_1))]^2 + [d((x_2, y_1), (x_2, y_2))]^2}$$
$$= \sqrt{|x_1 - x_2|^2 + |y_1 - y_2|^2}.$$

Since $|x_1 - x_2|^2 = (x_1 - x_2)^2$ and $|y_1 - y_2|^2 = (y_1 - y_2)^2$, we have the following result.

Notice the similarity between the distance between points in the plane and the distance between points on a line: $d(x_1, x_2) = |x_1 - x_2| = \sqrt{(x_1 - x_2)^2}$. In the latter parts of calculus you will see a similar formula for the distance between points (x_1, y_1, z_1) and (x_2, y_2, z_2) in space. You can probably guess what it will be.

The Distance Between Points in the Plane

$$d((x_1, y_1), (x_2, y_2)) = \sqrt{(x_1 - x_2)^2 + (y_1 - y_2)^2}.$$

For example, the distance between the points $(1, 2)$ and $(-2, 6)$, as shown in Figure 1.17, is given by

$$d((1, 2), (-2, 6)) = \sqrt{(1 - (-2))^2 + (2 - 6)^2} = \sqrt{9 + 16} = 5.$$

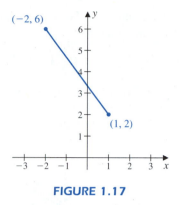

FIGURE 1.17

In Section 1.2 we used the distance formula between two numbers on a coordinate line to determine the midpoint of the numbers. We can use this result to determine the midpoint of a line segment in the plane.

Suppose that we want to find the x-coordinate of M, which is the midpoint of the line segment joining $P(x_1, y_1)$ and $Q(x_2, y_2)$. We drop lines from P, M, and Q perpendicular to the x-axis meeting the axis at the points labeled, respectively, A, C, and B. This situation is shown in Figure 1.18.

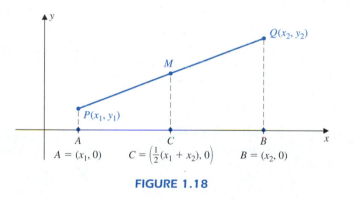

FIGURE 1.18

The quadrilaterals $APMC$ and $APQB$ are similar, and M is the midpoint of the line segment joining P and Q, so C is the midpoint of the line segment joining A and B. This implies that C has coordinates $(\frac{1}{2}(x_1 + x_2), 0)$ and that the x-coordinate of M is $\frac{1}{2}(x_1 + x_2)$. The y-coordinate of M can be found in a similar manner by looking at the y-coordinates of P, Q, and M.

The Midpoint Formula

The midpoint of the line segment with endpoints (x_1, y_1) and (x_2, y_2) has coordinates

$$\left(\frac{1}{2}(x_1 + x_2), \frac{1}{2}(y_1 + y_2) \right).$$

The midpoint of the line segment joining $(1, 2)$ and $(-2, 6)$, is shown in Figure 1.19. It has the coordinates

$$\left(\frac{1}{2}(1 + (-2)), \frac{1}{2}(2 + 6) \right) = \left(-\frac{1}{2}, 4 \right).$$

The distance formula for points in the plane can also be used to obtain the equation of a circle. A circle is the set of all points whose distance from a given point, its center, is a fixed distance, the radius.

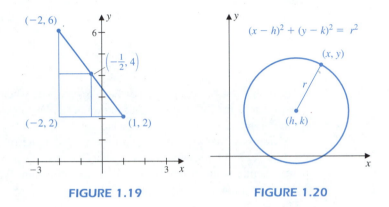

FIGURE 1.19   FIGURE 1.20

The circle with center (h, k) is just the unit circle scaled by the factor r and moved h units horizontally and k units vertically. The ability to quickly recognize the graph of an equation by shifting and scaling the graph of a familar equation is extremely important in calculus.

Figure 1.20 shows that a point (x, y) will be on the circle with center at (h, k) having radius r precisely when

$$r = d((x, y), (h, k)) = \sqrt{(x - h)^2 + (y - k)^2}.$$

Squaring both sides produces the formula in its most familiar, or *standard*, form.

Circles in the Plane

The point (x, y) lies on the circle with center (h, k) and radius r precisely when

$$(x - h)^2 + (y - k)^2 = r^2.$$

We call a circle whose radius is 1 a *unit circle*. The applications involving the unit circle whose center is at $(0, 0)$ are so extensive that this circle is frequently known as *the* unit circle. The points (x, y) on the unit circle centered at the origin, then, are those that satisfy the equation $x^2 + y^2 = 1$.

EXERCISE SET 1.3

In Exercises 1–4, sketch the listed points in the same coordinate plane.

1. $(1, 0), (0, 1), (-1, 0), (0, -1)$

2. $(2, 3), (3, 2), (-2, -3), (-3, -2)$

3. $(2, 3), (-2, -3), (2, -3), (-2, 3)$

4. $(5, -10), (10, 20), (-20, 10), (30, 40)$

In Exercises 5–8, find (a) the distance between the points and (b) the midpoints of the line segments joining the points.

5. $(2, 4), (-1, 3)$

6. $(-1, 5), (7, 9)$

7. $(\pi, 0), (-1, 2)$

8. $(\sqrt{3}, \sqrt{2}), (\sqrt{2}, \sqrt{3})$

In Exercises 9–18, indicate on an xy-plane those points (x, y) for which the statement holds.

9. $x = 3$

10. $y = -2$

11. $x > 1$

12. $x \geq 1$ and $y \geq 2$

13. $-1 \leq x \leq 2$

14. $-1 \leq x \leq 2$ and $2 < y < 3$

15. $4 \leq |x|$

16. $|y + 1| < 2$

17. $|x - 1| < 3$ and $|y + 1| < 2$

18. $|x| + |y| > 0$

In Exercises 19–22, find the standard form of the equation of each of the circles and sketch the graphs.

19. center $(0, 0)$; radius 1

20. center $(-2, 3)$; radius 2

21. center $(0, 2)$; radius 3

22. center $(1, -4)$; radius 4

In Exercises 23–26, (a) find the center and radius of each circle, and (b) sketch its graph.

23. $x^2 + y^2 = 9$

24. $x^2 + (y - 1)^2 = 1$

25. $(x - 2)^2 + (y - 1)^2 = 9$

26. $x^2 - 4x + y^2 - 2y = 4$

In Exercises 27–30, sketch the region in the xy-plane.

27. $\{(x, y) \mid x^2 + y^2 \leq 1\}$

28. $\{(x, y) \mid x^2 + y^2 > 2\}$

29. $\{(x, y) \mid 1 < x^2 + y^2 < 4\}$

30. $\{(x, y) \mid x^2 + y^2 \leq 4, y \geq x\}$

31. Find the distances between the points $(-1, 4)$, $(-3, -4)$, and $(2, -1)$, and show that they are vertices of a right triangle. At which vertex is the right angle?

32. Show that the points $(2, 1)$, $(-1, 2)$, and $(2, 6)$ are vertices of an isosceles triangle.

33. Find a fourth point that will form the vertices of a rectangle when added to the points in Exercise 31. Is the point unique?

34. Find a fourth point that will form the vertices of a parallelogram when added to the points in Exercise 32. Is the point unique?

35. Find an equation of the circle with center $(0, 0)$ that passes through $(2, 3)$.

36. Find an equation of the circle with center $(1, 3)$ that passes through $(-2, 4)$.

37. Find an equation of the circle with center $(3, 7)$ that is tangent to the y-axis.

38. Find a point on the y-axis that is equidistant from the points $(2, 1)$ and $(4, -3)$.

39. Which of the points $(-7, 2)$ and $(6, 3)$ is closer to the origin?

40. Find an equation of the circle whose center lies in the second quadrant, has radius 3, and is tangent to both the x-axis and the y-axis.

1.4 EQUATIONS AND GRAPHS

We use an *equation* to indicate that two mathematical expressions are equivalent. Sometimes the equation is an *identity*, which means that the equation is true for all values of the variable for which the equation is defined. For example, the equations

$$|x^2 - 4| = |x - 2| \, |x + 2| \quad \text{and} \quad \frac{1}{x^2 - 4} = \frac{1}{(x - 2)(x + 2)}$$

are identities. The first of these holds for all real numbers x. The second holds only when $x \neq \pm 2$, since for the expression to be defined, the denominator must be nonzero.

More often an equation is *conditional*, which means that it is true for some values of x, but not all. For conditional equations we need to determine and express the set of those values of the variable for which the equation is true. Equations such as

$$3x + 2 = 5, \quad x^2 - 3x + 2 = 0, \quad \text{and} \quad x^2 = -1$$

are conditional. The first has the single solution $x = 1$. The second is solved by factoring

$$0 = x^2 - 3x + 2 = (x - 1)(x - 2) \quad \text{to give the solutions} \quad x = 1 \quad \text{and} \quad x = 2.$$

The third conditional equation, $x^2 = -1$, has no real number solutions since the right side is negative and the left side cannot be.

Conditional equations commonly involve more than one variable, but the objective is the same. In the case of two variables the objective is

---

Conditional Equations in Two Variables

Determine which collection of ordered pairs satisfy the equation and give a representation of the solution.

For example, the conditional equation in the two variables x and y given as $y = 2x + 1$ has as its solutions those pairs of real numbers of the form $(x, 2x + 1)$, where x can be any real number. A few of the solutions to this equation, then, are $(0, 1)$, $(2, 5)$, and $(-3, -5)$.

Graphical representation can add significantly to our understanding of a problem. Always try to draw a picture before you begin to solve a problem. It tends to clear your mind and assure you that you really know what you are trying to determine.

Graphs of Equations

The set of ordered pairs (x, y) that satisfy an equation is called the *graph* of the equation. To give a visual representation of this set we *sketch the graph* of the equation, that is, we illustrate in the coordinate plane precisely those points (x, y) that satisfy the equation. Because the graph of an equation and the sketch of the graph are so frequently used, it is common not to distinguish explicitly between the graph of an equation and the sketch in the coordinate plane that represents the graph.

EXAMPLE 1   Sketch the graph of the equation $2x + 3y = 6$.

Solution   One way to obtain a rough sketch of the graph of an equation is to find the coordinates of several points, as we have done in Table 1.2. Then we plot the points in the coordinate plane, and connect those points, as best we can, with a curve. Figure 1.21 indicates this process produces a straight line for the graph of the equation $2x + 3y = 6$. In Section 1.7 we will see that equations of the form $Ax + By = C$ always have graphs that produce straight lines so this sketch is exact.   ■

TABLE 1.2

x	y
0	2
1	$\frac{4}{3}$
2	$\frac{2}{3}$
3	0
-1	$\frac{8}{3}$

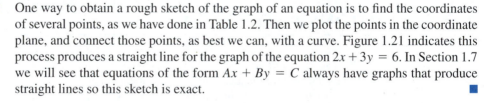

FIGURE 1.21

The points $(0, 2)$ and $(3, 0)$ shown on the graph of the line in Example 1 are the axis intercepts of the graph. Points of the form $(0, y)$ that satisfy an equation are the *y-intercepts* of its graph, and points of the form $(x, 0)$ that satisfy an equation are the *x-intercepts*. When these points can be easily determined they should be plotted on the graph.

EXAMPLE 2   Sketch the graph of the equation $y = \dfrac{x^2 + x - 2}{x - 1}$.

Solution   First notice that the numerator of the fraction can be factored to give

$$y = \frac{x^2 + x - 2}{x - 1} = \frac{(x - 1)(x + 2)}{x - 1} = x + 2, \quad \text{provided that } x \neq 1.$$

As in Example 1, we plot several points on the graph of the equation and connect the points with a curve. The result is the straight line shown in Figure 1.22. The open circle placed on the graph at the point $(1, 3)$ indicates that this point is missing from the graph. ■

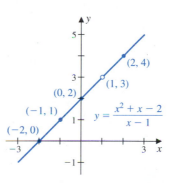

FIGURE 1.22

EXAMPLE 3   Sketch the following graphs.
  a. $y = x^2$ b. $x = y^2$ c. $x^2 + y^2 = 1$.

Solution   a. The graph of $y = x^2$ is the *parabola* shown in Figure 1.23(a). This graph was obtained by plotting representative points that satisfy the equation and then connecting

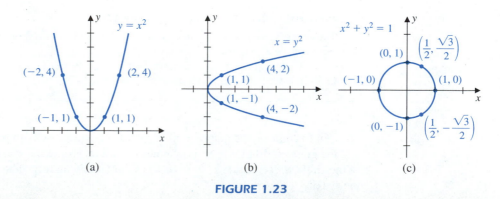

FIGURE 1.23

the points with a smooth curve. Its shape should be familiar from your previous mathematics courses.

b. Since this equation is the same as the equation in part (a), except that the roles of the two variables are interchanged, the graph of $x = y^2$ is the parabola shown in Figure 1.23(b).

c. We saw the graph of $x^2 + y^2 = 1$ in Section 1.3. It is the unit circle shown in Figure 1.23(c).   ■

Symmetry of a Graph

The graphs of the equations in Example 2 illustrate a feature known as *symmetry* with respect to a line. A graph is symmetric with respect to a line when the portion of the graph on one side of the line is the mirror image of the portion on the other side. Symmetry of a graph with respect to a line is often easy to determine when the line is one of the coordinate axes.

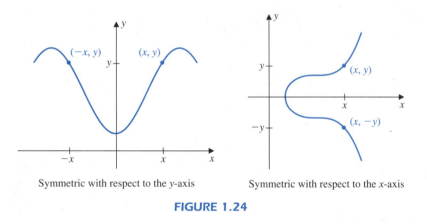

Symmetric with respect to the *y*-axis   Symmetric with respect to the *x*-axis

FIGURE 1.24

These axis symmetries are illustrated in Figure 1.24 and defined as follows.

Symmetry with Respect to the Coordinate Axes

The graph of an equation is *symmetric with respect to the y-axis* if whenever (x, y) is on the graph, $(-x, y)$ is also on the graph.

The graph of an equation is *symmetric with respect to the x-axis* if whenever (x, y) is on the graph, $(x, -y)$ is also on the graph.

In Example 3, the graph in (a) of $y = x^2$ is symmetric with respect to the *y*-axis, and the graph in (b) of $x = y^2$ is symmetric with respect to the *x*-axis. The graph in (c) of the circle with equation $x^2 + y^2 = 1$ is symmetric with respect to both the *x*-axis and the *y*-axis.

Symmetry is also defined with respect to a point in the plane. In this case, the graph has a mirror reflection property with respect to the point. This feature can be difficult to detect for arbitrary points in the plane, but it is easier when the point is the origin. A graph having this symmetry is shown in Figure 1.25. The unit circle shown in Figure 1.23(c) also has this property.

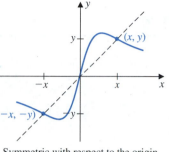

Symmetric with respect to the origin

FIGURE 1.25

Symmetry with Respect to the Origin

The graph of an equation is *symmetric with respect to the origin* if whenever (x, y) is on the graph, $(-x, -y)$ is also on the graph.

EXAMPLE 4   Determine any symmetry properties of the *cubing* function $y = x^3$, and sketch its graph.

Solution   Suppose a point (x, y) is on the graph of $y = x^3$, which means that (x, y) satisfies this equation. Since $(-x)^3 = -x^3$, the point $(-x, -y)$ also satisfies the equation, and the graph is symmetric with respect to the origin. For example, the point $(2, 8)$ is on the graph of $y = x^3$ and $(-2)^3 = -8$, so $(-2, -8)$ is also on the graph as shown in Figure 1.26.

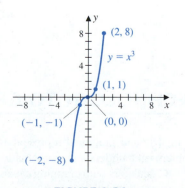

FIGURE 1.26

There is no axis symmetry since (x, y) satisfying the equation does not imply that either $(-x, y)$ or $(x, -y)$ satisfies the equation. For example, the equation is true for $(2, 8)$ but not true for $(-2, 8)$, since $(-2)^3 = -8 \neq 8$ or for $(2, -8)$, since $2^3 = 8 \neq -8$.

To determine a reasonable sketch of the graph, we plot a few points, and connect them in a simple and regular manner. Later in this chapter we will introduce some other aids to graphing, and you will learn other valuable techniques when studying calculus.

■

EXERCISE SET 1.4

In Exercises 1–4, specify any symmetries of the graphs that are shown.

1.

2.

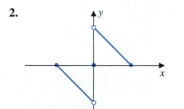

3.

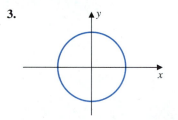

4.

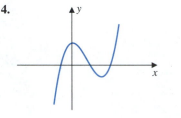

In Exercises 5–28, sketch the graph labeling any x- and y-intercepts and describing any axis or origin symmetry.

5. $y = x + 1$ **6.** $y = 2x - 3$

7. $x + y = 1$ **8.** $2x - y = 2$

9. $y = x^2 - 1$ **10.** $y = x^2 + 2$

11. $y = x^2 + 2x$ **12.** $3y = x^2$

13. $x = y^2$ **14.** $x + y^2 = 4$

15. $y = x^3 + 1$ **16.** $2y = x^3$

17. $y = \dfrac{(x + 2)(x - 2)}{x - 2}$ **18.** $y = \dfrac{x^2 - x - 6}{x + 2}$

19. $y = \sqrt{x} + 2$ **20.** $y = \sqrt{x - 1}$

21. $x^2 + y^2 = 4$ **22.** $(x - 1)^2 + y^2 = 1$

23. $y = \sqrt{9 - x^2}$ **24.** $y = -\sqrt{9 - x^2}$

25. $y = |x|$ **26.** $y = |x - 1|$

27. $y = |x| - 1$ **28.** $y = 2 - |x|$

29. Symmetry was discussed in the text with respect to the x-axis, the y-axis, and the origin. Show that if a graph has x-axis and y-axis symmetry, then it must also have origin symmetry.

1.5   USING TECHNOLOGY TO GRAPH EQUATIONS

Geometric and graphical representations can add significantly to our ability to solve problems and understand mathematics.

Calculators with graphing capabilities now cost little more than a standard textbook and are easily worth their price. In addition, powerful computer algebra systems such as Maple, Mathematica, and Derive are available on many campuses, and all these contain sophisticated graphing techniques. Technology is becoming more advanced, cheaper, and constantly growing. Ask your instructor for advice if you are interested in purchasing a graphing device for this course and for the calculus sequence you will be taking. Some calculus sequences are oriented toward a particular technological device. In this book, we will take a generic approach. Use any tool that is available. It will not be necessary for an understanding of the material in this book, but it will be helpful if you use it intelligently.

Graphing calculators and computer algebra systems can plot graphs of equations quickly by plotting as many points on the graph as the resolution of the screen will permit. When using a graphing device to plot an equation, we must be careful to ensure we have uncovered all the interesting aspects of the graph. Using technology without an understanding of the underlying concepts can result in accepting misleading information. This is particularly true in the case of plotting curves.

The plots in this section and throughout this book were generated using a computer algebra system. If you use a graphing calculator or a computer algebra system, you need to become familiar with the specific syntax of the device. A rectangular portion of the plane is called a *viewing rectangle* for a plot and is defined by specifying the range of values for x and the range of values for y. We denote a viewing rectangle specified by the inequalities

$$a \leq x \leq b \quad \text{and} \quad c \leq y \leq d$$

as $[a, b] \times [c, d]$.

Choosing an appropriate viewing rectangle is essential to obtaining a representative plot of an equation, and understanding the concepts of graphing is essential to recognizing when you have a good representation. The following examples show how to use technology to plot curves and how to choose an appropriate viewing rectangle to maximize the information.

EXAMPLE 1 Sketch the graph of the equation $y = x^2 + 5$ in the viewing rectangles $[-3, 3] \times [-3, 3]$, $[-6, 6] \times [-6, 6]$, $[-10, 10] \times [-3, 30]$, and $[-50, 50] \times [0, 1000]$.

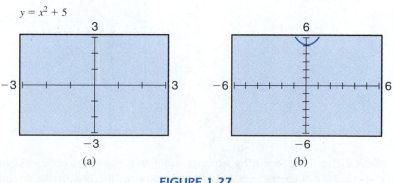

FIGURE 1.27

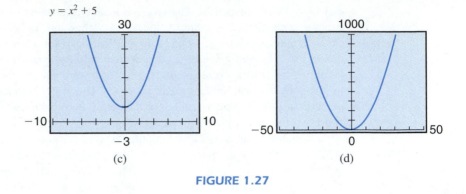

FIGURE 1.27

Solution   The graphs for these viewing rectangles are shown in Figure 1.27. In 1.27(a) the viewing rectangle does not contain any portion of the graph. This is because $x^2 \geq 0$ implies that $x^2 + 5 \geq 5$, and the range of y-values lies outside the viewing rectangle. We see better representations of the graph in parts 1.27(b), 1.27(c), and 1.27(d). More complete pictures appear in 1.27(c) and 1.27(d), although in part (d) the y-scale is so large that it appears the graph passes through the origin instead of through $(0, 5)$.   ■

EXAMPLE 2   Sketch the graph of the equation $y = x^3 - 25x$.

An appropriate viewing rectangle is needed if a graphing device is to give a good representation of the graph. Use all the information you know to determine an initial viewing window, and then modify the window to refine the view.

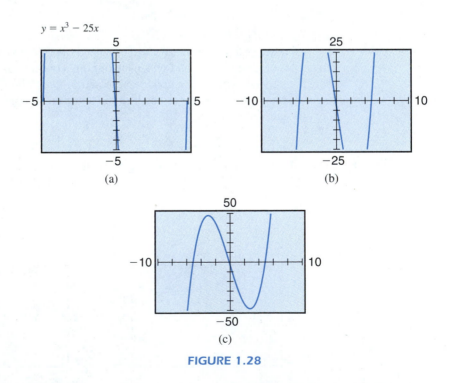

FIGURE 1.28

Solution   We begin by experimenting with viewing rectangles. In Figure 1.28(a), we chose a viewing rectangle $[-5, 5] \times [-5, 5]$, but we strongly suspect that the graph does not

Finding the location
of the points on a
graph that are locally
high or locally low is
one of the first
applications you will
see in calculus.

consist of the several straight lines that appear in the first plot. To obtain a more representative plot, we need to extend the range of points plotted to reveal points outside this viewing rectangle. In Figure 1.28(b) and 1.28(c), we have used viewing rectangles $[-10, 10] \times [-25, 25]$, and $[-10, 10] \times [-50, 50]$. With some confidence we accept the plot in Figure 1.28(c) as representative of the true curve. It appears from this figure that the graph has a local high point and a local low point somewhere near -3 and 3, respectively.

In Chapter 3 we will study these specific types of equations and see that the plot given here does reveal the important features of the curve. To completely analyze curves of this type requires concepts that you will study in calculus. ∎

EXAMPLE 3   Sketch the graph of $y = kx$ for $k = \pm1, \pm2, \pm3, \pm\frac{1}{2}, \pm\frac{1}{3}$, and discuss the effect of the constant k on the graph.

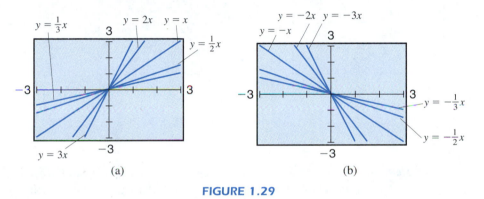

FIGURE 1.29

Solution   Figure 1.29(a) shows the plots for some positive values of k, and Figure 1.29(b) shows the plots for negative values of k. The figures indicate that the graph of the equation $y = kx$ is always a straight line, passing through the origin. For positive k, the graphs rise, or increase, from left to right, and for negative k they fall, or decrease, from left to right. The larger the magnitude of k, the steeper the incline or decline. ∎

EXAMPLE 4   Sketch the graph of $y = x$ in the viewing rectangle $[-3, 3] \times [-3, 3]$, and then sketch the graph of $y = 2x$ in the viewing rectangle $[-3, 3] \times [-6, 6]$.

When using a
graphing device, be
on guard for
problems like the one
shown in Example 4.
The scaling used on
the x- and y- axes
determines the shape
of the curve. It can be
deceptive if you are
not careful.

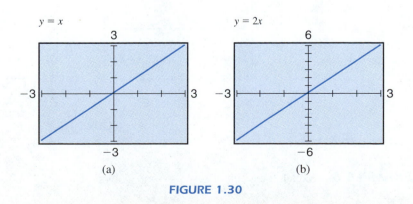

FIGURE 1.30

Solution   The graphs are shown in Figure 1.30(a) and 1.30(b). It appears the two graphs are the same, but we know this is not the case, since the graph of $y = 2x$ increases twice as fast as the graph of $y = x$. The visual problem occurs because the scale on the y-axis is twice as large for the plot of $y = 2x$ as it is for the plot of $y = x$. This makes the plots look the same, even though they are not.   ■

EXAMPLE 5   Sketch the graph of $y = x^3 + kx$ for several values of k and describe the effect of k on the curve.

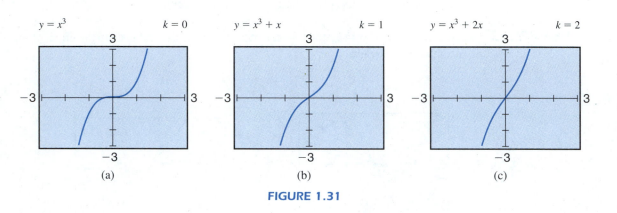

FIGURE 1.31

Solution   Figure 1.31 shows the graphs of $y = x^3 + kx$ for $k = 0, 1$, and 2 and Figure 1.32 shows the graphs for $k = -1$ and $k = -2$. For k positive the graphs appear to rise continually from left to right (the larger the value of k the steeper the increase) and the graphs do not have local high or low points like the graph in Example 2. For negative k, the graphs have both local high and low points. The more negative k becomes, the farther these points move away from the x-axis.   ■

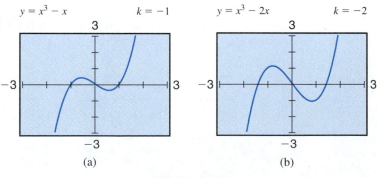

FIGURE 1.32

EXAMPLE 6   Sketch the graph of $y = \dfrac{x + 2}{x - 1}$.

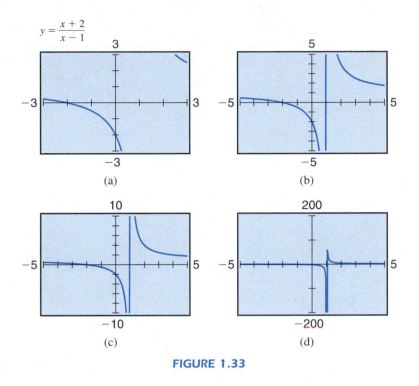

FIGURE 1.33

Solution   Figure 1.33 shows four views of the graph of the equation, each using a different viewing rectangle. The viewing rectangle in each case is determined by the largest and smallest values on the axes of the graph. In Figure 1.33(a), it is clear the viewing rectangle is not sufficient to show all the interesting parts of the graph. Figures 1.33(b) and 1.33(c) give successively better views. We should not be surprised that the graph seems to be broken near $x = 1$, since the original equation is not defined at $x = 1$. The vertical lines in Figure 1.33(b) and 1.33(c) should not appear on the graph. They are the result of the graphing device connecting by a straight line the closest points on either side of $x = 1$. Figure 1.33(d) shows this more clearly. Always be skeptical when using a graphing device. It simply plots points and connects them, which may lead to incorrect graphs. In this case, Figure 1.33(b) appears to be the best at representing the graph, even though it shows a vertical line at $x = 1$ that is not a part of the graph.   ■

EXAMPLE 7   Graphically solve each of the following inequalities.

a. $x^2 + x - 2 \geq 0$
b. $x^2 + 2 \geq 0$
c. $-x^2 + 3x + 1 \leq 0$

Solution   To solve an inequality graphically, we plot a good representative for the curve determined by the inequality and then locate, from the graph, the x-intercepts. This process allows us to find the intervals of x for which the inequality is satisfied by observing where the graph lies above or below the x-axis.

a. The graph of the equation $y = x^2 + x - 2$ is shown in Figure 1.34(a). The curve appears to cross the x-axis at $x = 1$ and $x = -2$. This can be seen algebraically by

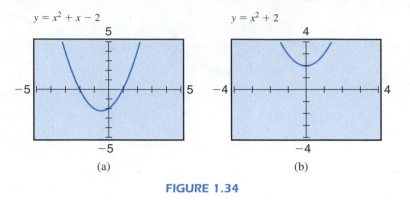

FIGURE 1.34

factoring the quadratic as

$$y = x^2 + x - 2 = (x - 1)(x + 2).$$

Then $x^2 + x - 2 = 0$ exactly when $x = 1$ or $x = -2$. By observing where the graph is on or above the x-axis, we have

$$x^2 + x - 2 \geq 0 \quad \text{for} \quad x \geq 1 \quad \text{or for} \quad x \leq -2.$$

b. The graph of the equation $y = x^2 + 2$, in Figure 1.34(b), shows that the expression $x^2 + 2$ is always greater than 2. This is evident without plotting the curve since $x^2 \geq 0$ for all x, which implies that $x^2 + 2 \geq 2$ for all x.

c. Figure 1.35(a) shows the graph of $y = -x^2 + 3x + 1$. In this example we cannot determine the x-intercepts exactly from the graph. We can, however, find an

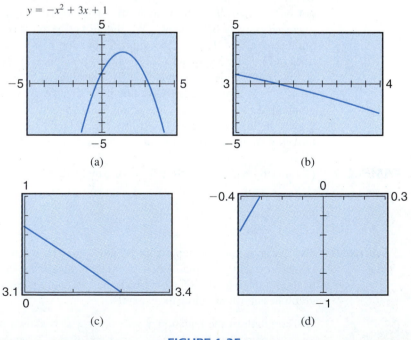

FIGURE 1.35

approximation by zooming in on the graph for x-values close to where it crosses the x-axis. Figure 1.35(b) shows the plot of the equation using a viewing rectangle $[3, 4] \times [-5, 5]$. It appears that one of the x-intercepts is near 3.3. Figure 1.35(c) zooms closer to this x-intercept. Using the trace feature of a graphing calculator or the mouse to click on the point when using a computer algebra system, we see that the graph crosses the x-axis at approximately 3.3. Figure 1.35(d) shows the graph close to the other x-intercept, which is approximately -0.3, often written using the notation ≈ -0.3. So an approximate solution to the original inequality is

$$-x^2 + 3x + 1 \le 0 \quad \text{when} \quad x \le -0.3 \quad \text{or when} \quad x \ge 3.3.$$

We would not be able to find the x-intercepts exactly in this example using this graphing technique since the intercepts turn out not to be rational numbers. ■

EXAMPLE 8   Approximate the point of intersection of the graphs of the equations $y = x^2 + x - 2$ and $y = 2x - 1$ for which x is largest.

Calculus applications require determining the boundary of a region between two curves. To determine this boundary, we need to find where the curves intersect.

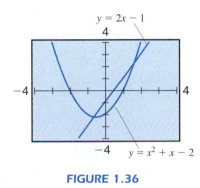

FIGURE 1.36

Solution   Figure 1.36 shows the graphs of the two equations on the same set of axes. The point of intersection for which x is largest appears to have its x-coordinate between 1 and 2.

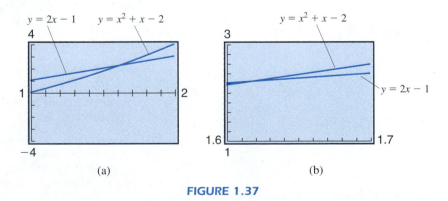

(a)

(b)

FIGURE 1.37

Figures 1.37(a) and 1.37(b) show close-ups of the two curves for $1 \le x \le 2$. In Figure 1.37(b) the y-range has also been reduced using a viewing rectangle $[1.6, 1.7] \times$

[1, 3] to get a better view of the curves. The point of intersection is approximately (1.62, 2.2).

■

We have presented only a few of the valuable ways that graphing devices can be used to help us graph equations and solve algebraic problems. We will mention more of these techniques as we proceed through the book.

EXERCISE SET 1.5

1. Use a graphing device to sketch a graph of $y = x^2 - 4x + 9$ with the following viewing rectangles, and determine which gives the best representation for the graph of the equation.

   a. $[-3, 3] \times [-3, 3]$

   b. $[-7, 7] \times [-7, 7]$

   c. $[-100, 100] \times [-100, 100]$

   d. $[-3, 20] \times [-3, 20]$

2. Use a graphing device to sketch a graph of $y = x^2 + 10x + 15$ with the following viewing rectangles, and determine which gives the best representation for the graph of the equation.

   a. $[-7, 7] \times [-7, 7]$

   b. $[-100, 100] \times [-500, 500]$

   c. $[-10, 10] \times [-10, 10]$

   d. $[0, 10] \times [-10, 300]$

3. Use a graphing device to sketch a graph of $y = x^3 - 20x + 25$ with the following viewing rectangles, and determine which gives the best representation for the graph of the equation.

   a. $[-2, 2] \times [-2, 2]$

   b. $[-5, 5] \times [-5, 5]$

   c. $[-10, 10] \times [-70, 70]$

   d. $[-100, 100] \times [-200, 200]$

4. Determine an appropriate viewing rectangle for the graph of each equation and use it to sketch the graph.

   a. $y = x^2 - 10x + 18$ b. $y = x^2 + 14x + 59$

   c. $y = \sqrt{3x - 8}$ d. $y = \sqrt{x^2 - 2x - 15}$

   e. $y = \dfrac{x + 3}{x - 2}$ f. $y = x^2 + \dfrac{1}{x}$

5. Use a graphing device to approximate all solutions to the following equations.

   a. $x^3 - 0.5x^2 - 7x + 5 = 0$

   b. $x^4 - 2 = x^3 + 2x^2 - x + 1$

6. Graphically approximate the solutions to the following inequalities.

   a. $x^2 + 3x - 2 \geq 0$ b. $x^3 - 2x^2 - 6x + 9 < 0$

7. Graph $y = x^4 + cx^2$ for different values of c, and describe the effect of the constant c on the curve.

8. Graph the equations $y = x^4$ and $y = x^4 - 4x^3 + 3x^2$ on the same set of axes using the viewing rectangles

   a. $[-5, 5] \times [-10, 10]$ b. $[-100, 100] \times [0, 10^8]$

   c. $[-500, 500] \times [0, 6 \times 10^{10}]$

   What can you conclude from the graphs?

9. Consider the family of curves given by

$$y = \frac{1}{x^n},$$

   where $x \neq 0$ and n is a positive integer. Plot the graph for $n = 2, 4, 6$, and 8 on the same set of axes. On another set of axes plot the graph for $n = 1, 3, 5$, and 7. For a given value of n, compare the sizes of $1/x^n$ and $1/x^{n+2}$ on each of the intervals.

   a. $0 < x < 1$ b. $x > 1$

   c. $-1 < x < 0$ d. $x < -1$

10. The number of bacteria in a culture at time t is given by $n = 10000 \left(\frac{3t^2 + 1}{t^2 + 1} \right)$. As the time t increases, does the size of the bacteria colony become stable? If so, what is the stabilizing level?

11. Use a graphing device to plot a variety of curves of the form $y = ax + b$, where a and b are real numbers. Describe the effect that both positive and negative values of a and b have on the graph.

12. Use a graphing device to plot a variety of curves of the form $y = (x - a)^2 + b$, where a and b are real numbers. Describe the effect that both positive and negative values of a and b have on the graph.

1.6 FUNCTIONS

There are a variety of ways for expressing a functional relationship between two quantities. One common way is through a formula expressed as an equation. For example, the formula for the area of a circle, $A = \pi r^2$, expresses how the area of the circle depends on the length of the radius. Another way of expressing a functional relationship between two quantities is through a table of values. Table 1.3 shows the rapid increase in the public debt in the United States after 1940.

TABLE 1.3

Public Debt (in millions)			
Year	Amount	Year	Amount
1940	42,968	1970	370,094
1945	258,682	1975	533,189
1950	256,087	1980	907,701
1955	272,087	1985	1,823,103
1960	284,093	1990	3,233,313
1965	313,819	1992	4,002,669

This information is shown graphically in Figure 1.38. Straight lines have been used to join the data points so that the increasing trend is clearer.

Calculus studies the properties, applications, and behavior of functions, one of the fundamental tools of mathematics. A function expresses a special type of relationship between two quantities, where one quantity depends on the other in a very specific way.

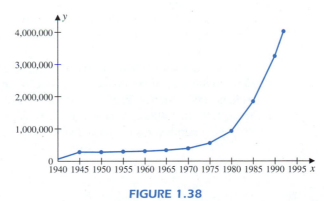

FIGURE 1.38

Computer algebra systems and graphing calculators are useful in analyzing tables of data since they can easily plot large numbers of data points and either simply connect the points with straight-line segments or use other techniques to fit a smooth curve to the data points. The curves can then be used to detect trends in the data.

In a functional relationship, whether given by a formula, a table, or a graph, the common property is that for each value of the first quantity, called the *independent variable*, there is associated a unique value of the second quantity, called the *dependent variable*.

Functions

A *function* from a set $\mathbb{X}$ into a set $\mathbb{Y}$ is a rule of correspondence that assigns each element in $\mathbb{X}$ to precisely one element in $\mathbb{Y}$.

Figure 1.39 illustrates the function concept, where $\mathbb{X}$ represents the first, or given, set and $\mathbb{Y}$ represents the second set. Notice that each value of x is assigned only one value of y, although x_3 and x_4 are assigned to the same value, y_3.

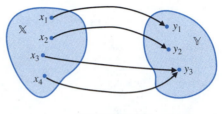

FIGURE 1.39

If f is used to denote a function from a set $\mathbb{X}$ into a set $\mathbb{Y}$, then $f(x)$ is called the *image* of x under f. The image $f(x)$ is the unique element in $\mathbb{Y}$ that corresponds to the element x in $\mathbb{X}$. The set $\mathbb{X}$ is called the *domain* of f. The *range* of f is the set of elements in $\mathbb{Y}$ that are associated with some element in $\mathbb{X}$. This means that the domain of the function must be the entire set $\mathbb{X}$, but the range need not be all of $\mathbb{Y}$. For example, the function $f(x) = x^2$ takes the set of real numbers, $\mathbb{R}$, into $\mathbb{R}$. This function has domain $\mathbb{R}$ but its range is the subset $[0, \infty)$ of $\mathbb{R}$.

To indicate that f is a function with domain $\mathbb{X}$ and range in $\mathbb{Y}$, we write $f\colon \mathbb{X} \to \mathbb{Y}$ and say f maps $\mathbb{X}$ into $\mathbb{Y}$. If x is in the domain of a function f, we say that "f is defined at x" or that "$f(x)$ exists."

Although each element in the domain of a function corresponds to a unique element in the range, more than one element in the domain can be associated with the same element of the range. The most extreme examples are constant functions, as shown in Figure 1.40. In 1.40(a), every number x is associated with the same value 2, and in 1.40(b), every number x is associated with -1.

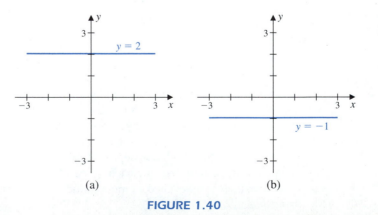

(a)   (b)

FIGURE 1.40

The study of calculus is concerned primarily with functions whose domain and range both consist of real numbers. Functions of this type are usually described by simply giving their rule of correspondence. For example, $f(x) = x^3 + 1$ describes a function f with domain and range both consisting of real numbers. Unless otherwise specified, the domain of the function is assumed to be the largest subset of real numbers for which the correspondence produces a real number. The domain of $f(x) = 1/x$, then, is $(-\infty, 0) \cup (0, \infty)$ since we can divide by anything but zero. The range is the set of real numbers associated with some number in the domain.

If you have access to a graphing device, we encourage you to use it throughout the text to help you to visualize problems using graphical representations.

In this section, we have given graphs of the functions we are using for the examples to better illustrate the concepts we wish to explore. If you have a graphing device, you should check that these graphs are correct. Later in this chapter we will see that these graphs were obtained by simple modifications of the graphs of some common functions.

EXAMPLE 1   Use the graph in Figure 1.41 to find the domain and range of the function f whose rule of correspondence is $f(x) = \sqrt{x} + 1$.

Solution   The square root of x, $\sqrt{x}$, is a real number if and only if $x \geq 0$. Consequently, the domain of f is the set of all nonnegative real numbers.

Since the symbol $\sqrt{}$ indicates the principal, or nonnegative, square root, we have $f(x) = \sqrt{x} + 1 \geq 1$ for all $x \geq 0$. The range of f is the set of real numbers greater than or equal to 1.   ■

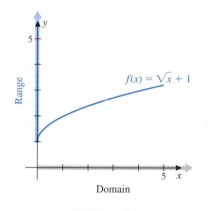

FIGURE 1.41

There is no special significance to the variable x; in fact, it might be better to describe the function in Example 1 by writing $f(\boxed{}) = \sqrt{\boxed{}} + 1$. This form indicates more clearly that whatever value is used to fill the box on the left side of the equation must also be used to fill the corresponding box on the right side. For example, 9 is in the domain of f and $f(\boxed{9}) = \sqrt{\boxed{9}} + 1 = 3 + 1 = 4$.

It is important to be completely comfortable with function notation since it is used extensively in calculus. The following example illustrates the first steps of a common calculus problem.

EXAMPLE 2   Consider the function f described by $f(x) = x^2 + x$, whose graph is shown in Figure 1.42. Determine, for arbitrary x and $h \neq 0$, simplified expressions for

$$f(x + h), \qquad f(x + h) - f(x), \qquad \text{and} \qquad \frac{f(x + h) - f(x)}{h}.$$

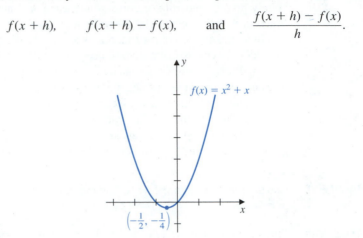

FIGURE 1.42

Solution   The solution to this problem is easier if we think of the function written as

$$f(\Box) = \Box^2 + \Box,$$

where replacing the box on the left by something in the domain of f has the same effect as placing that item in the boxes on the right. This implies, for example that

$$f(x + h) = (x + h)^2 + (x + h) = x^2 + 2xh + h^2 + x + h = x^2 + (2h + 1)x + h^2 + h.$$

As a consequence,

$$f(x + h) - f(x) = \left(x^2 + (2h + 1)x + h^2 + h\right) - \left(x^2 + x\right) = 2hx + h^2 + h,$$

and

$$\frac{f(x + h) - f(x)}{h} = \frac{2hx + h^2 + h}{h} = 2x + h + 1,$$

provided that $h \neq 0$.   ■

The difference quotient is the basis for the definition of the *derivative* in calculus. Be sure that the notation is clear, and that you are able to simplify this type of expression.

The quotient in Example 2 is called a *difference quotient* of the function f. It describes the average rate of change of the values of the function as the variable changes from x to $x + h$, as shown in Figure 1.43. One of the primary concepts of calculus, the derivative, considers the limiting case of this difference quotient as h tends to zero.

EXAMPLE 3   In economics, the profit function P describes the profit $P(x)$ when x units of a commodity are sold. From experience a manufacturer determines that the profit function for a product is approximately

$$P(x) = 200x - x^2.$$

a. What is the average rate of change in the profit as the number of units changes from x to $x + h$?

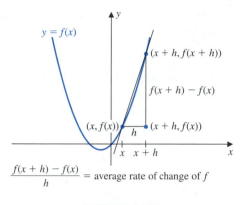

$$\frac{f(x + h) - f(x)}{h} = \text{average rate of change of } f$$

FIGURE 1.43

b. Use the result in part (a) to find the average rate of change in the profit if the number of units produced changes from 50 to 75.
c. Plot the graph of $y = P(x)$, and the line that passes through the points $(50, P(50))$ and $(75, P(75))$.

Solution   The graph of the profit function shown in Figure 1.44(a) seems like a reasonable representation. It indicates that there is no profit when no items are sold, a maximum profit is produced when 100 units are sold, and beyond that point profit begins to decline.

a. The average rate of change is determined by computing the difference quotient of the function P; that is,

$$\frac{P(x + h) - P(x)}{h} = \frac{200(x + h) - (x + h)^2 - (200x - x^2)}{h}$$

$$= \frac{200x + 200h - x^2 - 2xh - h^2 - 200x + x^2}{h}$$

$$= \frac{200h - 2xh - h^2}{h}$$

$$= 200 - 2x - h$$

In this part of the example we are finding the *average* rate of change in the profit when a specified number of units is produced. In calculus we will determine the *instantaneous* rate of change.

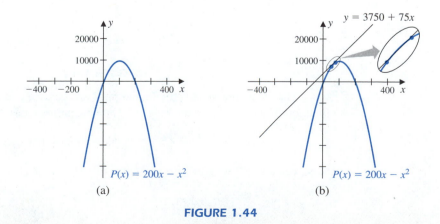

FIGURE 1.44

b. Since the number of units starts at 50 and changes by 25, the average rate of change
is given by

$$200 - 2(50) - 25 = 75.$$

Thus, for each additional unit manufactured and sold, the profit will increase on the
average by $75.

c. The plot is shown in Figure 1.44(b). The inclination of the line indicates the average
rate of change in the profit as the number of units produced increases from 50 to 75.

∎

EXAMPLE 4   Given $f(x) = \dfrac{1}{x-1} + 2$, find (a) the domain of f, and (b) a value of x for which
$f(x) = 5$.

Solution   a. The quotient $1/(x-1)$ is defined for all real numbers except $x = 1$ (written $x \neq 1$),
which makes the denominator 0. Since we can add 2 to any real number, the domain
of the function in interval notation is $(-\infty, 1) \cup (1, \infty)$.

b. The problem assumes that the number 5 is in the range of the function f, for unless
it did, no such x would exist. To determine the value x in the domain corresponding
to $f(x) = 5$, we solve the equation $f(x) = 5$. This gives

$$5 = \frac{1}{x-1} + 2, \qquad \text{which implies that} \qquad \frac{1}{x-1} = 3.$$

Inverting both sides of this equation, assuming $x \neq 1$, gives

$$x - 1 = \frac{1}{3}, \qquad \text{so} \qquad x = 1 + \frac{1}{3} = \frac{4}{3}.$$

The graph of f is shown in Figure 1.45.

∎

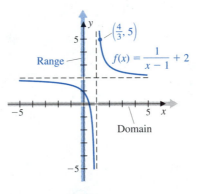

FIGURE 1.45

The procedure in part (b) of Example 4 can be used to determine the range of this
function. If y is a value in the range of f, then a number x, with $x \neq 1$, must exist with

$$y = f(x) = \frac{1}{x-1} + 2, \qquad \text{which implies that} \qquad y - 2 = \frac{1}{x-1}.$$

Now, when $y \neq 2$ we solve for x in terms of y to obtain

$$x - 1 = \frac{1}{y-2} \qquad \text{and} \qquad x = \frac{1}{y-2} + 1.$$

The range of f consists of the set of numbers y for which this relation is defined. This is the set of all real numbers y with $y \neq 2$, as shown in Figure 1.45. So, the range of f in interval notation is $(-\infty, 2) \cup (2, \infty)$.

We will always be interested in the domain of a given function since this tells us the values for which the function is defined. We do not always need to determine the range, which is fortunate, since it can be much more difficult to determine.

EXAMPLE 5   Find the domain of the function given by

$$g(x) = \sqrt{x^2 - 4x + 3}.$$

Solution   The domain of g is the set of all real numbers x for which the expression under the radical is nonnegative—that is, all values of x for which $x^2 - 4x + 3 \geq 0$. The quadratic factors as

$$x^2 - 4x + 3 = (x - 3)(x - 1),$$

so x is in the domain of g if the factors $(x - 3)$ and $(x - 1)$ are both nonnegative or the factors are both nonpositive.

This type of chart for analyzing inequalities is often used to determine the behavior of graphs of functions on intervals. It is more convenient than solving inequalities using algebra.

$$
\begin{array}{rl}
(x - 3) & - - - - - - - - - - \ 0 + + + + + + \\
(x - 1) & - - - - - - \ 0 + + + + + + + + + + \\
(x - 3)(x - 1) & + + + + + + \ 0 - - - \ 0 + + + + + +
\end{array}
$$

$$-2 \quad -1 \quad 0 \quad 1 \quad 2 \quad 3 \quad 4 \quad 5 \quad 6 \quad x$$

FIGURE 1.46

The top line of the sign graph in Figure 1.46 indicates that the factor $(x - 3)$ is nonnegative when $x \geq 3$ and is nonpositive when $x < 3$. The second line gives similar information about the factor $(x - 1)$. The third line in this figure shows that, as a consequence, the product $(x - 3)(x - 1)$ is nonnegative if either $x \leq 1$ or $x \geq 3$. Hence, the domain of g is $(-\infty, 1] \cup [3, \infty)$.   ■

In Section 1.4 we considered the graphs of equations and discussed some symmetry properties of graphs of equations. In Example 2 of that section we saw the graphs of the equations $y = x^2$, $x = y^2$, and $x^2 + y^2 = 1$. These graphs, reproduced in Figure 1.47, illustrate a test to distinguish equations that represent functions from those that do not.

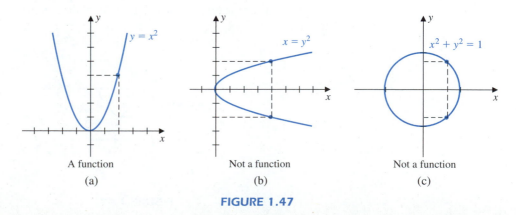

A function
(a)

Not a function
(b)

Not a function
(c)

FIGURE 1.47

Any vertical line intersects the graph of $y = x^2$ shown in Figure 1.47(a) in exactly one place, so for each real number x there is precisely one point $(x, y) = (x, x^2)$ on the graph. That is, for each real number on the x-axis, there corresponds precisely one number on the nonnegative y-axis. So the graph of $y = x^2$ is the graph of a function, and the domain of this function is $(-\infty, \infty)$. Since horizontal lines intersect the graph only on or above the x-axis, the range is the interval $[0, \infty)$.

On the other hand, $x = y^2$ and $x^2 + y^2 = 1$ both have the property that certain vertical lines intersect their graphs more than once. For example, the line $x = 4$ in Figure 1.47(b) intersects the graph of $x = y^2$ at both $y = 2$ and $y = -2$. The line $x = 1/2$ in Figure 1.47(c) intersects the graph of $x^2 + y^2 = 1$ at both $y = \sqrt{3}/2$ and $y = -\sqrt{3}/2$. Neither of these graphs can represent a function since for some values of x there corresponds more than one value of y.

Vertical Line Test for Functions

An equation describes y as a function of x if and only if every vertical line intersects the graph of the equation at most once.

A general illustration of this test is shown in Figure 1.48.

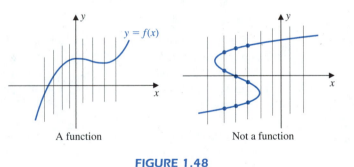

A function   Not a function

FIGURE 1.48

In addition, when the graph is that of a function, the following statements are true (see Figure 1.49).

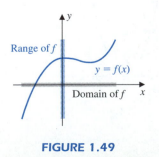

FIGURE 1.49

> **Finding the Domain and Range of a Function from a Graph**
>
> The domain of a function is described by those values on the horizontal axis through which a vertical line intersects the graph.
>
> The range of a function is described by those values on the vertical axis through which a horizontal line intersects the graph.

EXAMPLE 6   Find the domain and range of the function defined by

$$f(x) = \begin{cases} x - 1, & \text{if } x < 0 \\ x^2, & \text{if } 0 \le x \le 2. \end{cases}$$

Solution   A function that is defined by differing expressions on various portions of its domain is called a *piecewise* defined function. To sketch the graph of this piecewise defined function, we first sketch the graphs of $y = x - 1$ and $y = x^2$ as shown in Figure 1.50(a). The graph of $y = f(x)$ switches from the graph of $y = x - 1$ to the graph of $y = x^2$ when $x = 0$ and is shown in Figure 1.50(b).

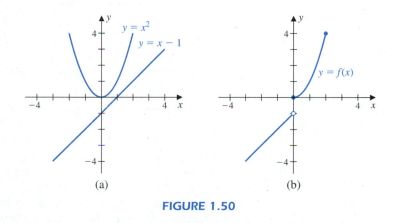

(a)   (b)

FIGURE 1.50

Vertical lines intersect the graph when $x \le 2$, so the domain of f is $(-\infty, 2]$. Horizontal lines intersect the graph when $y < -1$ or when $0 \le y \le 4$, so the range of f is $(-\infty, -1) \cup [0, 4]$.   ■

Graphing devices are helpful in determining the range of a function that has a complicated representation.

EXAMPLE 7   Use the graph of the function

$$f(x) = \frac{x^2 + 4}{x^2 - 4}$$

to determine its range.

Solution   The graph on the viewing rectangle $[-5, 5] \times [-5, 5]$ is shown in Figure 1.51. It appears from this sample that the range includes all values except those in the interval $(-1, 1]$. (Algebra can be used to show this conclusion is correct.)   ■

Finding the range of a function is often difficult, and a graphing device can be useful here. Once we have the graph, the range of the function can be found from the y-coordinates of points on the graph.

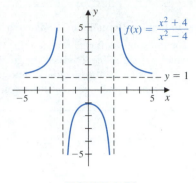

FIGURE 1.51

In Section 1.4 we discussed the symmetry properties of graphs. The following definition characterizes those functions whose graphs are symmetric with respect to the y-axis or with respect to the origin.

Knowing that a graph is symmetric with respect to the y-axis or to the origin cuts in half the work required to sketch the graph. We need to know the graph only when $x \geq 0$. The symmetry will give the graph when $x < 0$.

Odd and Even Functions

A function f is said to be *even* if whenever x is in the domain of f, $-x$ is also in the domain and $f(-x) = f(x)$.

  A function f is said to be *odd* if whenever x is in the domain of f, $-x$ is also in the domain and $f(-x) = -f(x)$.

This definition implies that the graph of an even function is symmetric with respect to the y-axis, for if $(x, f(x))$ is on the graph, then $(-x, f(x))$ is on the graph as well. The graph of an odd function is symmetric with respect to the origin since $(x, f(x))$ on the graph implies that $(-x, -f(x))$ is also on the graph.

  The graph of a *function* cannot be symmetric with respect to the x-axis unless the function is the constant zero function. This follows from the fact that (x, y) and $(x, -y)$ cannot both be on the graph of a function unless $y = -y$, which implies that $y = 0$.

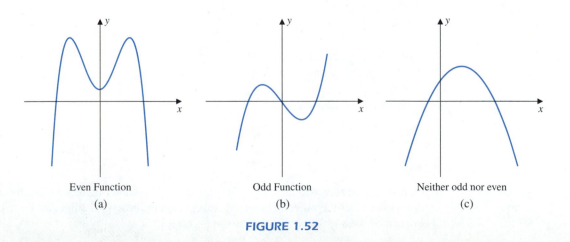

Even Function   Odd Function   Neither odd nor even

(a)   (b)   (c)

FIGURE 1.52

Figure 1.52(a) shows the graph of an even function and its symmetry with respect to the y-axis. Figure 1.52(b) shows the graph of an odd function and its symmetry with respect to the origin. The graph in Figure 1.52(c) has no axis or origin symmetry, so it represents the graph of a function that is neither odd nor even.

Our final examples illustrate the use of function notation to express real-life problems in a mathematical form. Generally, the most difficult part of a problem is translating the written description of the problem to a form using mathematical equations to construct a mathematical model. The "Know–Find" outline used in Example 8 can be helpful in making this translation.

EXAMPLE 8   Two ships sail from the same port. The first ship leaves port at 1:00 A.M. and travels eastward at a rate of 15 knots (nautical miles per hour). The second ship leaves port at 2:00 A.M. and travels northward at a rate of 10 knots. Find the distance between the ships as a function of the time after 2:00 A.M.

Solution   We first introduce a time variable into the problem so that we can express the distances using equations. Let t denote the time in hours after 2:00 A.M. Since the second ship travels at a rate of 10 nautical miles per hour, it is $10t$ nautical miles from port at time t.

The first ship is traveling at a rate of 15 nautical miles per hour and has traveled for $(1+t)$ hours when it is t hours after 2:00 A.M. Its distance from port is $15(1+t) = 15+15t$ nautical miles at time t. Figure 1.53 shows the position of the ships at a given time t.

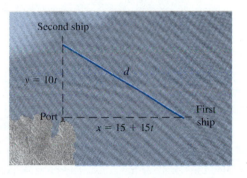

FIGURE 1.53

The following is a concise description of the problem.

KNOW	FIND
1. The distance of the first ship from the port: $x = 15 + 15t$, $t \geq 0$. 2. The distance of the second ship from the port: $y = 10t$, $t \geq 0$.	The distance d separating the ships as a function of t.

Since the paths of the ships are perpendicular, we can use the Pythagorean Theorem to determine the answer. The ships are a distance

$$d(t) = \sqrt{x^2 + y^2} = \sqrt{(15 + 15t)^2 + (10t)^2} = \sqrt{325t^2 + 450t + 225}$$

nautical miles apart t hours after 2:00 A.M.   ∎

Note that the domain of the function described in Example 8 has been restricted to the interval $[0, \infty)$ by the physical conditions of the problem. In fact, a reasonable approximation to the actual distance between the ships is given by $d(t)$ only for small values of t. This is due to variations in the actual speed of the ships, deviation from the assumed course, the curvature of the earth, and many other factors. Keep in mind that when mathematical expressions are used to solve physical problems, the answer to the mathematical problem can vary from the answer to the physical problem due to such neglected technicalities.

With the aid of a graphing device we can approximate maximum and minimum values of functions by using the zoom to isolate high and low points on the graph of the function.

EXAMPLE 9   A box with no top is to be constructed from a square piece of cardboard with side 3 ft by cutting out squares of equal size at each corner and bending up the flaps. Approximate the size of the square that should be removed in order to produce the maximum volume.

One of the basic tools of calculus is finding maximums and minimums of functions. Setting up an application problem is generally the difficult part of a calculus application; using calculus to solve the problem is often relatively easy.

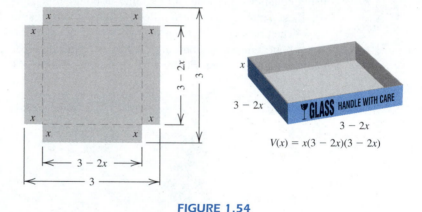

$$V(x) = x(3 - 2x)(3 - 2x)$$

FIGURE 1.54

Solution   We will write the volume as a function of x, the size of the square removed, which is also the height of the constructed box. Since the volume of a box is the area of the base times the height, we see from Figure 1.54 that the volume V is given by

$$V(x) = x(3 - 2x)(3 - 2x) = 9x - 12x^2 + 4x^3.$$

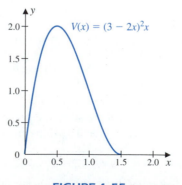

FIGURE 1.55

Notice the domain of V is $(0, 1.5)$ since the side of the flap must be positive and must be less than half the width of the side of the original cardboard piece.

To find the value of x that maximizes the volume, we first plot the graph of $y = V(x)$, as shown in Figure 1.55. Moving to the peak of the curve we see the maximum volume occurs when x is near 0.5, so the volume is approximately $V(0.5) = 0.5(3 - 2(0.5))^2 = 2.0 \text{ ft}^3$. Using calculus it can be shown these values are correct. ■

EXERCISE SET 1.6

1. If $f(x) = 4x^2 + 5$, find each of the following.

   a. $f(2)$ **b.** $f(\sqrt{3})$

   c. $f(2 + \sqrt{3})$ **d.** $f(2) + f(\sqrt{3})$

   e. $f(2x)$ **f.** $f(1 - x)$

2. If $f(t) = |t - 2|$, find each of the following.

   a. $f(4)$ **b.** $f(1)$

   c. $f(0)$ **d.** $f(t + 2)$

   e. $f(2 - t^2)$ **f.** $f(-t)$

In Exercises 3–8, determine which of the curves are graphs of functions.

3.

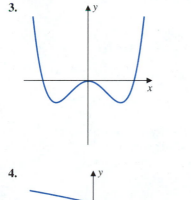

4.

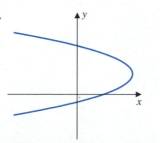

5.

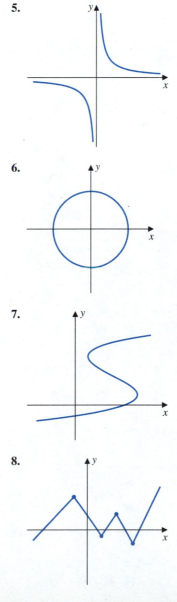

6.

7.

8.

In Exercises 9–12, use the graphs to determine the domain and range of the function.

9.

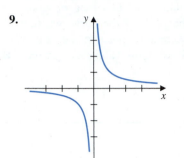

10.

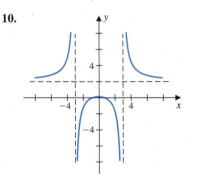

11.

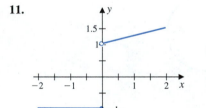

12.

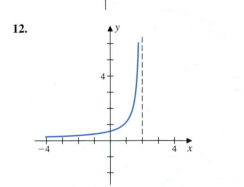

In Exercises 13–20, a function is described. Find the domain and range of the function.

13. $f(x) = x^2$

14. $f(x) = \dfrac{1}{x - 3}$

15. $f(x) = \dfrac{x^4 - x^2}{x^2 - 1}$

16. $f(x) = \dfrac{1}{\sqrt{x - 3}}$

17. $f(x) = \sqrt{x(x - 2)}$

18. $f(x) = \dfrac{1}{\sqrt{x^2 - x - 6}}$

19. $f(x) = \begin{cases} 1, & \text{if } x \geq 0 \\ -1, & \text{if } x < 0 \end{cases}$

20. $f(x) = \begin{cases} x^2, & \text{if } x \geq 0 \\ -x, & \text{if } x < 0 \end{cases}$

In Exercises 21–24, determine whether the equation describes y as a function of x.

21. $3x + 2y = 5$

22. $y = x^2 + 3$

23. $x = y^2 + 3$

24. $x^2 + y^2 = 16$

In Exercises 25–28, determine formulas for $f(-x)$, $-f(x)$, $f(1/x)$, $1/f(x)$, $f(\sqrt{x})$, and $\sqrt{f(x)}$.

25. $f(x) = x^2 + 2$

26. $f(x) = x^2 + 2x + 3$

27. $f(x) = 1/x$

28. $f(x) = \sqrt{x}$

In Exercises 29–32, determine a number a in the domain of the function f so that $f(a) = b$ for the given value of b.

29. $f(x) = x^2 - 1; b = 0$

30. $f(x) = x^2 - 1; b = 2$

31. $f(x) = \sqrt{x - 1}; b = 1/2$

32. $f(x) = \sqrt{x} - 1; b = 3/4$

In Exercises 33–38, find $f(x + h)$ and $\dfrac{f(x + h) - f(x)}{h}$, when $h \neq 0$.

33. $f(x) = 2x - 4$

34. $f(x) = \frac{3}{2}x - \frac{1}{4}$

35. $f(x) = x^2$

36. $f(x) = 4x^2 + 3x + 1$

37. $f(x) = \sqrt{x}$

38. $f(x) = \dfrac{x}{x - 3}$

In Exercises 39–42, classify the graph as that of a function that is even, odd, or neither even nor odd.

39.

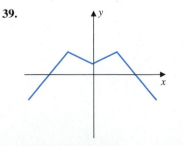

40.

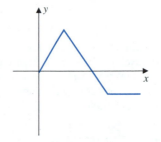

41.

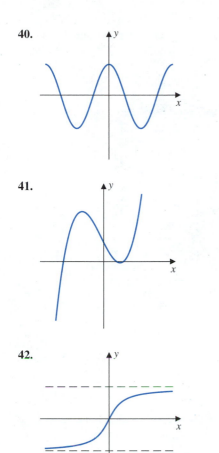

a. f is even.   **b.** f is odd.

46. Repeat Exercise 45 for the accompanying graph.

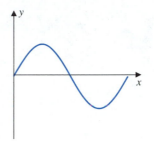

42.

47. Redraw the following graph, and on the same set of axes draw each graph.

a. $y = -f(x)$ **b.** $y = f(-x)$ **c.** $y = |f(x)|$

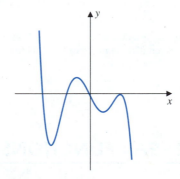

43. Determine whether each of the following functions is even, odd, or neither even nor odd.

a. $f(x) = x^2 + 1$ **b.** $f(x) = x^3 - 1$

c. $f(x) = x^3 + 3x$ **d.** $f(x) = \sqrt{x}$

e. $f(x) = x^2 + x$ **f.** $f(x) = x^2 + x^3$

g. $f(x) = x^4 - x^2$ **h.** $f(x) = 1/x$

44. Determine which of the functions described in Exercise 43 have graphs that are

(i) Symmetric with respect to the y-axis.

(ii) Symmetric with respect to the origin.

45. The graph of a function f is given for $x \geq 0$. Extend the graph for $x < 0$ if

48. a. Suppose f is an odd function whose domain contains 0. Explain why $f(0) = 0$.

b. Can a function be both even and odd?

49. Use algebra to show that the range of the function defined in Example 7 is $(-\infty, -1] \cup (1, \infty)$.

50. Sketch the graph of $y = f(x) = x(x - 1)^2$. Then use a graphing device to sketch the graphs of the following for $a = 1, 2, -1,$ and -2.

a. $y = f(x) + a$ **b.** $y = f(x + a)$

c. $y = af(x)$ **d.** $y = f(ax)$

51. Two ships sail from the same port. The first ship leaves at noon and travels eastward at 10 knots. The second ship leaves at 3:00 P.M. and travels southward at 15 knots. Find the distance d between the ships as a function of the time after 3:00 P.M.

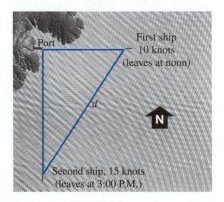

52. You are driving at 50 mi/h along a straight road at a constant speed between cities A and B. Along the way you pass through city C. The cities B and C are 100 and 25 mi, respectively, from city A. Draw a graph that indicates your distance from city C as a function of time, measured from the time you leave city A.

53. A rectangular plot of ground containing 432 ft^2 is to be fenced within a large lot.

a. Express the perimeter of the plot as a function of the width. What is the domain of this function?

b. Use a graphing device to approximate the dimensions of the plot that requires the least amount of fence.

54. A manufacturer estimates the profit on producing x units of their product at

$$P(x) = 300x - 2x^2$$

a. What is the average rate of change in the profit as the number of units changes from x to $x + h$?

b. Use the result in part (a) to find the average rate of change in the profit as the number of units produced changes from 25 to 50.

c. Sketch the graph of $y = P(x)$ and the line that passes through the points $(25, P(25))$ and $(50, P(50))$.

1.7 LINEAR FUNCTIONS

A linear relationship between two variables occurs when there is a steady increase or a steady decrease in one of the variables with respect to the other. Linear functions have the property that any change in the independent variable results in a proportional change in the dependent variable, as shown in Figure 1.56.

Many physical situations can be modeled using a linear relationship. Table 1.4 shows the increasing trend in the global atmospheric concentrations of carbon dioxide in parts per million (ppm) from 1960 to 1993. Many scientists are concerned that a continuation of this trend will lead to a "greenhouse effect," with the result that the average temperature of the earth will increase and produce catastrophic effects.

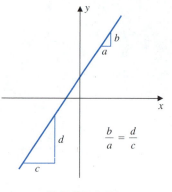

FIGURE 1.56

The data in Table 1.4 do not form an exact linear relationship since the carbon dioxide levels do not always increase by a fixed amount. However, a plot of the data called a *scatter plot*, shown in Figure 1.57(a), shows a steady, increasing pattern that is approximately linear.

TABLE 1.4

Year	Carbon Dioxide (ppm)
1960	317
1970	325
1980	339
1990	354
1993	357

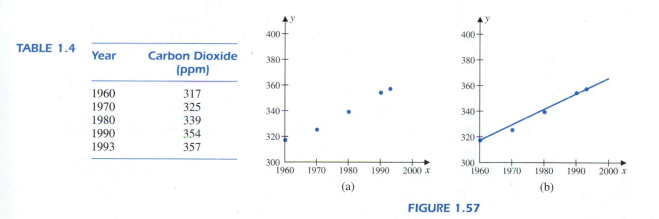

FIGURE 1.57

The line shown in Figure 1.57(b) passes through the two endpoints of the data and does a reasonable job of approximating the remaining points. Later in this section we will determine the equation of this line and use it to approximate the concentration in the year 2000.

Calculus permits us to find the equation of the line that "best" fits the data.

A function f defined by a linear equation of the form

$$y = f(x) = ax + b, \quad \text{where } a \text{ and } b \text{ are constants,}$$

is called a **linear function**. In this section we will see that linear functions have graphs that are straight lines and that any nonvertical straight line is the graph of a linear function.

Suppose that l is a nonvertical straight line and that $P(x_1, y_1)$ and $Q(x_2, y_2)$ are two distinct points lying on l, as illustrated in Figure 1.58. The quotient of the difference of the y-coordinates over the difference of the x-coordinates is called the *slope* of the line.

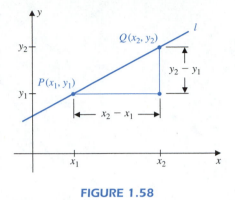

FIGURE 1.58

It is important to have a good feel for the slopes of lines. In calculus we will see that the slope of a line is used to describe the slope of the graph of a function. A good knowledge of the slopes of lines will keep you from making obvious errors.

The Slope of a Line

The *slope* of the line passing through $P(x_1, y_1)$ and $Q(x_2, y_2)$ is $m = \dfrac{y_2 - y_1}{x_2 - x_1}$.

Any two distinct points on the line will give the same value for the slope. So, if we replace one of the pairs of points, say, (x_2, y_2), with an arbitrary pair (x, y) lying on the line, we have what is called the *point-slope* form of the equation of the line.

Point-Slope Equation of a Line

The line with slope m that passes through the point $P(x_1, y_1)$ has the **point-slope** equation

$$y - y_1 = m(x - x_1).$$

Using a graphing device to plot a variety of lines quickly in the same viewing rectangle, as shown in Figure 1.59, provides a visualization of the slope of a line.

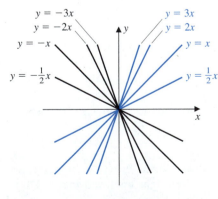

FIGURE 1.59

The slope of a line determines its direction, in the following sense. (See Figure 1.59.)

The Slope of a Line

(i) A line with a positive slope is directed upward (looking from left to right), and the values of y increase as the values of x increase. The values of y increase more rapidly on a line with a large positive slope than on a line with a small positive slope.

(ii) A line with a negative slope is directed downward (looking from left to right), and the values of y decrease as the values of x increase. The values of y decrease more rapidly on a line with a negative slope that is large in magnitude than on a line with a negative slope that is small in magnitude.

(iii) A line with zero slope is horizontal.

The point-slope equation can be rewritten as

$$y = m(x - x_1) + y_1 = mx + (y_1 - mx_1).$$

This gives the *slope-intercept* form of the equation of the line.

Slope-Intercept Equation of a Line

The line with slope m passing through the point $P(x_1, y_1)$ has the **slope-intercept** equation

$$y = mx + b, \qquad \text{where} \qquad b = y_1 - mx_1.$$

The name of this equation comes from the fact that it involves the slope m of the line and the point $(0, b)$ where the line crosses the y-axis. The number b is the y-*intercept* of the line. The linear function that is described by this equation has the form $f(x) = mx + b$.

When $m = 0$ the linear function $y = b$ is constant and the line is horizontal, since all the points on the line have the same y-values. Points lying on vertical lines have the same x-coordinates, so they have equations of the form $x = c$ for some constant c. Vertical lines cannot be graphs of functions. (See Figure 1.60.)

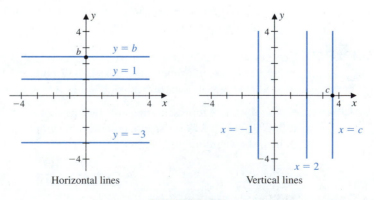

Horizontal lines   Vertical lines

FIGURE 1.60

EXAMPLE 1   The line passing through the points $(2, 3)$ and $(-4, 0)$ is shown in Figure 1.61.

a. Find a point-slope form of the equation of this line.
b. Find the slope-intercept form of the equation of this line.

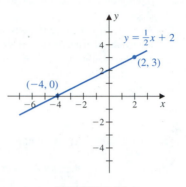

FIGURE 1.61

Solution   a. The slope of this line is

$$m = \frac{0 - 3}{-4 - 2} = \frac{-3}{-6} = \frac{1}{2}.$$

Using the point-slope equation with the point $(2, 3)$, we have

$$y - 3 = \frac{1}{2}(x - 2).$$

If we use the point-slope equation with the point $(-4, 0)$, we have

$$y - 0 = \frac{1}{2}(x - (-4)), \quad \text{or} \quad y = \frac{1}{2}(x + 4).$$

b. Both point-slope equations in part (a) reduce to the same slope-intercept form,

$$y = \frac{1}{2}x + 2. \qquad \blacksquare$$

In part (a) of Example 1 we found two different point-slope formulas, but these reduced to the same slope-intercept formula in part (b). Each point on a line with nonzero slope generates a different point-slope formula, but they all reduce to the same slope-intercept formula.

EXAMPLE 2   The carbon dioxide data given at the beginning of this section state that there were 317 ppm in the atmosphere in 1960 and 357 ppm in 1993. Determine a linear function to predict the level of carbon dioxide in the atmosphere in the year 2000.

Solution   Suppose we let t represent time and $f(t)$ the number of parts per million at time t. Then $(1960, 317)$ and $(1993, 357)$ are points on the graph, and the slope of the line joining the points, shown in Figure 1.62, is

$$m = \frac{357 - 317}{1993 - 1960} = \frac{40}{33}.$$

The line has equation

$$f(t) - 317 = \frac{40}{33}(t - 1960), \quad \text{or} \quad f(t) = 317 + \frac{40}{33}(t - 1960).$$

The predicted concentration in the year 2000 is

$$f(2000) = 317 + \frac{40}{33}(2000 - 1960) \approx 365 \text{ ppm}.$$

If different points on the graph were used to determine the equation of the line, we would expect to have a different approximation. ∎

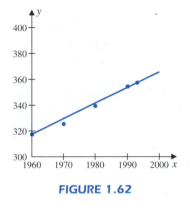

FIGURE 1.62

EXAMPLE 3 Sketch the graph of the linear function described by $f(x) = 2 - \frac{2}{3}x$.

Solution The equation

$$y = 2 - \frac{2}{3}x = -\frac{2}{3}x + 2$$

describes the line with slope $-\frac{2}{3}$ and y-intercept 2, so the point $(0, 2)$ lies on the graph of the line.

To find another point on the line we can set $y = 0$ to find the x-intercept of the line. Substituting $y = 0$ and solving for x gives

$$0 = 2 - \frac{2}{3}x, \quad \text{so} \quad \frac{2}{3}x = 2 \quad \text{and} \quad x = 3.$$

The straight line through $(0, 2)$ and $(3, 0)$ is the graph of f, as shown in Figure 1.63. ∎

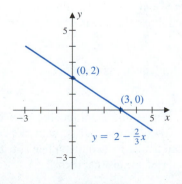

FIGURE 1.63

Parallel and Perpendicular Lines

You will frequently
need to find
equations of lines in
calculus as part of a
larger problem.
Particularly impor-
tant are tangent lines
to curves and the
lines that are perpen-
dicular to them.

Straight lines that never intersect are *parallel*, and lines that intersect at right angles are *perpendicular*. The slopes of the lines provide an easy way to determine when two lines are parallel or perpendicular.

Slopes of Parallel and Perpendicular Lines

Suppose the nonvertical lines l_1 and l_2 have the slopes m_1 and m_2.

(i) Lines l_1 and l_2 are perpendicular if and only if $m_1 m_2 = -1$.
(ii) Lines l_1 and l_2 are parallel if and only if $m_1 = m_2$.

To show the first of these statements, consider the lines shown in Figure 1.64(a), where we have placed the origin of the coordinate system at the point of intersection. Line l_1 has y-intercept 0 and slope m_1, so it has the equation $y = m_1 x$, and it passes through the point $(1, m_1)$. Similarly, line l_2 has the equation $y = m_2 x$ and passes through $(1, m_2)$.

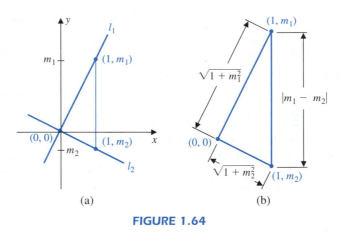

(a)   (b)

FIGURE 1.64

Lines perpendicular
to curves and
surfaces are called
normals and describe
points from which an
object striking the
surface will return to
the point. Radar, for
example, must be
emitted from a
location along a line
perpendicular to the
surface it is
attempting to locate.
An important
application of
calculus is the
determination of
normal lines to
surfaces.

Figure 1.64(b) shows the lengths of the sides of the triangle with vertices $(0, 0)$, $(1, m_1)$, and $(1, m_2)$. The lines are perpendicular if and only if this is a right triangle, for which the Pythagorean Theorem implies that

$$\left(1 + m_1^2\right) + \left(1 + m_2^2\right) = |m_1 - m_2|^2, \qquad \text{that is,} \qquad 2 + m_1^2 + m_2^2 = m_1^2 - 2m_1 m_2 + m_2^2.$$

The last equation is equivalent to

$$-2 = 2m_1 m_2, \qquad \text{or to} \qquad m_1 m_2 = -1.$$

The result about parallel lines in part (ii) follows from the perpendicular result, since two lines l_1 and l_2, with slopes m_1 and m_2 are parallel if and only if they are perpendicular to a common line l_3 with slope $m_3 \neq 0$, as shown in Figure 1.65.
    This means that both

$$m_1 m_3 = -1 \qquad \text{and} \qquad m_2 m_3 = -1,$$

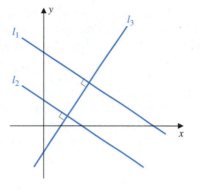

FIGURE 1.65

which implies that

$$m_1 m_3 = m_2 m_3.$$

Since $m_3 \neq 0$ we must have $m_1 = m_2$.

EXAMPLE 4   Find an equation of the line passing through (3, 1) and

a. Parallel to the line with equation $y = 2x - 1$;
b. Perpendicular to the line with equation $y = 2x - 1$.

Solution   a. The slope of the line with equation $y = 2x - 1$ is 2, which is also the slope of any line parallel to $y = 2x - 1$. Since the required line passes through (3, 1), as shown in Figure 1.66(a), it has equation

$$y - 1 = 2(x - 3), \qquad \text{or} \qquad y = 2x - 5.$$

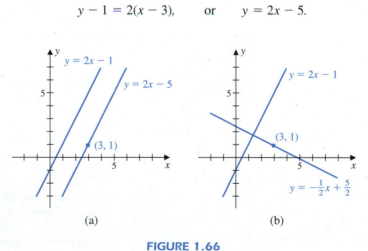

FIGURE 1.66

b. Since $y = 2x - 1$ has slope 2, any line perpendicular to $y = 2x - 1$ must have slope $-\frac{1}{2}$. As a consequence, the perpendicular line through (3, 1), which is shown in Figure 1.66(b), has equation

$$y - 1 = -\frac{1}{2}(x - 3), \qquad \text{or} \qquad y = -\frac{1}{2}x + \frac{5}{2}.$$

We have been careful to exclude vertical lines to this point in the section because these lines cannot be the graphs of linear *functions*. Vertical lines, are, however, included in the general form of linear equations, equations of the form

$$Ax + By + C = 0,$$

where A, B, and C are constants.

When $B \neq 0$, the equation describes a line with slope $-A/B$. The special case of a horizontal line occurs when $A = 0$, which implies that the slope is 0. A vertical line results when $B = 0$ and $A \neq 0$, and vertical lines have no slope.

EXERCISE SET 1.7

1. Plot the pair of points given in each of the following, sketch the straight line determined by these points and find the slope of the line and an equation of the line.

    a. $(0, 0), (1, 2)$ **b.** $(1, 2), (3, -2)$

    c. $(1, -3), (5, 1)$ **d.** $(-1, 3), (-5, -1)$

2. Find equations of the lines that pass through the point $(3, 2)$ and have the given slope. Sketch each equation.

    a. 1   **b.** -1

    c. 0   **d.** 3

3. Sketch the graph of the line associated with each of the following linear equations, and tell which pairs of lines are parallel.

    a. $y = 2$ **b.** $y = x + 1$

    c. $y = -x + 1$ **d.** $y = x + 4$

    e. $y = 2x - 5$ **f.** $y = -4$

    g. $x + y = 0$ **h.** $2x - y = 2$

    i. $x - y = 4$ **j.** $2y - 2x = 1$

4. Sketch the graph of the line associated with each of the following linear equations, and tell which lines are perpendicular.

    a. $y = -1$ **b.** $x = -2$

    c. $y = 2x + 3$ **d.** $y = 3x + 5$

    e. $y = \frac{1}{3}x - \frac{1}{3}$ **f.** $y = -x - 3$

    g. $y = -2x - 5$ **h.** $y = -3x - 7$

    i. $x + 3y = -5$ **j.** $x + 2y = -1$

In Exercises 5–8, the equation of a line is given, together with a point that is not on the line. Find the slope-intercept form of the equation of the line that passes through the given point and is (a) parallel to the given line and (b) perpendicular to the given line.

5. $y = 2x + 1; (0, 0)$ 6. $y = 3x - 2; (1, 2)$

7. $y = -2x + 3; (-1, 2)$ 8. $x + y = -1; (0, 0)$

In Exercises 9–16, find an equation of the line that satisfies the given conditions.

9. Passes through $(1, -2)$ with slope 3.

10. Passes through $(-1, -3)$ with slope -2.

11. Has slope -1 and y-intercept 2.

12. Has x-intercept 2 and y-intercept 4.

13. Passes through $(2, -1)$ and is parallel to the x-axis.

14. Passes through $(4, 3)$ and is parallel to $2x - 3y = 2$.

15. Is perpendicular to $y = 2 - x$ at the point $(1, 1)$.

16. Passes through $(-3, 5)$ and is perpendicular to $x - 2y = 4$.

17. Find an equation of the line that is tangent to the circle $x^2 + y^2 = 3$ at the point $(1, \sqrt{2})$. At what other point on the circle will the tangent line be parallel to this line?

18. Show that the line with x-intercept at $a \neq 0$ and y-intercept at $b \neq 0$ has the *intercept-form* equation

$$\frac{x}{a} + \frac{y}{b} = 1.$$

19. The function defined by $v(t) = -32t$ describes the velocity of a rock t seconds after it has been dropped from the top of a 784 ft high building. Sketch the graph of v, and determine the velocity of the rock when it hits the ground, 7 s after it has been dropped.

20. Determine the linear function that relates the temperature in degrees Celsius to the temperature in degrees Fahrenheit, and use this function to determine the Fahrenheit temperature corresponding to 30°C. (*Note:* At sea level, water freezes at 32°F (0°C) and boils at 212°F (100°C).)

21. The average weight W, in grams, of a fish in a particular pond depends on the total number n of fish in the pond according to the model

$$W(n) = 500 - 0.5n.$$

   a. Sketch the graph of the function W.

   b. Express the total fish weight production in grams as a function of the number of fish in the pond.

   c. What happens when $n \geq 1000$?

22. A new computer workstation costs $10,000. Its useful lifetime is 5 years, at which time it will be worth an estimated $2000. The company calculates its depreciation using the linear decline method that is an option in the tax laws.

   a. Find the linear equation that expresses the value V of the equipment as a function of time t, $0 \leq t \leq 5$.

   b. How much will the equipment be worth after 2.5 years?

   c. What is the average rate of change in the value of the equipment from 1 to 3 years?

1.8   QUADRATIC FUNCTIONS

In calculus you will see that the motion of a falling object is described using a quadratic function. This quadratic function can be used to determine information about the object's position, velocity, and acceleration.

A function f defined by a quadratic equation of the form

$$y = f(x) = ax^2 + bx + c, \quad \text{where} \quad a \neq 0,$$

is called a **quadratic function**, and its graph is a *parabola*. The most basic parabola is the graph of the *squaring function* $y = f(x) = x^2$, the quadratic equation with $a = 1, b = 0$, and $c = 0$. Its graph is shown in Figure 1.67.

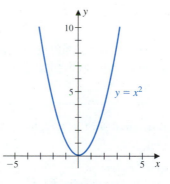

FIGURE 1.67

Understanding the notions of increasing and decreasing now will help you in calculus when you need to describe precisely how a curve increases and decreases.

For $x < 0$ the values of $f(x) = x^2$ decrease as x increases, and for $x > 0$ the values of $f(x)$ increase as x increases, so we say that f *decreases* for $x < 0$, and f *increases* for $x > 0$. Notice also that the function f has a minimum value of 0 at $x = 0$ but has no maximum value.

The graph of $f(x) = x^2$ can be used to determine the graphs of other quadratic functions by using a valuable shifting technique that can be applied to other functions as well. Graphing devices can plot a curve that represents the graph of a function, but care must be taken to be certain the graph is a true representative for the graph of the function. Knowing techniques for graphing functions will help to ensure that we can distinguish between a true representative of a graph and when the graphic information is incomplete.

EXAMPLE 1   Sketch the graph of the function given by $g(x) = (x - 1)^2$.

Solution   The correspondence described by $g(x) = (x - 1)^2$ is similar to that of $f(x) = x^2$ except that one unit is subtracted *before* the squaring operation is performed. Consequently,

$$g(1) = (1 - 1)^2 = 0^2 = f(0), \quad g(2) = 1^2 = f(1), \quad g(3) = f(2),$$

and so on. Because of this, the graph of $y = g(x)$ is the same as the graph of $y = x^2$ except that it is moved one unit to the right, as shown in Figure 1.68. The function g is decreasing when $x < 1$, is increasing when $x > 1$, and has a minimum of 0 at $x = 1$. ∎

This example shows how we can quickly sketch a good representation of the graph by using the graph of a simpler function. We will do this frequently throughout the book. The objective is one of the basic tenets of good problem solving: always try to reduce a complicated problem to a series of simpler or more familiar problems.

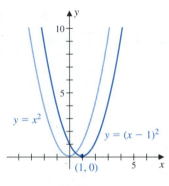

FIGURE 1.68

The horizontal shifting, or translation, technique described in Example 1 can be applied in general, as illustrated in Figure 1.69.

Horizontal Shifts of Graphs

Let $a > 0$.

(i) The graph of $y = f(x - a)$ is the graph of $y = f(x)$ shifted a units to the right.
(ii) The graph of $y = f(x + a)$ is the graph of $y = f(x)$ shifted a units to the left.

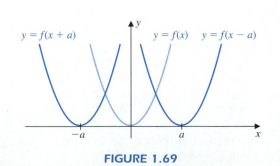

FIGURE 1.69

The Standard Form of a Quadratic Equation

Any quadratic term of the form

$$ax^2 + bx + c$$

can be written as the sum of a constant and a perfect square involving the variable x. Changing the quadratic into this *standard* form is called **completing the square**.

To complete the square, the coefficient of x^2 must be 1, so we first factor the a from the terms involving x to produce

$$ax^2 + bx + c = a\left(x^2 + \frac{b}{a}x\right) + c.$$

Now we add and subtract the term $(b/2a)^2$ inside the parentheses to obtain

$$ax^2 + bx + c = a\left(x^2 + \frac{b}{a}x + \left(\frac{b}{2a}\right)^2 - \left(\frac{b}{2a}\right)^2\right) + c$$

$$= a\left(x^2 + \frac{b}{a}x + \left(\frac{b}{2a}\right)^2\right) - a\left(\frac{b}{2a}\right)^2 + c.$$

Observing that the term inside the brackets is a perfect square and simplifying gives

$$ax^2 + bx + c = a\left(x + \frac{b}{2a}\right)^2 + \frac{4ac - b^2}{4a}.$$

The *vertex* of the parabola with equation $y = ax^2 + bx + c$ occurs at the point $(-b/2a, (4ac - b^2)/4a)$. This vertex gives the minimal y-value when $a > 0$ and the maximal y-value when $a < 0$. This result is illustrated in Figure 1.70.

A graphing device can be used to draw quickly the graphs of most functions you will see in calculus. Care must always be taken to ensure the graph is a good representation of the function. The graphing techniques introduced in this chapter will help you to visualize mentally the graph of a variety of functions, which in turn can help you to decide when it is appropriate to accept the graph given by a graphing device.

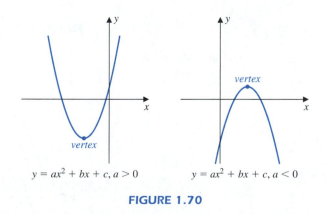

$$y = ax^2 + bx + c, a > 0 \qquad y = ax^2 + bx + c, a < 0$$

FIGURE 1.70

Completing the square on a quadratic function permits its graph to be determined by simply shifting the graph of a basic quadratic function of the form $f(x) = ax^2$. The following example illustrates the technique when the coefficient of the x^2 term is 1. Later in the section we consider the more general situation.

EXAMPLE 2   Sketch the graph of $h(x) = x^2 - 2x + 3$.

Solution   We first complete the square of this quadratic equation by adding and subtracting the term

$$\left(\frac{b}{2a}\right)^2 = \left(\frac{-2}{2 \cdot 1}\right)^2 = (-1)^2 = 1,$$

to obtain

$$h(x) = x^2 - 2x + 3 = (x^2 - 2x + 1) - 1 + 3 = (x - 1)^2 + 2.$$

When $h(x)$ is written in this form we see that it is the same as $g(x) + 2$, for $g(x) = (x-1)^2$ given in Example 1. The graph of $y = h(x) = g(x) + 2$ is the graph of $y = g(x)$ shifted 2 units upward. For a fixed x, the y-coordinate of the point on the graph of $y = h(x)$ is found by adding 2 units to the y-coordinate of the corresponding point on the graph of $y = g(x)$.

Consequently, the graph of $y = h(x) = x^2 - 2x + 3$, shown in Figure 1.71, is the graph of $y = x^2$ moved to the right 1 unit and upward 2 units. The function h is decreasing when $x < 1$, is increasing when $x > 1$, and has a minimum of 2 at $x = 1$. The vertex of the parabola is at $(1, 2)$. ∎

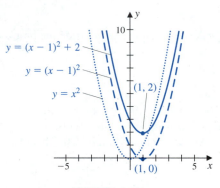

FIGURE 1.71

The general rule associated with the vertical shifting, or translation, technique described in Example 2 is illustrated in Figure 1.72.

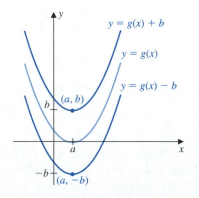

FIGURE 1.72

Vertical Shifts of Graphs

Let $b > 0$.

(i) The graph of $y = f(x) + b$ is the graph of $y = f(x)$ shifted b units upward.
(ii) The graph of $y = f(x) - b$ is the graph of $y = f(x)$ shifted b units downward.

This result combined with the horizontal shifting technique permits us to sketch quickly the graph of any quadratic function whose x^2 term has the coefficient 1, once we have completed the square.

Let us now consider the situation when the quadratic equation has an x^2-coefficient different from 1. This is done first for the most basic of these quadratics, those with $b = 0$ and $c = 0$. The graph of a quadratic function with equation $y = ax^2$, for an arbitrary constant $a \neq 0$, has a shape similar to the graph of $y = x^2$, but modified as follows. (See Figure 1.73.)

The Graph of the Quadratic Equation $y = ax^2$

(i) The graph opens upward when $a > 0$, and opens downward when $a < 0$.
(ii) The greater the magnitude of a, the narrower the opening of the graph.

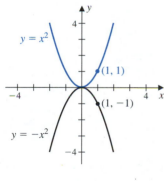

(a)

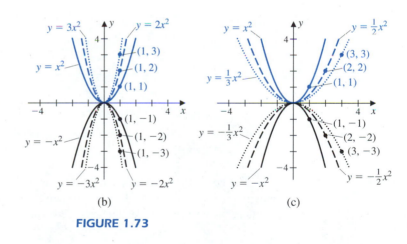

(b)   (c)

FIGURE 1.73

EXAMPLE 3 Sketch the graph of $f(x) = 2x^2 + 12x + 17$.

Solution We first complete the square on the quadratic. Then we use the horizontal and vertical shifting techniques to determine the graph.

Factor 2, the coefficient of x^2, from the first two terms of the expression, and then complete the square. This gives

$$f(x) = 2(x^2 + 6x) + 17 = 2(x^2 + 6x + 9) - 2 \cdot 9 + 17 = 2(x + 3)^2 - 1.$$

The graph of $y = 2x^2$ has the same shape as the graph of $y = x^2$, except that it is compressed by a factor of 2, as shown in Figure 1.73(b).

Applying a horizontal shift, the graph of $y = 2(x + 3)^2$ is the graph of $y = 2x^2$ shifted 3 units to the left. Then applying the vertical shifting technique, the graph of $y = 2(x + 3)^2 - 1$ is the graph of $y = 2(x + 3)^2$ shifted 1 unit downward.

Consequently, the graph of

$$y = 2x^2 + 12x + 17 = 2(x + 3)^2 - 1$$

has the same shape as the graph of $y = 2x^2$, but it is shifted 3 units to the left and 1 unit downward, as shown in Figure 1.74. The function f is decreasing when $x < -3$, is increasing when $x > -3$, and has a minimum of -1 at $x = -3$. The vertex of the parabola is at $(-3, -1)$.

To ensure that we have not made a translation error, we can check in the original equation that we indeed have

$$f(-3) = 2(-3)^2 + 12(-3) + 17 = 18 - 36 + 17 = -1.$$

We could still be in error, but it is unlikely.   ■

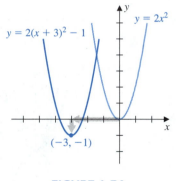

FIGURE 1.74

The next example shows the graph of a quadratic function with a negative coefficient of x^2. The analysis is the same as when the leading coefficient is positive, but care is required to ensure that the algebra is performed correctly.

EXAMPLE 4   Sketch the graph of $f(x) = -\dfrac{1}{3}x^2 + 2x + 3$.

Solution   First factor $-\frac{1}{3}$ from the x^2- and x-terms of the expression and then complete the square. This gives

$$f(x) = -\frac{1}{3}(x^2 - 6x) + 3$$

$$= -\frac{1}{3}(x^2 - 6x + 9 - 9) + 3$$

$$= -\frac{1}{3}(x^2 - 6x + 9) - \left(-\frac{1}{3}\right) \cdot 9 + 3$$

$$= -\frac{1}{3}(x - 3)^2 + 6.$$

We first sketch the graph of $y = \frac{1}{3}x^2$, which is shown with the graph of $y = x^2$ in Figure 1.75(a). Then we reflect the graph of $y = \frac{1}{3}x^2$ about the x-axis to produce the graph of $y = -\frac{1}{3}x^2$, which is also shown in Figure 1.75(a).

Notice that we have reduced the problem to a series of four easier problems. Never work a hard problem if you can avoid it. Reduce the problem to a few simpler problems. Not only will the work be easier, the reduction will often show you the heart of the problem.

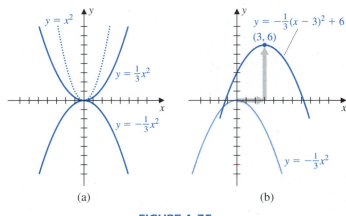

(a)   (b)

FIGURE 1.75

Applying the shifting techniques to $y = -\frac{1}{3}x^2$, we see that the graph of

$$y = -\frac{1}{3}x^2 + 2x + 3 = -\frac{1}{3}(x - 3)^2 + 6$$

has the same shape, except that it is shifted 3 units to the right and 6 units upward, as shown in Figure 1.75(b). The function f is increasing when $x < 3$, is decreasing when $x > 3$, and has a maximum of 6 at $x = 3$. The vertex of the parabola is at $(3, 6)$. Checking in the original equation gives

$$f(3) = -\frac{1}{3}(3)^2 + 2(3) + 3 = -3 + 6 + 3 = 6. \qquad \blacksquare$$

The graph in Figure 1.75(b) shows x-axis intercepts near -1 and 7. We can use the completed square of the quadratic to determine the exact values of these intercepts. Setting y to 0 and solving for x gives

$$0 = -\frac{1}{3}x^2 + 2x + 3 = -\frac{1}{3}(x - 3)^2 + 6.$$

So

$$\frac{1}{3}(x - 3)^2 = 6, \qquad \text{and} \qquad (x - 3)^2 = 18.$$

Taking the square root of both sides gives the two solutions

$$x - 3 = \pm\sqrt{18} = \pm3\sqrt{2},$$

which simplify to

$$x = 3 + 3\sqrt{2} \qquad \text{and} \qquad x = 3 - 3\sqrt{2}.$$

Since $\sqrt{2}$ is approximately 1.41 (written $\sqrt{2} \approx 1.41$), the solutions are

$$x = 3 - 3\sqrt{2} \approx 3 - 3(1.41) = -1.23 \qquad \text{and} \qquad x = 3 + 3\sqrt{2} \approx 7.23.$$

EXAMPLE 5   A company finds that on the average it can sell x units per day when the price is $p(x) = 100 - 0.05x$, for x between 250 and 800. The cost of operating the company, regardless of the number of items produced, is \$4000 per day, and the production cost per unit when x items are produced is $60 - 0.01x$. Determine the number of units that should be produced each day to maximize profit, and find this maximum profit.

Solution   The profit $P(x)$ is the revenue, $R(x)$, minus the cost, $C(x)$, where the revenue is the number of units sold times the price per unit. Since

$$R(x) = xp(x) = x(100 - 0.05x) = 100x - 0.05x^2$$

and

$$C(x) = 4000 + x(60 - 0.01x) = 4000 + 60x - 0.01x^2,$$

we have

$$P(x) = R(x) - C(x)$$
$$= (100x - 0.05x^2) - (4000 + 60x - 0.01x^2)$$
$$= -0.04x^2 + 40x - 4000.$$

Calculus will give you an alternative method for determining the vertex of this parabola, but completing the square of a quadratic is a tool you should master.

We have plotted $y = P(x)$ in Figure 1.76. It appears that the maximum profit is \$6000 per day, which is achieved when 500 items are produced and sold per day. To verify this algebraically, we find the vertex of the parabola. Completing the square gives

$$P(x) = -0.04(x^2 - 1000x) - 4000$$
$$= -0.04(x^2 - 1000x + (500)^2 - (500)^2) - 4000$$
$$= -0.04(x - 500)^2 + 0.04(500)^2 - 4000$$
$$= -0.04(x - 500)^2 + 6000.$$

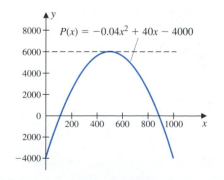

FIGURE 1.76

The vertex of the parabola is at $(500, 6000)$, so the maximum profit is \$6000 and it occurs when $(x - 500)^2 = 0$, that is, when 500 units are produced and sold.   ∎

EXAMPLE 6   Use a graphing device to approximate the point on the parabola $y = (x - 4)^2$ that is nearest the origin.

Solution   Figure 1.77(a) shows the parabola and the distance $d(x)$ that must be minimized. An arbitrary point on the parabola has coordinates $(x, (x - 4)^2)$, so, by the distance formula, the distance from $(0, 0)$ to $(x, (x - 4)^2)$ is

$$d(x) = \sqrt{x^2 + (x - 4)^4}.$$

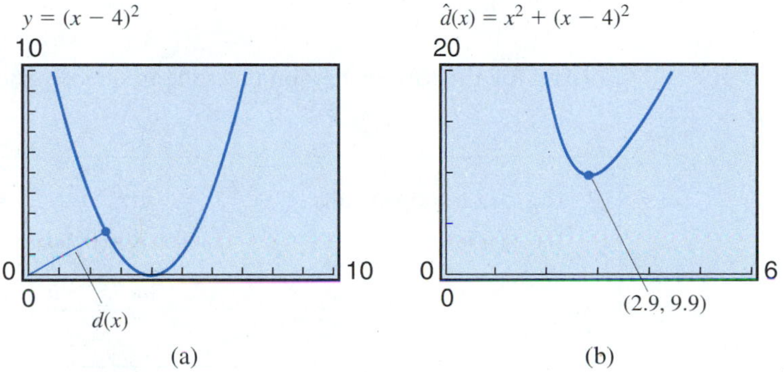

$y = (x - 4)^2$

$\hat{d}(x) = x^2 + (x - 4)^2$

(2.9, 9.9)

$d(x)$

(a)   (b)

FIGURE 1.77

This is a typical calculus problem. Although we do not have the power of calculus, a graphing calculator can be used to approximate the solution.

The value of x that minimizes $d(x)$ is the same as the value that minimizes its square $\hat{d}(x) = x^2 + (x - 4)^4$, and $\hat{d}$ is a simpler function than d. The plot of $y = \hat{d}(x)$ is shown in Figure 1.77(b) using a viewing rectangle $[0, 6] \times [0, 20]$. Clicking on the curve gives the minimum point on $y = \hat{d}(x)$ at approximately $(2.9, 9.9)$. The closest point to the origin occurs when $x \approx 2.9$, and $y \approx (2.9 - 4)^2 = 1.21$, and the minimal distance to the origin is

$$d(2.9) = \sqrt{(2.9)^2 + (2.9 - 4)^4} \approx 3.14.$$ ∎

The Quadratic Formula

We can determine the solutions of the general quadratic equation

$$ax^2 + bx + c = 0, \qquad \text{where } a \neq 0,$$

by completing the square, as we did to find the vertex of a parabola. Since

$$ax^2 + bx + c = a\left(x^2 + \frac{b}{a}x + \left(\frac{b}{2a}\right)^2 - \left(\frac{b}{2a}\right)^2\right) + c$$

$$= a\left(x^2 + \frac{b}{a}x + \left(\frac{b}{2a}\right)^2\right) - a\left(\frac{b}{2a}\right)^2 + c,$$

we have

$$0 = ax^2 + bx + c = a\left(x + \frac{b}{2a}\right)^2 + \frac{4ac - b^2}{4a}.$$

This result implies that

$$a\left(x + \frac{b}{2a}\right)^2 = -\frac{4ac - b^2}{4a} = \frac{b^2 - 4ac}{4a}.$$

So

$$\left(x + \frac{b}{2a}\right)^2 = \frac{b^2 - 4ac}{4a^2}, \qquad \text{and} \qquad x + \frac{b}{2a} = \pm\frac{\sqrt{b^2 - 4ac}}{2a}.$$

Solving for x in this expression gives the frequently used *quadratic formula*.

The Quadratic Formula

The solutions to $ax^2 + bx + c = 0$, when $a \neq 0$, are

$$x = -\frac{b}{2a} \pm \frac{\sqrt{b^2 - 4ac}}{2a} = \frac{-b \pm \sqrt{b^2 - 4ac}}{2a}.$$

The *discriminant* of the quadratic equation, $b^2 - 4ac$, determines the number of real solutions of the equation.

1. When $b^2 - 4ac > 0$, there are two distinct solutions.
2. When $b^2 - 4ac = 0$, there is only one solution.
3. When $b^2 - 4ac < 0$, there are no real solutions.

The linear and quadratic functions we have seen in this and the previous section are special cases of polynomial functions. A *polynomial of degree n* is an expression of the form

$$a_n x^n + a_{n-1} x^{n-1} + \cdots + a_1 x + a_0,$$

where n is a nonnegative integer and the $a_0, a_1, \ldots, a_n$ are all constants, with $a_n \neq 0$.

Linear functions that are not constant are first-degree polynomial functions since they can be expressed in the form

$$y = a_1 x + a_0,$$

where a_1 is the slope of the line and a_0 is the y-intercept. Quadratic functions are second-degree polynomial functions because they have the form

$$y = a_2 x^2 + a_1 x + a_0.$$

More general polynomial functions are considered in Chapter 2.

EXERCISE SET 1.8

In Exercises 1–18, sketch the graph of the quadratic equation.

1. $y = x^2 + 1$ **2.** $y = x^2 - 2$

3. $y = (x - 3)^2$ **4.** $y = (x + 3)^2$

5. $y = x^2 - 4x + 4$ **6.** $y = x^2 - 4x + 3$

7. $y = x^2 + 4x + 5$ **8.** $y = -(x - 1)^2$

9. $y = -x^2 + 6x - 5$ **10.** $y = 2x^2 + 1$

11. $y = 3x^2 + 6x$ **12.** $y = 2x^2 + 8x + 6$

13. $y = 2x^2 - 8x + 10$ **14.** $y = -4x^2 - 4x + 3$

15. $y = \frac{1}{2}x^2$ **16.** $y = \frac{1}{2}x^2 + 2$

17. $y = \frac{1}{2}x^2 - 2x + 2$ **18.** $y = \frac{1}{2}x^2 - 2x + 1$

19. Use the graph of the function shown in the accompanying figure to sketch the graph of each of the following.

  a. $y = f(x - 1)$ **b.** $y = f(x - 1) + 2$

  c. $y = f(x + 2)$ **d.** $y = f(x + 2) + 1$

  e. $y = f(x + 2) - 1$ **f.** $y = f(x - 1) - 2$

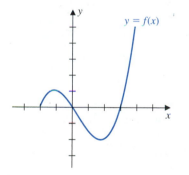

20. Let $f(x) = x^2 - 3x + 2$ and $g(x) = x^2 + 3x + 2$. Sketch the graphs of $y = f(x)$, $y = g(x)$, $y = |f(x)|$, and $y = f(|x|)$.

21. Sketch the graph of

$$f(x) = \begin{cases} x + 2, & \text{if } x < -1, \\ x^2, & \text{if } x \geq -1. \end{cases}$$

22. Find the domain of $\sqrt{f(x)}$ if

  a. $f(x) = x^2 - 3$.   **b.** $f(x) = x^2 - \frac{1}{2}x$.

23. What can you say about the values of a, b, and c in $f(x) = ax^2 + bx + c$ if

  a. $(1, 1)$ is on the graph of f?

  b. The y-intercept is 6?

  c. $(1, 1)$ is the vertex?

  d. Conditions (a), (b), and (c) are all satisfied?

24. Let $f(x) = 100x^2$ and $g(x) = 0.1x^3$. For what values of x is $f(x) \geq g(x)$?

25. The function defined by $s(t) = 576 + 144t - 16t^2$ describes the height, in feet, of a rock t seconds after it has been thrown upward at 144 ft/s from the top of a 50-story building, as shown in the figure.

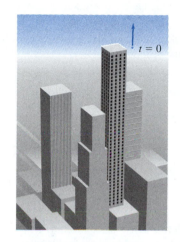

  a. Sketch the graph of s.

  b. How long does it take the rock to hit the ground?

  c. Determine the time it takes the rock to reach its maximum height.

  d. Determine physically reasonable definitions for the domain and range of s.

26. The function defined by $v(t) = 144 - 32t$ describes the velocity in feet per second of the rock t seconds after it is dropped from the building described in Exercise 25.

  a. Sketch the graph of v.

  b. What is the velocity of the rock when it strikes the ground?

  c. Determine physically reasonable definitions for the domain and range of v.

27. For a small manufacturing firm the unit cost $C(x)$ in dollars of producing x units per day is given by

$$C(x) = x^2 - 120x + 4000$$

How many items should be produced per day to minimize the unit cost, and what is the minimum unit cost?

28. A company that produces computer terminals analyzes production and finds that they should make a profit $P(x)$, in dollars, for selling x terminals per month, where

$$P(x) = -0.1x^2 + 160x - 20{,}000.$$

How many terminals should be sold per month to produce the maximum profit? What is the maximum profit?

29. National health-care spending, in billions of dollars, has taken the shape of a parabola over the last few decades, increasing at an alarming rate. (See the table.)

  a. Using only the 1965, 1980, and 1990 data, fit a parabola of the form $y = a(x - 1965)^2 + b(x - 1965) + c$ to the data.

  b. What does the parabola predict the population will be in the year 2000?

Year	Dollars (in billions)
1965	30
1970	80
1975	120
1980	250
1985	400
1990	690

30. The profit function of a manufacturer when x units of a commodity are produced and sold is given by

$$P(x) = 200x - x^2$$

  a. Sketch a graph of the profit function.

  b. How many units should be produced to yield the maximum profit?

  c. Compute the difference quotient $\dfrac{P(x + h) - P(x)}{h}$. What value does the difference quotient approach as h approaches 0? This quantity, in economics, is called the *marginal profit* when x units are sold. It approximates the change in the profit when one additional unit is produced.

31. A driver traveling on an interstate highway sees traffic coming to a stop ahead and applies the brakes. The car's speed, in feet per second, is given by $v(t) = -4t^2 - 4t + 80$.

  a. Make a sketch of the velocity curve for $t \geq 0$.

  b. How long does it take for the car to come to rest?

  c. Make a table of speeds starting from time $t = 0$, the instant when the driver applies the brakes, including every half s until the car comes to rest.

  d. What is the slowest the car is traveling in the first half s after the brakes are applied? What is the fastest the car is traveling in the first half s?

  e. Repeat part (d) for the second half s and the third half s.

  f. What is the minimum distance the car can travel in the first half s? In the second half s? What is the maximum distance the car can travel in the first half s? In the second half s?

  g. Use the table from part (c) to estimate a lower and an upper bound on the distance it takes for the car to come to rest.

32. Using a graphing device to approximate to one decimal place the point on the line $y = 2x - 3$ that is closest to the point $(4, 1)$.

1.9   OTHER COMMON FUNCTIONS

In the previous two sections we considered linear and quadratic functions, functions that have frequent applications in calculus. In this section we will consider some other commonly used functions.

Throughout this section we will use the graphical shifting techniques introduced in Section 1.8. These techniques are reviewed in Figure 1.78. If you have a graphing device, by all means use it, but be sure that you master these shifting techniques. They will be very valuable when you quickly need to see the general shape of a function.

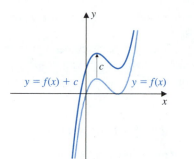

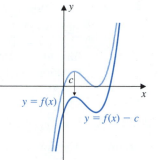

Vertical translation
c units upward

Vertical translation
c units downward

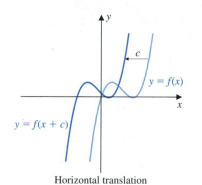

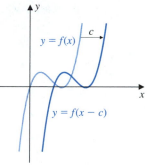

Horizontal translation
c units to the left

Horizontal translation
c units to the right

FIGURE 1.78

The Absolute Value Function

The absolute value function is the first example that we have seen that is not "smooth" at every number in its domain. At $(0, 0)$ the absolute value function turns to form a right angle. This nonsmooth behavior is of particular interest in calculus when the concept of the derivative is discussed.

We saw in Section 1.2 that the absolute value of a real number x,

$$|x| = \begin{cases} x, & \text{if } x \geq 0, \\ -x, & \text{if } x < 0, \end{cases}$$

describes the distance from x to the origin. The graph of the **absolute value function**, $f(x) = |x|$, is shown in Figure 1.79.

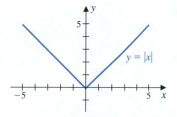

FIGURE 1.79

When $x \geq 0$ the graph coincides with the line $y = x$. When $x < 0$ the graph coincides with the line $y = -x$.

Since $|x| = |-x|$ for any real number x, the absolute value function is an even function, and its graph is symmetric with respect to the y-axis. The absolute value function is decreasing when $x < 0$, increasing when $x > 0$, and has a minimum of 0 at $x = 0$.

EXAMPLE 1   Use the graph of $y = |x|$ to sketch the graph of $g(x) = |x - 1| - 2$.

Solution   The horizontal shifting procedure implies that the graph of $y = |x - 1|$ is obtained from shifting the graph of $y = |x|$ one unit to the right. Then applying vertical shifting, the graph of $y = g(x) = |x - 1| - 2$ is the graph of $y = |x - 1|$ shifted two units downward. Putting these results together gives the graph shown in Figure 1.80. Notice that g is decreasing when $x < 1$, is increasing when $x > 1$, and has a minimum of -2 at $x = 1$. ∎

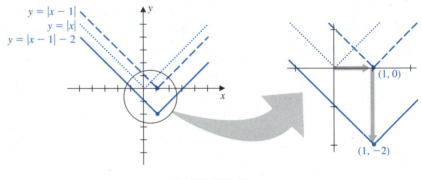

FIGURE 1.80

EXAMPLE 2   Sketch the graph of $h(x) = -|2x - 1| + 2$

Solution   The graph of $y = 2|x|$ is similar to the graph of $y = |x|$ but is narrower, just as the graph of $y = 2x^2$ is narrower than the graph of $y = x^2$. Since $|2x - 1| = 2|x - \frac{1}{2}|$, the graph of $y = |2x - 1|$ is the graph of $y = 2|x|$ shifted $\frac{1}{2}$ unit to the right. These graphs are shown in Figure 1.81(a).

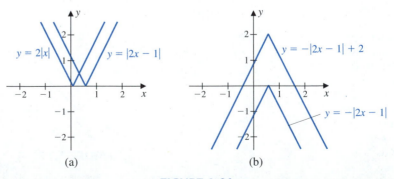

(a)   (b)

FIGURE 1.81

The graph of $y = -|2x - 1|$ is the reflection of the graph of $y = |2x - 1|$ about the x-axis. As a consequence, it opens downward instead of upward. Finally, the graph of $y = h(x) = -|2x - 1| + 2$ has the same shape as the graph of $y = -|2x - 1|$, but it is shifted upward 2 units, as shown in Figure 1.81(b). So h is increasing when $x < \frac{1}{2}$, is decreasing when $x > \frac{1}{2}$, and has a maximum of 2 at $x = \frac{1}{2}$. ∎

EXAMPLE 3   Sketch the graph of

$$f(x) = |x^2 - 4x + 3|.$$

Solution   Figure 1.82(a) shows the graph of the parabola $y = x^2 - 4x + 3$. Since the absolute value of a number gives the positive magnitude of the number, the portion of the graph in Figure 1.82(a) below the x-axis is reflected above the x-axis. The portion already above the x-axis remains the same. The graph of $f(x) = |x^2 - 4x + 3|$ is shown in Figure 1.82(b). ∎

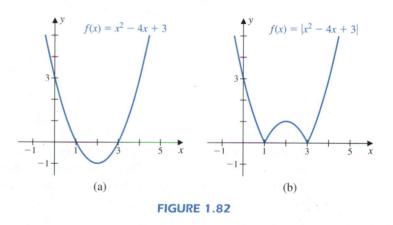

(a)   (b)

FIGURE 1.82

The Square Root Function

The graph of the **square root function**, $f(x) = \sqrt{x}$, is shown in Figure 1.83. The domain of the square root function is the set of all nonnegative real numbers, $[0, \infty)$. The range is also $[0, \infty)$, since $\sqrt{}$ is defined as the principal, or nonnegative, square root. The square root function is increasing on its entire domain and has a minimum of zero at $x = 0$.

Notice that the graph of $f(x) = \sqrt{x}$ appears to approach the y-axis vertically as x approaches zero. This feature of the graph will be important in calculus.

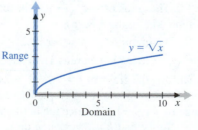

FIGURE 1.83

EXAMPLE 4   Sketch the graph of $g(x) = \sqrt{x-2} - 1$ and determine the domain and range of g.

Solution   The graph of $y = \sqrt{x-2}$ is the graph of $y = \sqrt{x}$ shifted 2 units to the right. As a consequence, the graph of $y = \sqrt{x-2} - 1$ is the graph of $y = \sqrt{x}$ shifted 2 units to the right and 1 unit downward, as shown in Figure 1.84. It is always increasing and has a minimum value of -1 at $x = 2$.

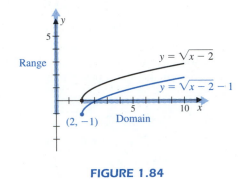

FIGURE 1.84

The domain and range of g can easily be determined by using the horizontal and vertical line tests on the graph. As illustrated in the figure, the domain of g is $[2, \infty)$ since the numbers in this interval of the x-axis are those through which a vertical line intersects the graph of the function. In a similar manner the range of g is $[-1, \infty)$ since the numbers in this interval of the y-axis are those through which a horizontal line intersects the graph. ∎

EXAMPLE 5   Sketch the graph of $h(x) = \sqrt{-x-2} - 1$ and determine the domain and range of h.

Solution   First notice that we can rewrite $h(x)$ as

$$h(x) = \sqrt{-x-2} - 1 = \sqrt{-(x+2)} - 1.$$

So the basic function whose graph we need is $y = \sqrt{-x}$, which is defined only when $x \le 0$. The graph of $y = \sqrt{-x}$ is the reflection about the y-axis of the graph of $y = \sqrt{x}$, as shown in Figure 1.85(a).

The graph of $y = \sqrt{-x-2} = \sqrt{-(x+2)}$ is obtained by shifting the graph of $y = \sqrt{-x}$ to the left 2 units. Then the final graph of $y = h(x) = \sqrt{-x-2} - 1$ is found by shifting the graph of $y = \sqrt{-x-2}$ downward by 1 unit, as shown in Figure 1.85(b). Vertical lines intersect the graph whenever $x \le -2$, so the domain of f is $(-\infty, -2]$. The range of f is $[-1, \infty)$, since the horizontal lines that intersect the graph have $y \ge -1$.

Notice that h is decreasing on its entire domain and has a minimum value of -1 at $x = -2$. ∎

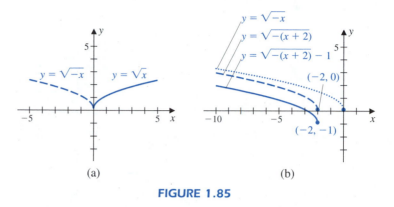

FIGURE 1.85

The Greatest Integer Function

The **greatest integer function**, traditionally denoted $f(x) = [x]$, is defined for a given real number x to be the largest integer that is less than or equal to x. Thus

$$[x] = m, \qquad \text{where } m \text{ is the integer satisfying } m \leq x < m + 1.$$

In essence, the greatest integer function "rounds down" a real number to the next lowest integer. For example,

$$[1.2] = 1, \qquad [-1.51] = -2, \qquad [2] = 2, \qquad \text{and} \qquad [0.33] = 0.$$

The greatest integer function has wide application in computer science, where it is known as the *floor* function and is denoted $f(x) = \lfloor x \rfloor$. Complementary to the floor function is the *ceiling* function, denoted $f(x) = \lceil x \rceil$, which represents the least integer that is greater than or equal to x. This influence of computer science on the notation is so strong that we will subsequently use the notation $\lfloor \ \rfloor$ in place of $[\ ]$ when discussing the greatest integer function since this is the notation that will most likely be used in the future.

The greatest integer function assumes the constant integer value m on the interval $[m, m + 1)$, so its graph consists of horizontal line segments beginning at each of the integers. For example, if x is in the interval $[-2, -1)$, then $-2 \leq x < -1$ and $f(x) = \lfloor x \rfloor = -2$; if $0 \leq x < 1$, then $f(x) = \lfloor x \rfloor = 0$; and so on, as shown in Table 1.5 and Figure 1.86.

TABLE 1.5

When	$-2 \leq x < -1$	$-1 \leq x < 0$	$0 \leq x < 1$	$1 \leq x < 2$	$2 \leq x < 3$
$\lfloor x \rfloor$ is	-2	-1	0	1	2

The domain of the greatest integer function is the set of real numbers, and its range is the set of integers. Because the function is constant between integers, it is not increasing, nor is it decreasing. In addition, it has no maximum or minimum.

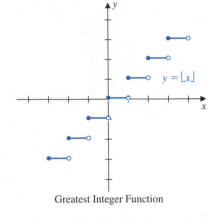

Greatest Integer Function

FIGURE 1.86

The greatest integer function is the first function we have seen whose graph "jumps", or has gaps. In calculus you will find that this behavior implies the function is not *continuous* at each of the integers.

EXAMPLE 6   Sketch the graph of $g(x) = \lfloor x + 3 \rfloor - 2$.

Solution   First the graph of $y = \lfloor x \rfloor$ is shifted three units to the left to produce the graph of $y = \lfloor x + 3 \rfloor$, as shown in Figure 1.87(a). Then the graph of $y = \lfloor x + 3 \rfloor$ is shifted downward two units to give the final graph of $y = g(x) = \lfloor x + 3 \rfloor - 2$, as shown in Figure 1.87(b). Notice that the graph is the same as the graph obtained by shifting the greatest integer function one unit to the left. You should consider what this says algebraically. ∎

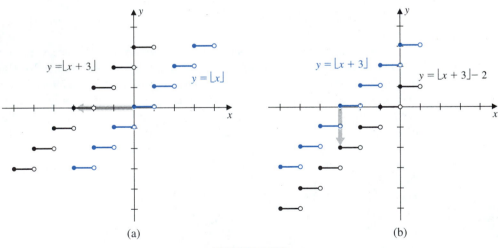

(a)   (b)

FIGURE 1.87

Since the greatest integer function jumps at each integer, it can be used to describe and approximate certain real-world situations, as illustrated in the following example.

EXAMPLE 7   Calculators commonly permit you to enter the number of decimal places you would like to display and round all displayed results when appropriate. As a first step in examining this procedure, we construct a function that displays no decimal places, that is, it rounds a number to the nearest integer.

Solution To round to the nearest integer, we need a function such as

$$f(x) = \begin{cases} \lfloor x \rfloor, & \text{if } \lfloor x \rfloor \leq x < \lfloor x \rfloor + 0.5, \\ \lfloor x \rfloor + 1, & \text{if } \lfloor x \rfloor + 0.5 \leq x < \lfloor x \rfloor + 1. \end{cases}$$

For example, $f(2.2)$ will be $\lfloor 2.2 \rfloor = 2$, since

$$2 \leq 2.2 < 2.5,$$

but $f(2.534)$ will be $\lfloor 2.534 \rfloor + 1 = 2 + 1 = 3$ since

$$2.5 \leq 2.534 < 3.$$

A simpler representation can be given as

$$g(x) = \lfloor x + 0.5 \rfloor.$$

When

$$\lfloor x \rfloor \leq x < \lfloor x \rfloor + 0.5, \qquad \text{we have} \qquad \lfloor x \rfloor + 0.5 \leq x + 0.5 < \lfloor x \rfloor + 1,$$

and

$$\lfloor x + 0.5 \rfloor = \lfloor x \rfloor.$$

On the other hand, when

$$\lfloor x \rfloor + 0.5 \leq x < \lfloor x \rfloor + 1, \qquad \text{we have} \qquad \lfloor x \rfloor + 1 \leq x + 0.5 < \lfloor x \rfloor + 1.5$$

and

$$\lfloor x + 0.5 \rfloor = \lfloor x \rfloor + 1.$$

So g and f give the same values, but $g(x)$ gives a simpler representation that rounds numbers to the nearest integer. ■

EXERCISE SET 1.9

In Exercises 1–6, use the graph of $y = |x|$ to sketch the graph of each of the functions.

1. $f(x) = |x - 3|$ **2.** $f(x) = |x + 1| - 1$

3. $f(x) = |x - 2| + 2$ **4.** $f(x) = |x + 2| - 2$

5. $f(x) = |2x|$ **6.** $f(x) = |2x - 5|$

In Exercises 7–10, (a) use the graph of $y = \sqrt{x}$ to sketch the graph of each of the functions, and (b) find the domain and range of each of the functions.

7. $g(x) = \sqrt{x} + 2$ **8.** $g(x) = \sqrt{x + 2} - 2$

9. $g(x) = -\sqrt{x + 2}$ **10.** $g(x) = 2 - \sqrt{x + 2}$

11. Use the graph of $f(x) = x^3$, shown in the following figure, to sketch the graph of g defined by each equation.

     a. $g(x) = x^3 + 1$ **b.** $g(x) = (x + 1)^3$

c. $g(x) = x^3 - 1$ **d.** $g(x) = (x - 1)^3$

e. $g(x) = 2x^3$ **f.** $g(x) = -2x^3$

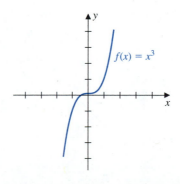

12. Use the graph of the *cube root* function $f(x) = \sqrt[3]{x}$, shown in the following figure, to sketch the graph of g defined by each equation.

   a. $g(x) = \sqrt[3]{x} + 1$ **b.** $g(x) = \sqrt[3]{x + 1}$

   c. $g(x) = \sqrt[3]{x} - 1$ **d.** $g(x) = \sqrt[3]{x - 1}$

   e. $g(x) = 2\sqrt[3]{x}$ **f.** $g(x) = -2\sqrt[3]{x}$

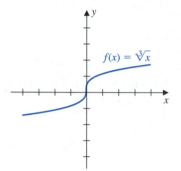

$$f(x) = \sqrt[3]{x}$$

In Exercises 13–16, sketch $y = f(x)$ and $y = |f(x)|$.

13. $f(x) = 2x - 3$ **14.** $f(x) = -x + 2$

15. $f(x) = -x^2 + 2$ **16.** $f(x) = x^2 - 5x + 6$

In Exercises 17–20, use the graph of $y = \lfloor x \rfloor$ to sketch the graph of each function.

17. $f(x) = \lfloor x - 2 \rfloor$ **18.** $f(x) = \lfloor x \rfloor - 2$

19. $f(x) = \lfloor x + 1 \rfloor - 2$ **20.** $f(x) = \lfloor x - 2 \rfloor + 1$

In Exercises 21 and 22, the graph of $y = f(x)$ is given. From it obtain the graph of each of the following:

   a. $y = f(x) + 1$ **b.** $y = f(x - 2)$

   c. $y = 2f(x)$ **d.** $y = f(2x)$

   e. $y = -f(x)$ **f.** $y = f(-x)$

   g. $y = |f(x)|$ **h.** $y = f(|x|)$

21.

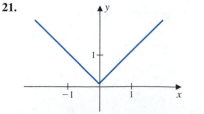

22.

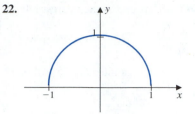

23. A function f satisfies $f(1) = -2$, the graph has slope 1 for $x < 1$, and the graph has slope -1 for $x > 1$.

   a. Sketch the graph of $y = f(x)$.

   b. Sketch $y = f(x - 2)$.

   c. Sketch $y = f(x + 1) + 2$.

   d. Sketch $y = f(3x)$.

24. Let $f(x) = x + 1$.

   a. Sketch the graph of $y = f(x)$ on the interval $[0, 4]$.

   b. Find an expression $d(x)$ for the distance from the origin to the point $(x, f(x))$. Use a graphing device to sketch the graph of $y = d(x)$ on the interval $[0, 4]$.

   c. Let $A(t)$ denote the area of the region bounded by the x-axis, the y-axis, the curve $y = f(x)$, and the vertical line $x = t$. Find an expression for $A(t)$ and sketch the graph of $y = A(t)$ on the interval $[0, 4]$.

25. Sketch a possible graph for a function that satisfies all the following conditions.

   a. $f(0) = 1$.

   b. $f(x)$ is increasing on the interval $(0, 2)$.

   c. $f(x)$ is decreasing on the interval $(2, 5)$.

   d. $f(x)$ is increasing on the interval $(5, \infty)$.

   e. $f(x)$ approaches 3 as x becomes large and approaches -2 as x approaches $-\infty$.

26. In 1996 the cost of mailing a first class letter in the United States was 32 cents up to the first ounce and an additional 23 cents for each ounce beyond the first ounce, up to a total of 12 oz. We can use the greatest integer function to express this rate as

$$P(w) = \begin{cases} 0.32 + 0.23\lfloor w \rfloor, & \text{if } w \text{ is not a positive integer,} \\ 0.32 + 0.23(w - 1), & \text{if } w \text{ is a positive integer.} \end{cases}$$

   a. Sketch the graph of $y = P(w)$ and determine the domain and range of P.

   b. Show that, alternatively, we can express the postal rate in the form

$$P(w) = 0.09 - 0.23\lfloor -w \rfloor.$$

27. The Ohio Turnpike is 241 mi in length and has service plazas located 75 and 160 mi from Eastgate, the entrance to the turnpike at the Pennsylvania line. Express the distance of a car from the nearest service plaza as a function of the car's distance from Eastgate, and sketch the graph of this function.

28. The ancient Babylonians discovered a rule for approximating the square root of a positive number p. Begin by making a reasonable estimate a of $\sqrt{p}$ and construct the improved estimate

$$b = \frac{1}{2}\left(a + \frac{p}{a}\right).$$

Continue the process with b as the new value of a. Use the method to approximate $\sqrt{2}$ to three decimal places. Repeat the process to approximate $\sqrt{13}$ to this accuracy.

1.10 ARITHMETIC COMBINATIONS OF FUNCTIONS

Functions are routinely combined in calculus to form more complex functions. It is important to understand and feel comfortable with the notation we introduce in this section since you will see it on a daily basis in your calculus course.

We have seen that the profit for a business is the difference between the revenue, (the amount of money taken in) and the cost of operating the business. Suppose that x represents the number of items sold by a business, and $R(x)$ and $C(x)$ describe, respectively, the revenue and cost associated with selling x items. Then the profit function P for this business is described by the difference

$$P(x) = R(x) - C(x).$$

This is an example of a function defined by combining two functions using the arithmetic operation of subtraction. In this section we will show how combinations of this type are used to construct complex functions by using just a few elementary functions and operations.

The rules involving the arithmetic combinations of functions are quite natural, the only minor difficulty involves the domain of the quotient of two functions.

Arithmetic Combinations of Functions

If f and g are functions, then the functions $f + g$, $f - g$, $f \cdot g$, and f/g are defined by

$$(f + g)(x) = f(x) + g(x), \qquad (f - g)(x) = f(x) - g(x),$$

$$(f \cdot g)(x) = f(x) \cdot g(x), \qquad (f/g)(x) = \left(\frac{f}{g}\right)(x) = \frac{f(x)}{g(x)}.$$

Both $f(x)$ and $g(x)$ must be defined for any of these operations to be defined at x. This implies that the domains of $f + g$, $f - g$, and $f \cdot g$ consists of those real numbers that are common to both the domain of f and the domain of g. The domain of the quotient f/g consists of those real numbers x that are in both the domain of f and the domain of g and that also satisfy $g(x) \neq 0$.

EXAMPLE 1 Let f and g be defined by Table 1.6. Show that f and g are functions, and then find $f + g$, $f - g$, $f \cdot g$, and f/g.

TABLE 1.6

x	-3	-2	-1	0	1	2	3
$f(x)$	-1	1	2	-3	0	7	5
$g(x)$	0	-2	4	8	-1	3	5

Solution Both f and g are functions, since for each value of x there are unique image values. The new functions are shown in Table 1.7. Notice that since $g(-3) = 0$, the value $x = -3$ is not in the domain of the function f/g. ∎

TABLE 1.7

x	$f(x)$	$g(x)$	$(f + g)(x)$	$(f - g)(x)$	$(f \cdot g)(x)$	$(f/g)(x)$
-3	-1	0	-1	-1	0	Not defined
-2	1	-2	-1	3	-2	$-\frac{1}{2}$
-1	2	4	6	-2	8	$\frac{1}{2}$
0	-3	8	5	-11	-24	$-\frac{3}{8}$
1	0	-1	-1	1	0	0
2	7	3	10	4	21	$\frac{7}{3}$
3	5	5	10	0	25	1

EXAMPLE 2 Let $f(x) = 1/(x^2 - 1)$ and $g(x) = x/\sqrt{x + 2}$. Find $f + g$, $f - g$, $f \cdot g$, and f/g, and give their domains.

Solution By the definitions we have

$$(f + g)(x) = f(x) + g(x) = \frac{1}{x^2 - 1} + \frac{x}{\sqrt{x + 2}},$$

$$(f - g)(x) = f(x) - g(x) = \frac{1}{x^2 - 1} - \frac{x}{\sqrt{x + 2}},$$

$$(f \cdot g)(x) = f(x) \cdot g(x) = \frac{1}{x^2 - 1} \cdot \frac{x}{\sqrt{x + 2}},$$

$$\left(\frac{f}{g}\right)(x) = \frac{f(x)}{g(x)} = \frac{\dfrac{1}{x^2 - 1}}{\dfrac{x}{\sqrt{x + 2}}} = \frac{\sqrt{x + 2}}{x(x^2 - 1)}.$$

The domain of f is the set of all real numbers except 1 and -1, which would produce a 0 in the denominator. The domain of g is the interval $(-2, \infty)$ since the square root in the denominator requires that $x \geq -2$, but -2 would produce a 0 in the denominator. (See Figure 1.88(a).) Thus, the common domain of $f + g$, $f - g$, and $f \cdot g$ is the set of real numbers x that satisfy $x > -2$, excluding $x = \pm 1$; in interval notation we have $(-2, -1) \cup (-1, 1) \cup (1, \infty)$.

For x to be in the domain of the quotient f/g, it must also be true that $g(x) = x/\sqrt{x+2} \neq 0$, that is, $x \neq 0$. Therefore, the domain of f/g is the set of real numbers x that satisfy $x > -2$, excluding $x = \pm 1$ and $x = 0$. (See Figure 1.88(b).)

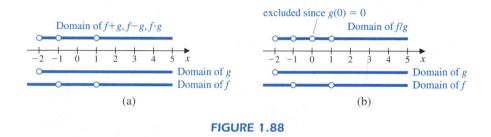

FIGURE 1.88

The simplified form on the right side of the equation describing the function f/g in this example,

$$\left(\frac{f}{g}\right)(x) = \frac{f(x)}{g(x)} = \frac{\dfrac{1}{x^2-1}}{\dfrac{x}{\sqrt{x+2}}} = \frac{\sqrt{x+2}}{x(x^2-1)},$$

seems to imply that $x = -2$ is in the domain of f/g. This is not the case, however. Since $x = -2$ is not in the domain of g, it is not a candidate for the domain of any arithmetic combination involving the function g. When computing the quotient or product of two functions, always be careful to exclude from the domain those values that are not in the common domain of the original functions, particularly when the final form has been simplified. ∎

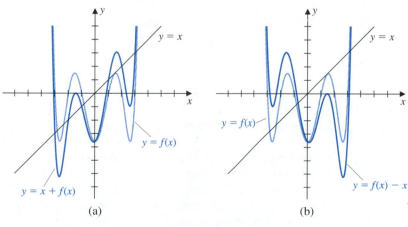

FIGURE 1.89

The graph of $y = (f + g)(x)$ is the graphical addition of the curves $y = f(x)$ and $y = g(x)$. For each x, the y-coordinate of the point on the graph of the sum is the sum of the corresponding y-coordinates of the points on $y = f(x)$ and $y = g(x)$.

Similarly the graph of $y = (f - g)(x)$ is the graphical subtraction of the graphs of $y = f(x)$ and $y = g(x)$. Figure 1.89(a) shows the graph of the sum of the arbitrary function $y = f(x)$ and $y = x$, and Figure 1.89(b) shows the difference. The x-intercepts for the graph of $y = (f + g)(x)$ are those x for which $f(x) = -g(x)$, and the x-intercepts for the graph of $y = f(x) - g(x)$ are those x for which $f(x) = g(x)$.

EXAMPLE 3   Sketch the graphs of $f(x) = \lfloor x \rfloor + |x - 1|$ and $g(x) = \lfloor x \rfloor - |x - 1|$.

Solution   The graph of $y = |x - 1|$ is the graph of $y = |x|$ shifted one unit to the right, as shown in Figure 1.90(a). Summing the vertical components of this graph with that of the greatest integer function, $y = \lfloor x \rfloor$, produces the graph of $f(x) = \lfloor x \rfloor + |x - 1|$ shown in Figure 1.90(b). By contrast, the graph of $g(x) = \lfloor x \rfloor - |x - 1|$ is shown in Figure 1.90(c).

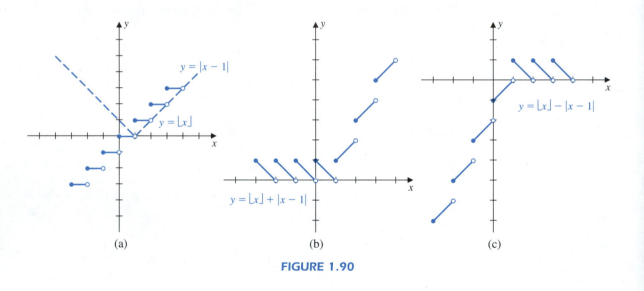

(a)   (b)   (c)

FIGURE 1.90

It is generally more difficult to determine the graphs for the product $f \cdot g$ and quotient f/g then for the sum and difference, but some valuable information about these graphs can often be obtained. The graph of $y = f(x)g(x)$ crosses the x-axis when $f(x) = 0$ or $g(x) = 0$, and the graph of $y = f(x)/g(x)$ crosses the x-axis when both $f(x) = 0$ and $g(x) \neq 0$. The function $y = f(x)/g(x)$ is undefined when $g(x) = 0$ and can go to ∞ or $-\infty$. This situation is illustrated for $g(x) = x$ in Figure 1.91.

When one of the functions making up the product $f \cdot g$ is a constant, say $f(x) = c$, the graph of the product is more easily analyzed. We encountered this problem in Section 1.8, where we obtained the graph of $y = ax^2$ from the graph of $y = x^2$. Each y-coordinate of a point on the graph of $y = x^2$ is multiplied by a to obtain the graph of $y = ax^2$. This resulted in the graph opening more narrow, when $a > 1$, and widening when $0 < a < 1$. Some modifications of those guidelines are needed to reflect the general nature of the graph of $y = c \cdot g(x)$.

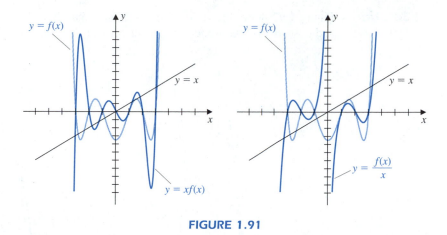

FIGURE 1.91

Vertical Compression and Elongation of Graphs

(i) For any constant $c > 0$, the basic shape of the graph of $y = c \cdot g(x)$ is the same as the graph of $y = g(x)$ but with a change in the vertical scale. The domains of cg and g are the same and the graphs have the same x-intercepts.

(ii) For $c > 1$, the graph of $y = c \cdot g(x)$ is a vertical *elongation* of the graph of $y = g(x)$.

(iii) For $0 < c < 1$, the graph of $y = c \cdot g(x)$ is a vertical *compression* of the graph of $y = g(x)$.

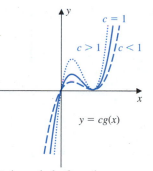

$c > 1$: vertical enlongation
$0 < c < 1$: vertical compression

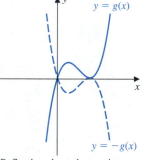

Reflection about the x-axis

(iv) The graph of $y = -g(x)$ is the reflection about the x-axis of the graph of $y = g(x)$. So, for $c < -1$, the graph of $y = c \cdot g(x)$ is a vertical elongation and reflection of the graph of $y = g(x)$. For $-1 < c < 0$, the graph of $y = c \cdot g(x)$ is a vertical compression and reflection of the graph of $y = g(x)$.

EXAMPLE 4   Compare the graphs of $g(x)$, $2g(x)$, $-2g(x)$, $\frac{1}{2}g(x)$, and $-\frac{1}{2}g(x)$ if

$$g(x) = (x - 1)(x - 2)(x + 1) = x^3 - 2x^2 - x + 2.$$

Solution Figure 1.92 shows the various graphs. Notice that they all have the same x-intercepts, and observe that the y-intercepts are directly proportional to the scaling factor c. ■

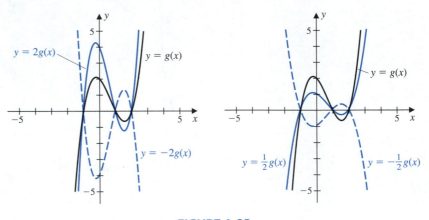

FIGURE 1.92

There is also a special case of the quotient of two functions that is both useful and quite elementary. The *reciprocal* of a function g, defined by $(1/g)(x) = 1/g(x)$, is a quotient that occurs when the function in the numerator is the constant 1 for all values of x. The following relationships between a function and its reciprocal can be used to produce a reasonably accurate graph of the reciprocal function. (See Figure 1.93.)

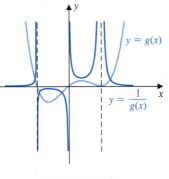

FIGURE 1.93

The Graph of the Reciprocal of a Function

Let $f(x) = \frac{1}{g(x)}$ be the reciprocal of $g(x)$.

(i) $f(x)$ is undefined when $g(x) = 0$.
(ii) $f(x)$ and $g(x)$ have the same value when $g(x) = 1$ or $g(x) = -1$.
(iii) $f(x)$ and $g(x)$ have the same sign.
(iv) The magnitude of $f(x)$ is large when the magnitude of $g(x)$ is small.
(v)   The magnitude of $f(x)$ is small when the magnitude of $g(x)$ is large.

EXAMPLE 5   Use the graph of $g(x) = x - 2$ to determine the graph of $f(x) = \frac{1}{x-2}$.

Solution   The graph of g is the straight line through $(2, 0)$ with slope 1. Consequently, the function f is undefined at $x = 2$, and its values increase in magnitude as x approaches 2. In addition, $f(x) = 1/(x - 2)$ approaches zero as x increases in magnitude. Noting that $f(x)$ and $g(x)$ always have the same sign and that both graphs pass through the points $(1, -1)$ and $(3, 1)$ leads to the graph of f shown in Figure 1.94.   ∎

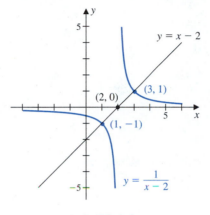

FIGURE 1.94

Some other examples of the reciprocal graphing technique are shown in Figure 1.95.

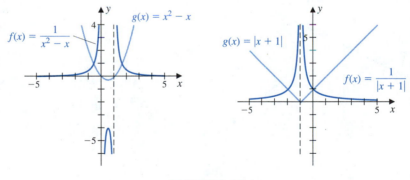

FIGURE 1.95

EXERCISE SET 1.10

In Exercises 1–8, find $f + g$, $f - g$, $f \cdot g$, and f/g, and give the domain of each new function.

1. $f(x) = x$; $g(x) = 3x - 2$

2. $f(x) = x^2$; $g(x) = x + 1$

3. $f(x) = \frac{1}{x}$; $g(x) = \sqrt{x - 1}$

4. $f(x) = \frac{1}{x - 1}$; $g(x) = \frac{1}{x + 1}$

5. $f(x) = \sqrt{x + 2}$; $g(x) = \sqrt{2 - x}$

6. $f(x) = \dfrac{1}{x};\ g(x) = \dfrac{x}{x-2}$

7. $f(x) = \begin{cases} -1, & \text{if } x < 0, \\ 1, & \text{if } x \geq 0, \end{cases}$

$g(x) = \begin{cases} 1, & \text{if } x < 0, \\ 0, & \text{if } x \geq 0 \end{cases}$

8. $f(x) = \begin{cases} 0, & \text{if } x < 0, \\ x, & \text{if } x \geq 0, \end{cases}$

$g(x) = \begin{cases} -x, & \text{if } x < 0, \\ 0, & \text{if } x \geq 0 \end{cases}$

In Exercises 9 and 10, the graphs of functions f and g are given. From the graphs sketch the graphs of $f + g$ and $f - g$.

9.

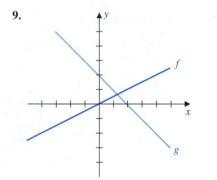

10.

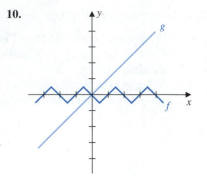

11. Functions f and g are defined by $f(x) = \dfrac{x^2-4}{x-2}$ and $g(x) = x + 2$. How do the graphs of f and g differ?

12. Use the results about the graph of the reciprocal of a function to sketch the graph of $f(x) = 1/g(x)$ in each case.

    a. $g(x) = 2x - 1$ **b.** $g(x) = |x|$

    c. $g(x) = x^2 - 1$ **d.** $g(x) = x^2 - 4x + 3$

13. Sketch the graphs of the given functions in the order given, and observe the difference in the graph that each successive complication introduces.

    a. $f_1(x) = x - 1$ **b.** $f_2(x) = (x-1)^2$

    c. $f_3(x) = x^2 - 2x$ **d.** $f_4(x) = |x^2 - 2x|$

    e. $f_5(x) = \dfrac{1}{|x^2 - 2x|}$ **f.** $f_6(x) = \dfrac{-1}{|x^2 - 2x|}$

14. Sketch the graphs of the given functions in the order given, and observe the difference in the graph that each successive complication introduces.

    a. $f_1(x) = x - 2$ **b.** $f_2(x) = (x-2)^2$

    c. $f_3(x) = x^2 - 4x + 2$ **d.** $f_4(x) = |x^2 - 4x + 2|$

    e. $f_5(x) = \dfrac{1}{|x^2 - 4x + 2|}$ **f.** $f_6(x) = \dfrac{2}{|x^2 - 4x + 2|}$

1.11   COMPOSITION OF FUNCTIONS

In Section 1.10 we saw how functions can be combined using the arithmetic operations of addition, subtraction, multiplication and division. *Composition* is another method of combining functions that takes the output of one function and applies it as input to another function. For example, suppose that $f(x) = x^2 + 1$ and $g(x) = x - 2$. If the input to g is x, then the output is $g(x) = x - 2$. If we now use $x - 2$ as the input to f, we have $f(g(x)) = f(x - 2) = (x - 2)^2 + 1$. This final result is called the *composition* of f with g and is denoted $f \circ g$. It results in a building process that is used to make successively more complicated functions from a sequence of elementary functions.

Composition of Functions

The **composition** of the function f with the function g, denoted $f \circ g$, is defined by

$$(f \circ g)(x) = f(g(x)).$$

The domain of $f \circ g$ consists of all x in the domain of g for which $g(x)$ is in the domain of f.

The composition of a pair of functions f and g is illustrated in Figure 1.96.

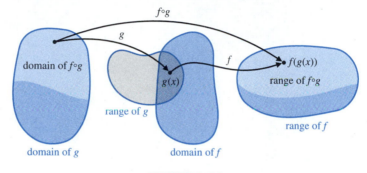

FIGURE 1.96

EXAMPLE 1   Find $(f \circ g)(2)$ and $(g \circ f)(2)$ for $f(x) = x - 2$ and $g(x) = x^2 - 1$.

Solution   As shown in Figure 1.97,

$$(f \circ g)(2) = f(g(2)) = f(3) = 1$$

and

$$(g \circ f)(2) = g(f(2)) = g(0) = -1.$$

Notice that

$$(f \circ g)(2) \neq (g \circ f)(2). \qquad \blacksquare$$

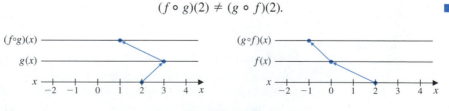

FIGURE 1.97

It is generally true about compositions that even when $f \circ g$ and $g \circ f$ have a common value x in their domains,

$$(f \circ g)(x) \neq (g \circ f)(x).$$

EXAMPLE 2   Let $f(x) = \sqrt{x - 1}$ and $g(x) = 1/x^2$.

a. Find $(f \circ g)(x)$ and the domain of $f \circ g$.
b. Find $(g \circ f)(x)$ and the domain of $g \circ f$.

Solution   a. We have

$$(f \circ g)(x) = f(g(x)) = f\left(\frac{1}{x^2}\right) = \sqrt{\frac{1}{x^2} - 1} = \sqrt{\frac{1 - x^2}{x^2}}$$

The domain of g is the set of all nonzero real numbers, and the domain of f is the interval $[1, \infty)$. To find the domain of $f \circ g$, we need to determine the numbers x in the domain of g for which $g(x)$ is in the domain of f.

First we must have $x \neq 0$, so that it will be in the domain of g. Then we must have $g(x) = 1/x^2 \geq 1$, that is, $x^2 \leq 1$. As a consequence, the domain of $f \circ g$ is $[-1, 0) \cup (0, 1]$, and the graph of $f \circ g$ is as shown in Figure 1.98(a).

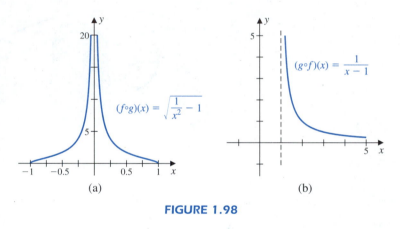

FIGURE 1.98

b. First we find that

$$(g \circ f)(x) = g(f(x)) = g\left(\sqrt{x - 1}\right) = \frac{1}{\left(\sqrt{x - 1}\right)^2} = \frac{1}{x - 1}.$$

The domain of f is the interval $[1, \infty)$ since the term under the square root cannot be negative. We can apply g to any of the resulting values of $f(x)$ except when $x = 1$ since $f(1) = 0$ is not in the domain of g. Hence, the domain of $g \circ f$ is the interval $(1, \infty)$, and the graph of $g \circ f$ is shown in Figure 1.98(b).   ■

Notice that the simplified form of $(g \circ f)(x) = 1/(x - 1)$ in the previous example cannot be used to determine the domain of $g \circ f$ since $\left(\sqrt{x - 1}\right)^2 = x - 1$ only when both of these terms are defined. The left side of the equation is not defined when $x < 1$.

In Section 1.10 we found that the graph of $y = c \cdot f(x)$ is a vertical compression or elongation of the graph of $y = f(x)$, together with a reflection about the x-axis in the case the constant is negative. The composition of a function f with $g(x) = cx$ is $(f \circ g)(x) = f(g(x)) = f(cx)$. Multiplying the *argument* of a function by a constant produces a similar effect to multiplying the function by a constant, but the scaling of the graph occurs in the horizontal rather than the vertical direction.

Horizontal Compression and Elongation of Graphs

(i) For any constant $c > 0$, the basic shape of the graph of $y = f(cx)$ is the same as the graph of $y = f(x)$ but with a change in the horizontal scale.

(ii) For $c > 1$, the graph of $y = f(cx)$ is a horizontal *compression* of the graph of $y = f(x)$.

(iii) For $0 < c < 1$, the graph of $y = f(cx)$ is a horizontal *elongation* of the graph of $y = f(x)$.

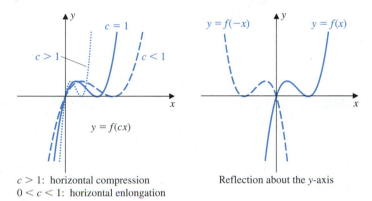

$c > 1$: horizontal compression
$0 < c < 1$: horizontal enlongation

Reflection about the y-axis

(iv) The graph of $y = f(-x)$ is the reflection about the y-axis of the graph of $y = f(x)$. So, for $c < -1$, the graph of $y = f(cx)$ is a horizontal compression and reflection of the graph of $y = f(x)$. For $-1 < c < 0$, the graph of $y = f(cx)$ is a horizontal elongation and reflection of the graph of $y = f(x)$.

EXAMPLE 3   Compare the graphs of $f(x)$, $f(2x)$, $f(-2x)$, $f(\frac{1}{2}x)$, and $f(-\frac{1}{2}x)$ if

$$f(x) = (x - 1)(x - 2)(x + 1) = x^3 - 2x^2 - x + 2.$$

Solution   Figure 1.99 shows the various graphs. Notice that the y-intercepts of the graphs agree. Since the horizontal scale has changed, the graphs do not have the same x-intercepts, they are inversely proportional to the scaling factor.   ∎

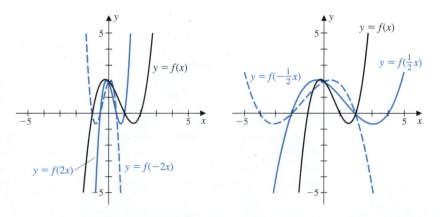

FIGURE 1.99

Decomposing a complicated function into a sequence of more familiar functions is a common technique in calculus. The following examples illustrate how this can be done.

EXAMPLE 4   a. Write the function

$$f(x) = \sqrt{x^2 + 1}$$

as the composition of two functions.

b. Write the function

$$k(x) = \frac{2}{\sqrt{x^2 + 1}}$$

as the composition of three functions.

Solution   a. One way to decompose the function f is to observe the following chain of transformations:

$$x \rightarrow x^2 + 1 \rightarrow \sqrt{x^2 + 1} = f(x).$$

This leads us to define a pair of functions g and h as

Being able to unravel a complex composition of functions is something you will need to do in calculus. It is essential to the understanding of the Chain Rule for derivatives and the Substitution Rule for integration.

$$g(x) = x^2 + 1 \qquad \text{and} \qquad h(x) = \sqrt{x}$$

so that

$$f(x) = h(g(x)) = \sqrt{g(x)} = \sqrt{x^2 + 1}.$$

This representation in terms of a composition is not unique. We could alternatively define, for example,

$$\hat{g}(x) = x^2 \qquad \text{and} \qquad \hat{h}(x) = \sqrt{x + 1},$$

so that

$$f(x) = \hat{h}(\hat{g}(x)) = \sqrt{\hat{g}(x) + 1} = \sqrt{x^2 + 1}.$$

This second decomposition does not, however, seem to be quite as natural as the first.

b. Most of the work has already been done in part (a). Consider the chain

$$x \rightarrow x^2 + 1 \rightarrow \sqrt{x^2 + 1} \rightarrow \frac{2}{\sqrt{x^2 + 1}} = k(x).$$

If we define $i(x) = 2/x$ and use the same functions, $g(x) = x^2 + 1$ and $h(x) = \sqrt{x}$ as in the first decomposition in part (a), we have

$$k(x) = i(f(x)) = i(h(g(x))) = \frac{2}{h(g(x))} = \frac{2}{\sqrt{x^2 + 1}}. \qquad \blacksquare$$

Composition of functions will cause no difficulty if you remember that function notation such as $f(x) = x^2 - 1$ means simply that for any element x in the domain, the value $f(x)$ is obtained by squaring x and subtracting 1. In particular, keep in mind that there is nothing special about the variable used in the description of f. Notation of the type

$$f(\square) = (\square)^2 - 1$$

is helpful when considering the composition since it better emphasizes that

$$f\left(\boxed{g(x)}\right) = \left(\boxed{g(x)}\right)^2 - 1.$$

When decomposing a complicated function into a sequence of more familiar functions, it is often useful to first illustrate a composition by a chain, as we did in Example 4. The next example shows how decomposition of a complicated function can be used to help sketch its graph.

EXAMPLE 5 Write $f(x) = \dfrac{1}{\left|x^2 - 4x + 1\right|}$ as a composition of functions, and then sketch the graph of f.

Solution First let us reemphasize that there is no single method of writing a function as a composition. There are, however, ways that are more natural than others, and we will help you discover these.

Since $f(x)$ includes a quadratic term in the denominator and we are now comfortable with graphing quadratics, let us first complete the square to write the quadratic in standard form. This gives

$$x^2 - 4x + 1 = x^2 - 4x + 4 - 3 = (x - 2)^2 - 3,$$

which allows us to write

$$f(x) = \frac{1}{\left|(x - 2)^2 - 3\right|}.$$

To construct $f(x)$ in a systematic manner involving only common operations, we use the "chain"

$$x \rightarrow (x - 2)^2 \rightarrow (x - 2)^2 - 3 \rightarrow |(x - 2)^2 - 3| \rightarrow \frac{1}{\left|(x - 2)^2 - 3\right|}.$$

The sequence of functions that produces each of these operations is

$$g_1(x) = (x - 2)^2, \qquad g_2(x) = x - 3, \qquad g_3(x) = |x|, \qquad \text{and} \qquad g_4(x) = \frac{1}{x}.$$

Then

$$g_4(g_3(g_2(g_1(x)))) = g_4\left(g_3\left(g_2\left((x - 2)^2\right)\right)\right)$$
$$= g_4\left(g_3\left((x - 2)^2 - 3\right)\right)$$
$$= g_4\left(\left|(x - 2)^2 - 3\right|\right)$$
$$= \frac{1}{\left|(x - 2)^2 - 3\right|} = f(x),$$

and $f(x)$ is written as a composition of familiar functions.

To use the decomposition of $f(x)$ to sketch its graph, we sketch in sequence the graphs of

$$y = g_1(x) = (x - 2)^2,$$
$$y = g_2(g_1(x)) = (x - 2)^2 - 3, \qquad \text{and}$$
$$y = g_3\left(g_2(g_1(x))\right) = \left|(x - 2)^2 - 3\right|.$$

These are shown in Figure 1.100.

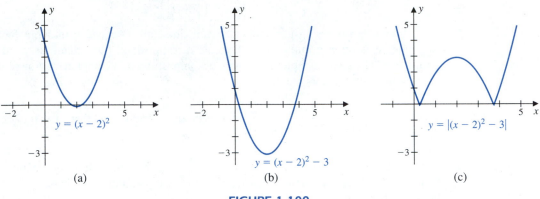

(a)   (b)   (c)

FIGURE 1.100

Since $f(x)$ is the reciprocal of the last function in the sequence,

$$f(x) = \frac{1}{g_3(g_2(g_1(x)))} = \frac{1}{|(x-2)^2 - 3|},$$

we can use the reciprocal graphing technique applied to the graph in Figure 1.100(c) to produce the graph of $y = f(x)$, as shown in Figure 1.101. ∎

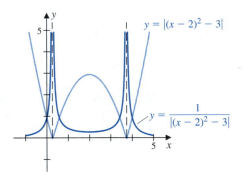

FIGURE 1.101

This example is the first, and most difficult, step in a standard calculus problem. In calculus you will want to know how the area of the circle changes as the radius changes relative to time.

The final example in this section illustrates the use of the composition of functions to describe a problem involving variation in time. These applications form an important part of any calculus course.

EXAMPLE 6   Oil is leaking from a tanker into a lake. Suppose the shape of the oil spill is approximately circular, and at any time t minutes after the leak has begun, the radius of the circle is $r(t) = \sqrt{t} + \sqrt[3]{t}$. Find the area of the spill at any time t after the leak has begun.

Solution   The area A of a circle of radius r is $A(r) = \pi r^2$. The area of the spill depends on the radius, and the radius depends on time. Thus A is the function of t given by the composition

$$(A \circ r)(t) = A(r(t)) = A(\sqrt{t} + \sqrt[3]{t}) = \pi(\sqrt{t} + \sqrt[3]{t})^2.$$ ∎

EXERCISE SET 1.11

In Exercises 1–6, let $f(x) = 2x - 3$ and $g(x) = x^2 + 2$ and evaluate each of the following.

1. $(f \circ g)(2)$

2. $(g \circ f)(-1)$

3. $f(g(-3))$

4. $g(f(5))$

5. $(f \circ f)(-2)$

6. $g(g(-1/2))$

In Exercises 7–10, find $f \circ g$, $g \circ f$, $f \circ f$, and $g \circ g$, and give the domain of each composition.

7. $f(x) = 2x + 1$; $g(x) = 3x - 1$

8. $f(x) = x^2 + 1$; $g(x) = x - 1$

9. $f(x) = \dfrac{1}{x}$; $g(x) = x^2 + 2x$

10. $f(x) = \sqrt{x - 1}$; $g(x) = x^2 - 3$

In Exercises 11–14, find functions f and g such that $h = f \circ g$.

11. $h(x) = (2 - 3x^2)^4$

12. $h(x) = \sqrt{x - 2}$

13. $h(x) = \dfrac{1}{x + 2}$

14. $h(x) = |x^2 + x + 1|$

15. Use the graphs of f and g in the figure to evaluate each expression.

a. $(f \circ g)(1)$

b. $(g \circ f)(-1)$

c. $(g \circ f)(0)$

d. $(f \circ g)(-2)$

x	$f(x)$	$g(x)$	$(f \circ g)(x)$	$(g \circ f)(x)$
-3	2	2		
-2	-1	-1		
-1	3	3		
0	0	0		
1				
2				
3				

17. We have seen that, in general, $f \circ g$ and $g \circ f$ are different. Find examples of f and g when these two compositions are the same.

18. Show that the composition of two odd functions is an odd function.

19. Show that the composition of an odd and an even function, in either order, is an even function.

20. Find a linear function f such that $(f \circ f)(x) = 4x + 3$.

21. Let f and g be linear functions with $f(x) = ax + b$ and $g(x) = cx + d$. When does

a. $(f \circ g)(x) = (g \circ f)(x)$?

b. $(f \circ g)(x) = f(x)$?

c. $(f \circ g)(x) = g(x)$?

22. What type of function is (a) the composition of two linear functions? (b) The composition of a linear and a quadratic function? (c) The composition of two quadratic functions?

23. A metal sphere is heated so that t seconds after the heat has been applied, the radius $r(t)$ is given by $r(t) = 3 + 0.01t$ cm. Express the volume of the sphere as a function of t.

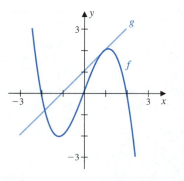

16. In the following table f is even and g is odd. Complete the table.

24. Sand is poured onto a conical pile whose radius and height are always equal, although both increase with time. The height of the pile t sec after pouring begins is given by $h(t) = 10 + 0.25t$ ft. Express the volume of the pile as a function of t.

25. A spherical balloon is inflated so that its radius at the end of t seconds is $r(t) = 3\sqrt{t} + 5$ cm, $0 \le t \le 4$. Express the volume and the surface area as a function of time. What are the units of these quantities?

26. In Exercise 25 of Exercise Set 1.8, the height of a rock above the ground t sec after it has been thrown upward from the top of a building was given as $s(t) = 576 + 144t - 16t^2$. The domain of the function s is the closed interval whose left endpoint is zero and whose right endpoint is the time it takes the rock to reach the ground. Define a function $\bar{s}$ that describes the height of the rock above the ground, assuming that the rock is

thrown at $t = 2$ instead of $t = 0$. What are the domain and range of $\bar{s}$?

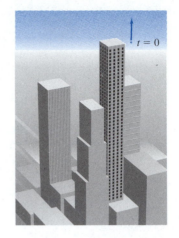

27. Newton's law of gravitation states that the attraction between an object of mass m_1 and an object of mass m_2 is directly proportional to the product of the masses m_1 and m_2 of the objects and inversely proportional to the square of the distance r between the centers of mass of the objects. Write a functional relationship expressing this force in terms of the distance r, assuming that the masses remain constant. What restrictions must be put on the domain of this function if it is to describe the physical situation? Sketch the graph of the function.

1.12   INVERSE FUNCTIONS

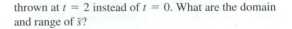

The study of function–inverse function pairs is an important topic in calculus since some of the most important functions cannot be approached directly. Learning the notation and concepts now will let you concentrate on the calculus concepts when you see them later.

The word *inverse* brings to mind a reversal of some process or operation. In the case of functions, the reversal involves the interchange of the domain with the range and a corresponding reversal of the operation describing the function.

Consider, as an example, the linear function that converts Celsius temperature to Fahrenheit temperature. The relationship between the scales can be determined by two facts:

1. Water freezes at $0°C$ and at $32°F$;
2. Water boils at $100°C$ and at $212°F$.

So the graph of the function that converts Celsius temperature into Fahrenheit temperature is a line that passes through the points $(0, 32)$ and $(100, 212)$.

The linear function converting Celsius, C, into Fahrenheit, $F(C)$, must consequently have slope

$$\frac{F(100) - F(0)}{100 - 0} = \frac{212 - 32}{100} = \frac{9}{5},$$

and, since $F(0) = 32$, its equation is

$$F \equiv F(C) = \frac{9}{5}C + 32.$$

It is equally useful to know the *inverse* relationship, the one that converts Fahrenheit temperature to Celsius. This is found by solving for C in terms of F in the Celsius-to-Fahrenheit equation,

$$F - 32 = \frac{9}{5}C,$$

so

$$C \equiv C(F) = \frac{5}{9}(F - 32).$$

Graphs of these equations are shown in Figure 1.102.

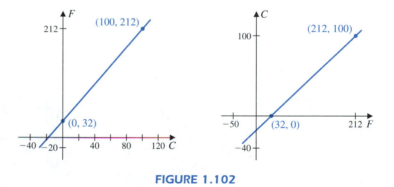

FIGURE 1.102

It is easily verified that if we start with a temperature given in Celsius, convert to Fahrenheit, and then convert back to Celsius, we arrive back at the starting temperature. A similar result occurs for a Fahrenheit temperature converted to Celsius and then back to Fahrenheit. So

$$F \equiv F(C) = \frac{9}{5}C + 32 \quad \text{if and only if} \quad C \equiv C(F) = \frac{5}{9}(F - 32).$$

This is the essence of the inverse function relationship: being able to find a function that reverses the process and brings you back unambiguously to your starting point.

Not all functions are reversible. The function $f(x) = x^2$ cannot be reversed since we have no way of knowing, for example, whether the number 4 in the range of f originated at $x = 2$ or at $x = -2$. The same quandary occurs at every number in the range of f except at 0, as shown in Figure 1.103.

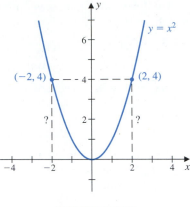

FIGURE 1.103

Before we can study inverse functions, then, we need to consider precisely which functions we can invert. These functions are called *one-to-one*. (See Figure 1.104.)

One-to-One Functions

A function f is **one-to-one** if for all x_1 and x_2 in its domain,

$$f(x_1) = f(x_2) \qquad \text{implies that} \qquad x_1 = x_2.$$

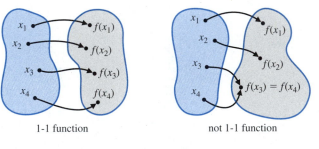

1-1 function   not 1-1 function

FIGURE 1.104

The stipulation for a function being one-to-one is often alternatively described by specifying that

$$f(x_1) \neq f(x_2), \qquad \text{whenever} \qquad x_1 \neq x_2,$$

which is logically equivalent to the statement in the definition. The terminology one-to-one refers to the fact that we want these functions to have the property that:

Each *one* of the range elements corresponds *to* precisely *one* domain element.

The Celsius-to-Fahrenheit conversion function is one-to-one since different Celsius temperatures always produce different Fahrenheit temperatures. The squaring function is not one-to-one since, for example, both $2^2 = 4$ and $(-2)^2 = 4$.

If we have the graph of a function, it is easy to determine if it is one-to-one.

Horizontal Line Test for One-to-One Functions

A function is one-to-one precisely when every horizontal line intersects its graph at most once.

Figure 1.105 illustrates that the graphs of the one-to-one functions $f(x) = x^3$ and $f(x) = 1/x$ satisfy the Horizontal Line Test. But the graphs of the functions $f(x) = x^2$ and $f(x) = |x|$ fail the Horizontal Line Test, and these functions are not one-to-one.

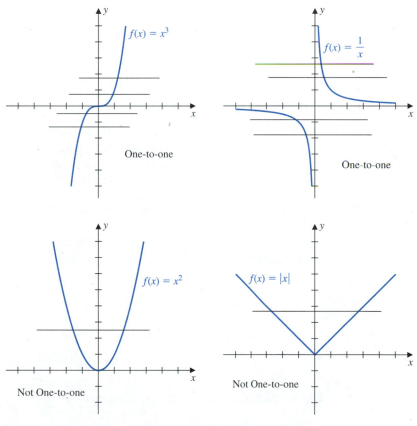

FIGURE 1.105

Although the Horizontal Line Test is a valuable tool when the graph of a function is known, it is generally necessary to use algebra to verify that a function is one-to-one. On the other hand, to verify that a function f is not one-to-one, we need only find a pair of numbers $x_1 \neq x_2$ with $f(x_1) = f(x_2)$. The next example illustrates a typical situation.

EXAMPLE 1   Show that

a. $f(x) = -|2x - 1| + 2$ is not one-to-one.
b. $f(x) = \sqrt{x - 2} - 1$ is one-to-one.

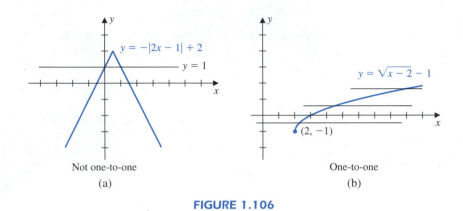

FIGURE 1.106

Solution   The graphs of these functions were considered in Section 1.9 and are reproduced in Figure 1.106.

a. The graph of $f(x) = -|2x - 1| + 2$ in Figure 1.106(a) shows quite clearly that the function is not one-to-one and gives us a good indication of how we should choose numbers to demonstrate it. For example, it appears that the horizontal line $y = 1$ crosses the graph at $(0, 1)$ and at $(1, 1)$. To verify this, note that

$$f(0) = -|0 - 1| + 2 = 1 \quad \text{and} \quad f(1) = -|2 - 1| + 2 = 1.$$

This is all that is necessary to show that the function is not one-to-one.

b. The graph in Figure 1.106(b) seems to imply that the function $f(x) = \sqrt{x - 2} - 1$ is one-to-one. But perhaps we have the graph slightly incorrect, and it dips down at some point instead of always going upward. To show that the function is truly one-to-one, we apply the definition.

Suppose that we have $f(x_1) = f(x_2)$. Then

$$\sqrt{x_1 - 2} - 1 = \sqrt{x_2 - 2} - 1, \quad \text{so} \quad \sqrt{x_1 - 2} = \sqrt{x_2 - 2}.$$

Squaring both sides of this last equation gives

$$x_1 - 2 = x_2 - 2, \quad \text{and we must have} \quad x_1 = x_2.$$

This shows conclusively that the function is one-to-one.   ∎

When a function is one-to-one, we can define its inverse function. (See Figure 1.107.)

Inverse Functions

Suppose that f is a one-to-one function with domain $\mathbb{X}$ and range $\mathbb{Y}$. The *inverse function* for the function f is the function f^{-1} with domain $\mathbb{Y}$ and range $\mathbb{X}$ defined for all values of $x \in \mathbb{X}$ by

$$f^{-1}(f(x)) = x.$$

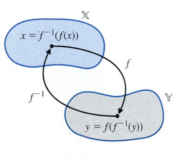

FIGURE 1.107

The inverse function of f in the definition just given is denoted f^{-1}, and we will use this notation in the remainder of the section. Many of the inverse functions that are particularly important, however, have special notation, which we will introduce when we study these functions.

There are some important properties that follow directly from the definition of an inverse function and the fact that the original function must be one-to-one for this inverse function to exist.

Properties of Inverse Functions

Suppose that f is a one-to-one function with domain $\mathbb{X}$ and range $\mathbb{Y}$. Then

(i) The inverse function f^{-1} is unique.
(ii) The domain of f^{-1} is $\mathbb{Y}$, the range of f.
(iii) The range of f^{-1} is $\mathbb{X}$, the domain of f.
(iv) For all $x \in \mathbb{X}$ and $y \in \mathbb{Y}$,

$$f(x) = y \qquad \text{if and only if} \qquad f^{-1}(y) = x.$$

(v) For all $x \in \mathbb{X}$ and $y \in \mathbb{Y}$,

$$f^{-1}(f(x)) = x \qquad \text{and} \qquad f(f^{-1}(y)) = y.$$

EXAMPLE 2   Show that the function defined by $f(x) = 2x - 3$ with the domain restricted to the set $[0, 4]$ is one-to-one and determine its inverse.

Solution   We have restricted the domain of the function in this example to better distinguish the domains and ranges of the function and its inverse. The graph of this function is a straight-line segment whose slope is positive, so the range of f is the interval

$[f(0), f(4)] = [-3, 5]$. The function f is one-to-one since

$$f(x_1) = 2x_1 - 3 = 2x_2 - 3 = f(x_2) \qquad \text{if and only if} \qquad 2x_1 = 2x_2,$$

which holds precisely when $x_1 = x_2$. This can also be seen from the graph shown in Figure 1.108.

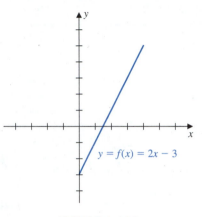

FIGURE 1.108

The domain of the inverse function, f^{-1}, is $[-3, 5]$, the range of f. The range of f^{-1} is $[0, 4]$, the domain of f.

To determine the correspondence relation for f^{-1}, we let $y = f(x) = 2x - 3$ and solve for the variable y in terms of x. Since

$$y = 2x - 3, \qquad \text{we have} \qquad f^{-1}(y) = x = \frac{1}{2}(y + 3).$$

We can verify that this relationship is correct by noting that indeed

$$f^{-1}(f(x)) = f^{-1}(2x - 3) = \frac{1}{2}((2x - 3) + 3) = x.$$

Although the relationship for f^{-1} is expressed here as x in terms of y, there is no significance concerning the defining variable for a function, it is the relationship that is of importance. The particular defining variable is simply a matter of convenience.

It is customary to reexpress the relationship for f^{-1} using the variable x to represent the values in the domain of f^{-1}. In this way we can give the graph of f^{-1} in the same framework as the graph of f, that is, with its domain along the x-axis. Hence, we have

$$f^{-1}(x) = \frac{1}{2}(x + 3).$$

The graph of f^{-1} is the straight-line segment with slope $1/2$ shown with the graph of f in Figure 1.109. ∎

Notice the steps we used to determine the inverse function:

1. Set $y = f(x)$.
2. Solve for x in terms of y.
3. Interchange the variables x and y so that the domain of f^{-1} is represented by numbers on the x-axis.

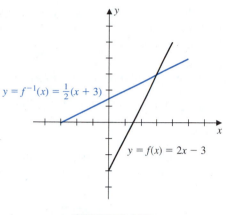

$y = f^{-1}(x) = \frac{1}{2}(x + 3)$

$y = f(x) = 2x - 3$

FIGURE 1.109

We can find the inverse for the function in Example 2 because it is linear and we can easily solve for x in terms of y. This is the critical step in determining inverse functions, one that is generally difficult and often impossible. Since we cannot always find explicit representations for inverse functions, we need to gain information about them in other ways. The next observation helps us considerably by describing the relationship between the graph of a one-to-one function and the graph of its inverse.

You might have noticed in Figure 1.109 that the graphs of f and f^{-1} are quite similar. There is a symmetry property that holds for the graph of a function and its inverse. If the point (a, b) is on the graph of the one-to-one function f, then $f(a) = b$ and $f^{-1}(b) = a$. This means that (b, a) is on the graph of f^{-1}. Geometrically, this implies the following. (See Figure 1.110.)

The Graph of an Inverse Function

The graph of $y = f^{-1}(x)$ is the reflection about the line $y = x$ of the graph of $y = f(x)$.

Notice that the graphs of a function and its inverse both increase or decrease together but that the steepness of the graph of the inverse function seems to be inversely related to the steepness of the graph of the function. This fact will be made precise in calculus.

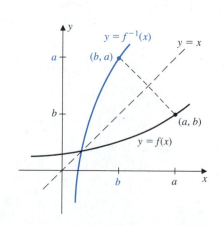

$y = f^{-1}(x)$

$y = x$

(b, a)

$y = f(x)$

(a, b)

FIGURE 1.110

This result will frequently permit us to determine many important features of the graph of an inverse function, even when we do not have an explicit expression for the inverse function.

EXAMPLE 3   Sketch the graph of the inverse function f^{-1} of the function defined by $f(x) = x + 2x^3$.

Solution   Although it is possible to solve for x in terms of y in the equation

$$y = x + 2x^3,$$

it is not easily done. Instead, we will first show that the inverse function exists by showing that f is one-to-one and then use the graph of f to produce the graph of f^{-1}.

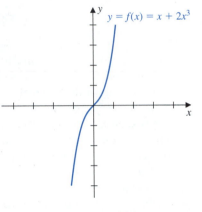

$y = f(x) = x + 2x^3$

FIGURE 1.111

The graph of f in Figure 1.111 appears to satisfy the horizontal line test, so we suspect that f is one-to-one. To show that this is true, suppose that $f(x_1) = f(x_2)$. Then

$$x_1 + 2x_1^3 = x_2 + 2x_2^3 \quad \text{and} \quad 0 = x_2 - x_1 + 2x_2^3 - 2x_1^3.$$

Performing some algebraic magic, we can reexpress this relation as

$$0 = (x_2 - x_1) + 2(x_2^3 - x_1^3)$$
$$= (x_2 - x_1) + 2(x_2 - x_1)(x_2^2 + x_2 x_1 + x_1^2)$$
$$= (x_2 - x_1)\left(1 + 2(x_2^2 + x_1 x_2 + x_1^2)\right)$$
$$= (x_2 - x_1)\left(1 + x_2^2 + (x_2^2 + 2x_2 x_1 + x_1^2) + x_1^2\right)$$
$$= (x_2 - x_1)\left(1 + x_2^2 + (x_2 + x_1)^2 + x_1^2\right).$$

Since all the terms within $\left(1 + x_2^2 + (x_2 + x_1)^2 + x_1^2\right)$ are positive, the sum is greater than or equal to 1, so the product in the last line of the right side of the equation is zero precisely when $x_2 - x_1 = 0$. As a consequence,

In calculus you will see that the derivative can be used to show more easily that the function f is one-to-one.

$$f(x_1) = f(x_2) \quad \text{if and only if} \quad x_1 = x_2,$$

and f is one-to-one.

The graph of f is shown in Figure 1.112 together with its reflection about the line $y = x$, which produces the graph of f^{-1}. ∎

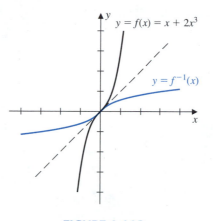

FIGURE 1.112

We admit that the function in the previous example was a bit contrived and that the factoring technique used here will not work in many situations. However, the functions whose inverses are of fundamental importance are not available to us yet, and this example does illustrate that by being sufficiently ingenious we might be able to determine a great deal about an inverse function based on very little explicit information.

The technique in Example 3 is sufficiently useful that many graphing devices have a built-in function to produce the graph of the reflection of a function about the line $y = x$. This is done whether or not the function is one-to-one. When it is not, the resulting reflected graph will not be the graph of a function. An example of this is shown in Figure 1.113, where the graph of $y = x^3 - x$ is reflected about the line $y = x$ using the $\boxed{\text{Dr Inv}}$ feature on a TI 85 graphing calculator. The reflected graph is certainly not the graph of a function, since $(0, 0)$, $(0, 1)$, and $(0, -1)$ are all on the graph.

Technology can often be used as a means of visualizing a graph or concept that we do not yet have the tools to analyze.

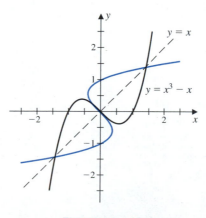

FIGURE 1.113

EXERCISE SET 1.12

In Exercises 1–6, determine if the function whose graph is given is one-to-one.

1.

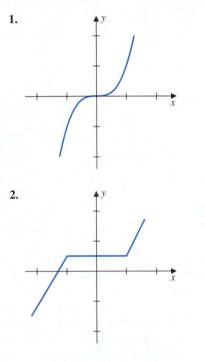

2.

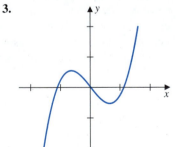

3.

4.

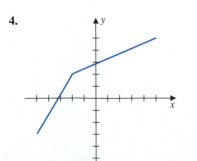

5.

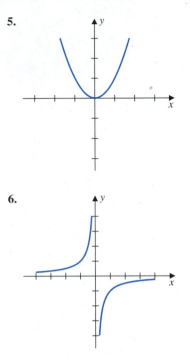

6.

In Exercises 7–12, determine if the given function is one-to-one.

7. $f(x) = 3x - 4$

8. $f(x) = |x|$

9. $f(x) = \sqrt{x - 1}$

10. $f(x) = x^2 - 3x + 2$

11. $f(x) = x^4 + 1$

12. $f(x) = \begin{cases} x^2, & \text{if } x \geq 0 \\ x - 1, & \text{if } x < 0 \end{cases}$

In Exercises 13–16, the graphs of one-to-one functions are given. Sketch the graph of the inverse of each function.

13.

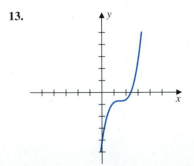

14.

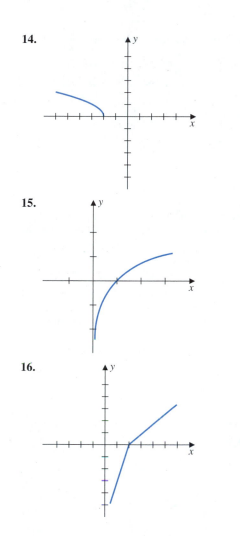

15.

16.

In Exercises 17–22, show algebraically that f is one-to-one, and find its inverse. Graph f and f^{-1} on the same coordinate axes.

17. $f(x) = 2x - 1$

18. $f(x) = \dfrac{4x - 3}{2}$

19. $f(x) = \sqrt{x - 3}$

20. $f(x) = \dfrac{x}{x + 1}$

21. $f(x) = \dfrac{1}{\sqrt{x}}$

22. $f(x) = x^2 + 1, \quad x \geq 0$

In Exercises 23–26, show that the given function is not one-to-one. Determine a subset of the domain of the function on which it is one-to-one, find its inverse on this restricted domain, and specify the domain of the inverse function.

23. $f(x) = |2 - 3x|$

24. $f(x) = x^2 - 2x$

25. $f(x) = \dfrac{1}{x^2 + 4}$

26. $f(x) = x^4 + 3$

27. For which values of m is the linear function $f(x) = mx + b$ one-to-one? For these values of m find $f^{-1}(x)$.

28. No object can assume a temperature lower than $-273°\,$C, a temperature that is called *absolute zero*. The Kelvin temperature scale, named for the Irish physicist William Thomson (Lord Kelvin), is defined as a translation of the Celsius scale so that $-273°\,$C corresponds to $0\,$K. (The degree symbol is generally omitted when expressing temperature in kelvins.) Determine a function that expresses temperature in Fahrenheit in terms of temperature in kelvins, show that this function is one-to-one, and find its inverse.

REVIEW EXERCISES FOR CHAPTER 1

In Exercises 1–4, express the given interval using inequalities and sketch the numbers in the interval.

1. $[-1, 7]$

2. $(0, \sqrt{5})$

3. $(-\infty, 7)$

4. $[-5, \infty)$

In Exercises 5–8, express the given inequalities using interval notation and sketch the numbers in the interval.

5. $x < -5$

6. $-3 < x \leq 3$

7. $2 \leq x < 10$

8. $x \leq 3$

In Exercises 9–14, find all values of x that satisfy the inequality.

9. $2x + 3 \geq 4$

10. $-(3x + 4) \geq 5$

11. $x^2 + 2x + 1 \geq 1$

12. $x^2 - 4x > -3$

13. $\dfrac{2x - 1}{x + 1} \leq -2$

14. $\dfrac{x^2 - 9}{x^2 - 4} \leq 0$

In Exercises 15–18, solve the given inequality and show the solution graphically.

15. $|2x - 3| < 5$

16. $|4x - 2| < 0.01$

17. $|2 - x| \geq 2$

18. $|x^2 - 4| > 1$

In Exercises 19–24, indicate in an xy-plane those points (x, y) for which the statement holds.

19. $2 < y \leq 3$ **20.** $|y| < 2$

21. $|x - 1| < 2$ **22.** $|x| \leq 1$ and $|y| \leq 2$

23. $|x| + |y| = 1$ **24.** $|x| > |y|$

In Exercises 25–28, a function is described. Find the domain and range of the function.

25. $f(x) = x^2 - 3$ **26.** $f(x) = \dfrac{1}{2x - 5}$

27. $f(x) = \dfrac{1}{\sqrt{x^2 - 6x + 8}}$

28. $f(x) = \begin{cases} -2x + 1, & \text{if } x < 0 \\ -x^2, & \text{if } x \geq 0 \end{cases}$

In Exercises 29–32, find $f(x + h)$ and $\dfrac{f(x + h) - f(x)}{h}$, where $h \neq 0$.

29. $f(x) = 7x + 4$ **30.** $f(x) = -\frac{3}{2}x + 5$

31. $f(x) = x^2 - 1$ **32.** $f(x) = 2x^2 - x$

In Exercises 33–36, (a) plot the pair of points given in each of the following, (b) determine the distance between the points, (c) sketch the straight line determined by these points, (d) find the slope-intercept equation of the line.

33. $(3, -1), (1, 2)$ **34.** $(3, 1), (-1, -2)$

35. $(-1, -2), (2, -3)$ **36.** $(1, -1), (4, -1)$

In Exercises 37–40, the equation of a line is given together with a point that is not on the line. Find the slope-intercept form of the equation of the line that passes through the given point and is (a) parallel to the given line, and (b) perpendicular to the given line.

37. $y = 4x + 3; (0, 0)$ **38.** $y = \dfrac{1}{2}x + 2; (1, 1)$

39. $-7x - 5y = -1; (-1, -3)$

40. $2x - 3y = 4; (1, -1)$

In Exercises 41–46, a quadratic equation is given. Sketch the graph of the equation, and find the range of the function described by the equation.

41. $y = x^2 - 4x$ **42.** $y = -x^2 + 4x - 5$

43. $y = 2x^2 - 12x + 18$ **44.** $y = 2x^2 + 5x + 10$

45. $y = -\dfrac{1}{2}x^2 + 3x - 3$ **46.** $y = -\dfrac{1}{3}x^2 + 2x - 1$

In Exercises 47–58, sketch the graph of each of the functions.

47. $f(x) = |x + 1|$ **48.** $f(x) = |x - 3| + 3$

49. $f(x) = |x + 2| - 2$ **50.** $f(x) = |2x + 4| - 3$

51. $g(x) = \sqrt{x + 2}$ **52.** $g(x) = \sqrt{x - 2} + 2$

53. $g(x) = -\sqrt{x} - 2$ **54.** $g(x) = 2 - \sqrt{x - 2}$

55. $f(x) = \lfloor x + 2 \rfloor$ **56.** $f(x) = \lfloor x \rfloor + 2$

57. $f(x) = -\lfloor x \rfloor$ **58.** $f(x) = 2\lfloor x \rfloor + 1$

In Exercises 59–62, the equation of a circle is given. (a) Find the center C and radius r of each circle, and (b) sketch its graph.

59. $x^2 + y^2 = 16$ **60.** $(x - 1)^2 + y^2 = 1$

61. $(x + 2)^2 + (y - 1)^2 = 9$ **62.** $x^2 + y^2 + 4x + 2y = 4$

In Exercises 63–66, sketch $y = f(x)$ and $y = |f(x)|$.

63. $f(x) = 3x - 2$ **64.** $f(x) = -2x + 1$

65. $f(x) = -x^2 + 3$ **66.** $f(x) = x^2 - 6x + 8$

67. Use the graph of the function shown in the figure to sketch the graph of each of the following.

  a. $y = f(x + 1)$ **b.** $y = f(x + 1) + 1$

  c. $y = f(x - 1)$ **d.** $y = f(x - 1) + 2$

  e. $y = f(x + 1) - 1$ **f.** $y = f(x - 1) - 2$

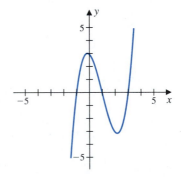

68. Determine formulas for $f(-x), -f(x), f(1/x), 1/f(x), f(\sqrt{x})$, and $\sqrt{f(x)}$ for the given $f(x)$.

  a. $f(x) = 2x^2 - 3$ **b.** $f(x) = 1/x^2$

69. Show that the points $(-3, 3), (3, -5)$, and $(7, -2)$ are vertices of a rectangle, and find the fourth vertex.

70. Find an equation of a circle with center $(-3, -1)$ that passes through $(-5, -3)$.

71. Find an equation of a circle with center $(4, 2)$ that is tangent to the x-axis.

72. Find a point on the x-axis equidistant from the points $(-2, -3)$ and $(3, 5)$.

73. Sketch the graph of

$$g(x) = \begin{cases} 0, & \text{if } x < 0 \\ x^2, & \text{if } 0 \le x \le 2 \\ -x + 6, & \text{if } x > 2 \end{cases}$$

74. From each of the data sets given, sketch a likely graph of f and determine a likely formula for $f(x)$.

a.

x	$f(x)$
0	-26
1	-6
2	6
3	10
4	6
5	-6
6	-26

b.

x	$f(x)$
-3	7
-2	5
-1	3
0	1
1	-1
2	-3
3	-5

75. Determine whether the given function is one-to-one.

a. $f(x) = 2 + x^3$ **b.** $f(x) = x^2 + 2x - 2$

c. $f(x) = 1/x^2$ **d.** $f(x) = \sqrt{x - 1} + 2$

76. Let $f(x) = x^2 - 2$.

a. Sketch the graph of $y = f(x)$ for $x \ge 0$.

b. Use the graph from a) to sketch the graph of $y = f^{-1}(x)$.

c. Find an equation for f^{-1}.

77. Let $f(x) = x^2 - 4x$.

a. Determine a subset of the function domain on which the function is one-to-one. Sketch the function restricted to the subset.

b. Use the graph in a) to draw the graph of f^{-1}

c. Find an equation for f^{-1} and specify the domain of f^{-1}.

78. A baseball is thrown upward from the ground with an initial velocity of 19.2 m/s. If air resistance is neglected, its distance $s(t)$, in meters, above the ground t seconds later is given by $s(t) = -4.8t^2 + 19.2t$. Estimate the maximum height of the ball.

79. A restaurant has a fixed price of \$36 for a complete dinner. The average number of customers per night is 200. The owner estimates that for each dollar increase in the price of the dinner, there will be, on average, four fewer customers per night. Estimate the price for the dinner that will produce the maximum amount the owner can expect to receive.

80. The daily truck rental for agency A is \$21, plus 21 cents per mile. For agency B the rate is \$32, plus 18 cents per mile. A driver plans to rent from one of these agencies and take a 320-mi trip. From which agency should the driver rent to get the better rate? Estimate the minimum number of miles beyond which agency B's cost is less than that of agency A.

81. A rectangular plot of ground containing 432 ft^2 is to be fenced off within a large lot, and a fence is to be constructed down the middle. Express the amount of fence required as a function of the length of the dividing fence. What is the domain of this function?

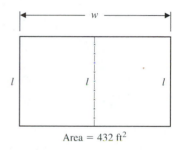

Area = 432 ft^2

82. An open rectangular box is to be made from a sheet of metal 5 cm wide by 9 cm long by cutting a square from each corner, bending up the sides, and welding the edges. Estimate the size of the square that should be removed in order to produce the maximum volume for the box.

83. A public works department is to construct a new road between towns A and B. Town A lies on an abandoned road running east-west. Town B lies 20 mi north of this road and 40 mi east of town A. An engineer proposes that the road be constructed by restoring a section of the old road leaving town A and joining it to an entirely new section of road at a point x to be determined, connecting that point with town B (see the figure).

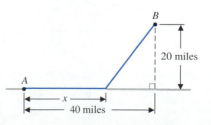

a. If the cost of restoring the old road is $200,000 per mile and the cost of new construction is $400,000 per mile, determine the function describing the total cost.

b. Estimate how much of the old road should be restored in order to minimize the department's cost.

84. Use a graphing device with each of the following viewing rectangles to sketch a graph of $y = x^2 - 2x + 5$. Which gives the best representation for the graph of the equation?

   a. $[-2, 2] \times [-2, 2]$ **b.** $[-5, 5] \times [-5, 5]$

   c. $[-10, 10] \times [-100, 100]$ **d.** $[-5, 20] \times [-5, 20]$

85. Use a graphing device with each of the following viewing rectangles to sketch a graph of $y = x^3 - 25x + 50$. Which gives the best representation for the graph of the equation?

   a. $[-5, 5] \times [-5, 5]$ **b.** $[-20, 20] \times [-20, 20]$

   c. $[-100, 100] \times [-100, 100]$

   d. $[-10, 10] \times [-100, 100]$

86. Determine an approximate viewing rectangle for the graph of the equation and use it to sketch the graph.

   a. $y = x^2 - 16x + 61$ **b.** $y = x^2 + 10x + 38$

   c. $y = \sqrt{x^2 - 10x + 29}$ **d.** $y = \dfrac{x + 4}{x - 3}$

87. Use the figure to answer the questions about the function f.

   a. Between which pairs of points labeled on the graph is the function increasing?

   b. Between which pairs of points labeled on the graph is the function decreasing?

   c. Between which pairs of points labeled on the graph does the function have a local maximum?

   d. Between which pairs of points labeled on the graph does the function have a local minimum?

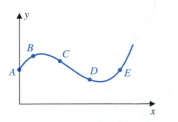

88. Match the equations with the graphs in the figures.

   a. $y = x + 3$ **b.** $y = -2x + 1$

   c. $y = 5$ **d.** $y = -4x - 3$

   e. $x = -2$ **f.** $y = 2x/3$

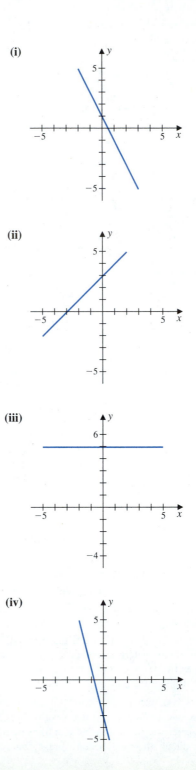

(v)

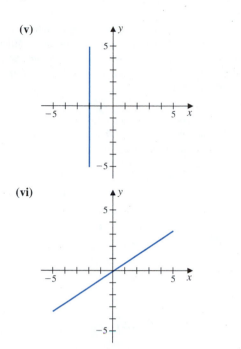

(vi)

89. Sketch the graph of a function that satisfies all of the conditions.

  a. $f(0) = 1$

  b. $f(x)$ is decreasing on the interval $(0, 2)$

  c. $f(x)$ is increasing on the interval $(2, 3)$

  d. $f(x)$ is decreasing on the interval $(3, \infty)$

  e. $f(x)$ approaches 0 as x becomes large and approaches 2 as x becomes small.

90. Sketch the graph of a function that satisfies all the conditions.

  a. $f(0) = -1$

  b. $f(x)$ is increasing on the interval $(0, 2)$

  c. $f(x)$ is decreasing on the interval $(2, 3)$

  d. $f(x)$ is increasing on the interval $(3, \infty)$

  e. $f(x)$ approaches 0 as x becomes large and approaches -2 as x becomes small.

CHAPTER 1: CALCULUS PREVIEW EXERCISES

1. One of the functions f or g in the accompanying table is linear and one is quadratic, of the form ax^2. Find equations for $f(x)$ and $g(x)$.

x	$f(x)$	$g(x)$
-2	2	2
2	2	14
6	18	26

2. The equation $3x + ay = -2a$ is the equation of a straight line. Determine any values of the constant a so that the line satisfies the specified condition.

  a. The line has slope 2.

  b. The line is horizontal.

  c. The line is vertical.

  d. The line passes through the point $(0, -2)$.

3. The sum of the first n consecutive natural numbers is given by

$$1 + 2 + 3 + \cdots + n = \frac{n(n+1)}{2}.$$

Find all values of n with the property that the sum is greater than 100 and less than 150.

4. Determine which of the following statements are always true, always false, and sometimes (but not always) true. Prove the statements that are always true, and give examples to justify your assertions for the other statements.

  a. The sum of two even functions is even, and the sum of two odd functions is odd.

  b. The product of two even functions is even, and the product of two odd functions is odd.

  c. The sum of an even function and an odd function is odd.

  d. The product of an even function and an odd function is odd.

5. Consider the family of functions given by

$$f_n(x) = x^n$$

for n a positive integer. Use a graphing device to sketch the graphs of f_n for $n = 2, 4$, and 6 on the same set of axes. On another set of axes draw f_n for $n = 1, 3$, and 5. Discuss and generalize your results. For a given positive integer n, compare the sizes of $f_n(x)$ and $f_{n+1}(x)$ on the intervals $0 < x < 1$ and $x > 1$.

6. Redraw the graph shown in the accompanying figure, and on the same set of axes draw the graph of each of the following.

   a. $y = -f(x)$ **b.** $y = |f(x)|$

   c. $y = f(-x)$ **d.** $y = f(|x|)$

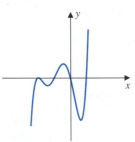

7. Sketch the graph of $y = f(x) = x^2(x - 1)$. Then use a graphing device to sketch the graphs of the following, for $a = 1, 2, -1,$ and -2.

   a. $y = f(x) + a$ **b.** $y = f(x + a)$

   c. $y = af(x)$ **d.** $y = f(ax)$

8. Let $y = \dfrac{ax + b}{cx + d}$, where $a, b, c,$ and d are positive constants. Plot a variety of graphs of the equation by varying $a, b, c,$ and d. Make some general comments about graphs of this type.

9. a. Sketch the graph of the equations $y = ax^2 + bx$, where a and b are arbitrary, and $a \neq 0$.

   b. Show that the graph of $y = ax^2 + bx$ is symmetric about the vertical line $x = -b/(2a)$.

   c. Show that for any number c, the graph of $y = ax^2 + bx + c$ is also symmetric about $x = -b/(2a)$.

10. An airplane makes one stop before reaching its final destination. This stop is 400 mi away and takes 1 hr to reach. The layover time at this stop is 1 hr and 30 min. It then completes its flight, landing at its final destination 800 mi from the first stop, taking another 2 hr. Let t be time in hours after the flight starts, $s(t)$ be the horizontal distance traveled, and $a(t)$ be the altitude of the plane.

   a. Sketch a plausible graph of $y = s(t)$.

   b. Sketch a plausible graph of $y = a(t)$.

   c. Sketch a plausible graph of the ground speed.

   d. Sketch a plausible graph of the vertical velocity.

11. Ice cubes are added to a glass of room temperature ($20°C$) water, and the water is allowed to stand until all the ice cubes melt 30 min later. Sketch a possible graph of the temperature of the water in the glass with respect to the elapsed time.

12. A child with a fever is given an antibiotic. The fever continues to rise for another hour, but not as rapidly as it did for the hour before the antibiotic was given. The child's temperature then begins to slowly drop until it stabilizes. Sketch a possible graph of the child's temperature as a function of time.

13. A standard can in the shape of a right circular cylinder with a top and a bottom contains a volume of 900 cm^3.

   a. Express the surface area of the can as a function of the radius of the bottom.

   b. Estimate the dimensions of the can that will minimize the metal needed to manufacture the can.

14. A small shoe company can produce between 100 and 1000 pairs of shoes per day. The cost of producing x pairs of shoes, $100 \leq x \leq 1000$, is given by

$$C(x) = 295 + 3.28x + 0.003x^2 \text{ dollars}$$

If x pairs of shoes are sold, then the price per pair that the company will receive is given by

$$p(x) = 7.47 + \frac{321}{x}$$

   a. Estimate the number of pairs of shoes that should be produced if the average cost per unit is to be a minimum.

   b. Estimate the number of pairs of shoes that should be produced in order to maximize the profit.

15. When a solid rod is heated, its length increases by a certain amount depending on its coefficient of linear expansion. This coefficient, a, is assumed to be constant, depending only on the material of the rod. The amount of increase in length is the product of the length, the change in temperature, and a. Suppose a steel rod has a length of 2 m at $0°C$ and that the coefficient of linear expansion for this material is $a = 11 \times 10^{-6}$.

a. Find a function that describes the length of the rod in terms of its temperature above $0°C$.

b. Determine its length when the temperature is $1000°C$.

16. According to Boyle's Law, if the temperature of a confined gas is held fixed, then the product of the pressure P (in $lb/in.^2$) and the volume V (in $in.^3$) is a constant. Suppose for a certain gas, $PV = 8000$. Determine the average rate of change of P as V increases from 200 in.^{3} to 250 in.^{3}.

17. In depreciating equipment it is common to assume that certain items will not depreciate to 0 but will have some residual value. Suppose a car originally valued at \$15,000 depreciates linearly to a residual value of \$2000 in 10 years. Find its value $V(t)$, t years after purchase, where $0 \leq t \leq 10$.

18. For an arbitrary function f whose domain contains $-x$ whenever it contains x, define g and h as follows:

$$g(x) = \frac{1}{2}[f(x) + f(-x)]$$

and

$$h(x) = \frac{1}{2}[f(x) - f(-x)]$$

a. Show that g is even and h is odd.

b. Show that $f = g + h$. This shows that f can be expressed as the sum of an even and an odd function.

19. Use the result of Exercise 18 to write each of the following functions as the sum of an even and an odd function.

a. $f(x) = x^2 + x$ **b.** $f(x) = \frac{1}{x} + 1$

c. $f(x) = |x| + x$ **d.** $f(x) = |x + 1|$

20. Let $f(x) = \frac{1}{x}$. Show that for any functions g and h,

$$f \circ \left(\frac{g}{h}\right) = \frac{f \circ g}{f \circ h}.$$

What restrictions must be placed on $g(x)$ and $h(x)$?

21. Let $f(x) = \frac{x+a}{x+b}$, where $a \neq b$. Show that f^{-1} exists and find $f^{-1}(x)$. Give the domain and range of f and f^{-1}. Verify that $(f^{-1} \circ f)(x) = x$ for all x in the domain of f and $(f \circ f^{-1})(x) = x$ for all x in the domain of f^{-1}.

ALGEBRAIC FUNCTIONS

2

CALCULUS CONNECTIONS

You may be wondering what makes the particular functions we have been studying valuable. One reason is that they can often be used to *model* real-world problems. A mathematical model is an equation, or system of equations, that reasonably describes a real-world situation. The phenomena we are modeling might be physical, biological, economic, or any other quantitative problem. Simplifying assumptions are generally made to make the modeling possible, and the resulting representation is not expected to describe the physical situation exactly, but instead give an indication of the important features of the problem. The fewer simplifying assumptions we make, the more likely the model is to describe the actual situation, but the more difficult the problem will be to solve.

Take, as an example, the problem of determining the path of water coming out of a hose. From experience we know that the path the water takes depends on both the velocity of the water and the angle at which we hold the nozzle of the hose. Place an *xy*-coordinate system with the end of the nozzle at the origin and the positive *y*-axis pointing upward, as shown in the figure.

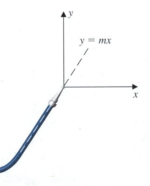

Suppose the velocity of the water as it leaves the origin is the constant v_0 and that the nozzle is pointed in the direction of the line with equation $y = mx$. Wind, the friction of the water in the air, the spread of the water as it moves through the air, and various other things will effect the path of the water, but our model will ignore all these (we hope) minor influences. We will assume that the only significant influence on the water is due to the gravity of the earth, the constant -32 ft/s^2. Physics (actually applied calculus) tells us that the path of the water is described by the equation

$$y = mx - \frac{16}{v_0^2}(1 + m^2)x^2,$$

whose graph is a parabola.

Setting y to zero in this polynomial equation and solving for x tells us where the water hits the ground. Determining zeros of polynomials will be just one of the topics we will consider in this chapter. In this case it is easy to do, and we have $y = 0$ precisely when

$$x = 0 \qquad \text{and} \qquad x = \frac{mv_0^2}{16(m^2 + 1)}.$$

The figure shows how the trajectory of the water can vary as the angle of the nozzle is changed. We can determine from the given equation the distance the water will travel, how high it will go, and what its velocity will be when it hits the ground.

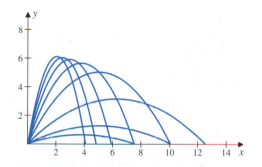

If we try to modify the mathematical model to take into consideration some of the other factors that influence the path of the water, we will have a much more difficult problem to solve. In Section 1.8 we considered the quadratic equation that the simple model provides. The quadratic equation involves a polynomial of degree two. In this chapter we will look at general polynomials and some other functions that are related to them.

2.1   INTRODUCTION

In Chapter 1 we discussed constant, linear, and quadratic functions—functions that have equations of the form $f(x) = c$, $f(x) = mx + b$, and $f(x) = ax^2 + bx + c$, respectively. These are special cases of *polynomial functions*, which have the form

$$P(x) = a_n x^n + a_{n-1}x^{n-1} + \cdots + a_1 x + a_0,$$

where $a_0, a_1, \ldots, a_n$ are constants and $a_n \neq 0$. The class of polynomials has the particularly nice property that the sum, difference, and product of a pair of polynomials always produces another polynomial.

The applications in these areas generally require calculus, but for an understanding of the calculus application you need to recognize the behavior of the polynomial.

Functions that arise in applications in the sciences, economics, and other areas are often polynomial functions or are approximated using them. This is due in part to the fact that the values of polynomial functions can be computed using only the operations of addition, subtraction, and multiplication, so their evaluation can be done quickly by computer.

In this chapter we will see how systematically to sketch the graphs of polynomial functions. We will also look at some common *rational functions*, which are those obtained by dividing polynomials, and more general *algebraic functions*. Algebraic functions are obtained from polynomials by any finite combination of the operations of addition, subtraction, multiplication, division, and extracting integral roots. For example,

$$f(x) = \sqrt{x^4 + 2x + 1} \qquad \text{and} \qquad g(x) = \frac{x^3 + 2x^2 - x + 5}{x^4 + 3x + 2}$$

are both algebraic functions. The function g is also rational since $g(x)$ is the quotient of the polynomials $P(x) = x^3 + 2x^2 - x + 5$ and $Q(x) = x^4 + 3x + 2$. The trigonometric functions that we will study in Chapter 3 and the exponential and logarithm functions we will consider in Chapter 4, on the other hand, are not algebraic functions. Functions that are not algebraic are called *transcendental*.

2.2   POLYNOMIAL FUNCTIONS

A **polynomial of degree n**, where n is a nonnegative integer, has the form

$$P(x) = a_n x^n + a_{n-1} x^{n-1} + \cdots + a_1 x + a_0,$$

where $a_0, a_1, \ldots, a_n$ are constants, and $a_n \neq 0$. The numbers $a_0, a_1, \ldots, a_n$ are called the *coefficients* of the polynomial; a_n is the *leading coefficient* and a_0 is the *constant term*.

From this definition we see, for example, that

(i)   A constant function $f(x) = c$, with $c \neq 0$, is a polynomial function of degree 0;
(ii) A linear function $f(x) = mx + b$, with $m \neq 0$, is a polynomial function of degree 1;
(iii) A quadratic function $f(x) = ax^2 + bx + c$, with $a \neq 0$, is a polynomial function of degree 2.

Some other examples of polynomials are shown in Table 2.1.

TABLE 2.1

Polynomial	Degree	Leading Coefficient	Constant Term
$-3x^6 + 2x^4 + x^2 - x + 1$	6	-3	1
17	0	17	17
$2x^3 - 2x^2 + x - 5$	3	2	-5
x^{10}	10	1	0

Graphing devices become a powerful tool when we have a general knowledge of the possibilities for the shape of the graph of a function or class of functions.

We saw in Section 1.7 that the graphs of polynomials of degree 0 and 1 are lines and in Section 1.8 that the graph of a polynomial of degree 2 is a parabola. (See Figure 2.1.)

Graphing devices, such as graphing calculators and computer algebra systems, can be very helpful for generating the graphs of polynomials, but the graph produced by these devices is found by simply sketching a large number of points. We want to develop

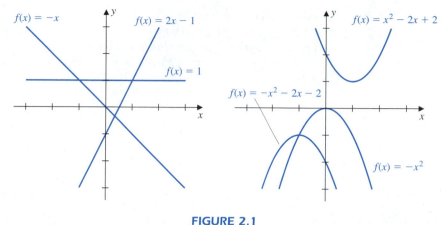

FIGURE 2.1

a logical way to approach graphing problems, one that relies only minimally on plotting points. We will be concentrating more on the overall behavior of the graph than on the fine detail. Remember that a primary objective of this book is to have you be familar with the graphs and behavior of the functions you will regularly see in calculus.

EXAMPLE 1 Use the graphing techniques of Chapter 1 to sketch the graph of $y = -3(x - 2)^3 + 1$.

Solution The graph of $y = 3x^3$ is found by vertically stretching the graph of the *cubing* function $y = x^3$, as shown in Figure 2.2(a). We obtain the graph of $y = -3x^3$ by reflecting the graph of $y = 3x^3$ about the x-axis, as shown in Figure 2.2(b).
    Then we shift the graph of $y = -3x^3$ two units to the right and one unit upward to obtain the graph of $y = -3(x - 2)^3 + 1$, shown in Figure 2.2(c).   ■

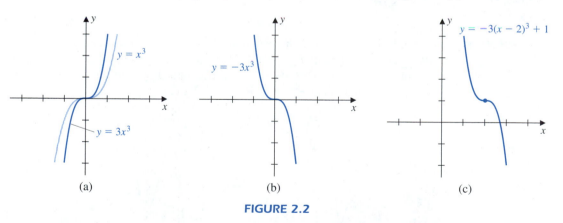

FIGURE 2.2

EXAMPLE 2 Sketch the graph of $f(x) = x^3 - 4x = x(x - 2)(x + 2)$.

Solution First notice that $f(0) = f(2) = f(-2) = 0$, so the graph has x-intercepts at 0, 2, and -2. In addition, the graph approaches the graph of $y = x^3$ as x becomes large in magnitude, so the graph is likely to have the features shown in Figure 2.3(a).
    Moreover, f is an odd function since

Comparing the rates of growth of different functions is an important topic in calculus.

$$f(-x) = (-x)^3 - 4(-x) = -x^3 + 4x = -(x^3 - 4x) = -f(x),$$

and the graph is symmetric with respect to the origin. To determine the behavior of the graph near the origin, we could plot some points, such as $f(1) = -3$ and $f(3) = 15$, and produce a graph like that shown in Figure 2.3(b). Alternatively, we could add the vertical components of the graphs of $y = x^3$ and $y = -4x$ to find the y-values on the graph of $y = x^3 - 4x$, as shown in Figure 2.3(c).

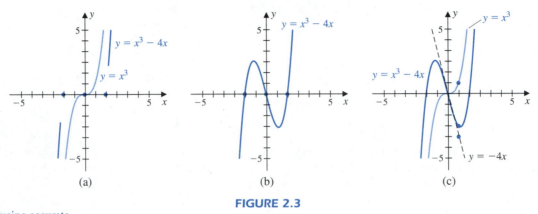

(a)   (b)   (c)

FIGURE 2.3

Producing accurate graphs of functions is an application of the derivative that you will study in your first calculus course.

To fine-tune the graph we would need to know the precise location of the local maximum and minimum points, but this requires techniques of calculus. To get an approximation to these points we could use a graphing device, but for most purposes the graph in Figure 2.3(b) or (c) should suffice.   ■

EXAMPLE 3   Sketch the graph of

$$y = f(x) = (x - 2)^3 - 4(x - 2) + 5.$$

Solution   The graph of $y = x^3 - 4x$ was found in Example 2. Shifting this graph to the right 2 units produces the graph of $y = (x - 2)^3 - 4(x - 2)$ shown in Figure 2.4(a), and the graph of $f(x) = (x - 2)^3 - 4(x - 2) + 5$ is obtained by shifting this graph five units upward, as shown in Figure 2.4(b).

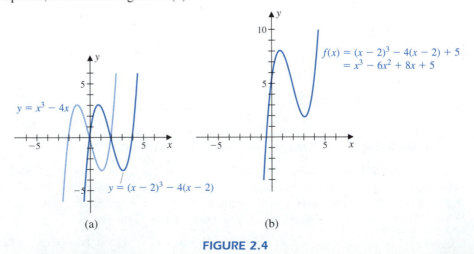

(a)   (b)

FIGURE 2.4

Notice that by shifting the graph to the right 2 units and then upward 5 units, we have lost the symmetry with the respect to the origin that we had in Example 2. It has been replaced by a symmetry with respect to the point $(2, 5)$. ■

These basic polynomials are frequently used for examples in calculus. Analyzing the properties of a graph is one of the basic applications of calculus.

The most elementary nth-degree polynomial is $f(x) = x^n$. When n is even, the graph of f is symmetric with respect to the y-axis and has a form similar to the graph of $y = x^2$, as shown in Figure 2.5(a). Notice, though, that as the value of n gets larger, the graph gets flatter near the origin and rises more rapidly when $x > 1$.

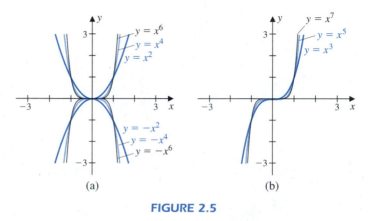

FIGURE 2.5

When n is odd, the graph of $f(x) = x^n$ is symmetric with respect to the origin, as shown in Figure 2.5(b). For $n > 3$ the graph is similar to the graph of $y = x^3$, and the larger the value of n the flatter the graph near the origin and the more rapidly it rises when $x > 1$.

EXAMPLE 4 Sketch the graphs.

a. $f(x) = \dfrac{1}{2}x^5$ and b. $f(x) = -2(x - 1)^4$

Solution a. The graph of $y = x^5$ is shown in Figure 2.6(a). By shrinking this vertically by a factor of $1/2$ we obtain the graph shown in Figure 2.6(b).

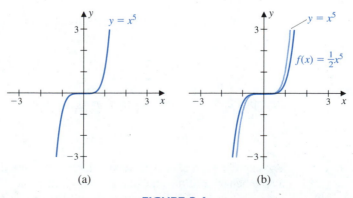

FIGURE 2.6

b. The first step in sketching the graph of $f(x) = -2(x - 1)^4$ is to sketch the graph of $y = x^4$ and then stretch this vertically by a factor of 2 to obtain the graph of $y = 2x^4$, as shown in Figure 2.7(a). Next we reflect this graph about the x-axis to produce the graph of $y = -2x^4$, and finally we shift this graph to the right 1 unit, as shown in Figure 2.7(b), to obtain the graph of $f(x) = -2(x - 1)^4$. ∎

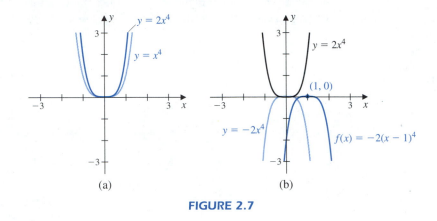

(a)   (b)

FIGURE 2.7

One of the most important features of the graph of a polynomial is the behavior of the y-values of points on the curve for large x-values, as well as for x-values that are negative but with large magnitude (that is, for x large in absolute value). This gives what we call the *end behavior* of the graph.

The end behavior of any polynomial function

$$P(x) = a_n x^n + a_{n-1} x^{n-1} + a_{n-2} x^{n-2} + \cdots + a_1 x + a_0$$

of degree n depends solely on the leading coefficient, a_n. To see this, let us factor x^n from each term to rewrite $P(x)$ in the form

$$P(x) = a_n x^n + a_{n-1} x^{n-1} + a_{n-2} x^{n-2} + \cdots + a_1 x + a_0$$
$$= x^n \left(a_n + \frac{a_{n-1}}{x} + \frac{a_{n-2}}{x^2} + \cdots + \frac{a_1}{x^{n-1}} + \frac{a_0}{x^n} \right).$$

As the values of x become large in magnitude, either positively or negatively, the terms

$$\frac{1}{x}, \quad \frac{1}{x^2}, \quad \cdots, \quad \frac{1}{x^{n-1}}, \quad \text{and} \quad \frac{1}{x^n}$$

all approach zero. Since $a_0, a_1, \ldots, a_{n-2}, a_{n-1}$ are all constants,

$$\frac{a_{n-1}}{x}, \quad \frac{a_{n-2}}{x^2}, \quad \cdots, \quad \frac{a_1}{x^{n-1}} \quad \text{and} \quad \frac{a_0}{x^n}$$

also approach zero. So, as x becomes large in magnitude, the behavior of $P(x)$ is modeled by the behavior of the equation $y = a_n x^n$. In Table 2.2 and Figure 2.8 we see how this equation behaves for the various possibilities of n and a_n.

The intuitive notion of approaching zero or getting close is made precise with the notion of *limit*, a fundamental concept underlying all of calculus.

TABLE 2.2

n	$a_n > 0$	$a_n < 0$
Even	$x \to +\infty, \quad P(x) \to +\infty$ $x \to -\infty, \quad P(x) \to +\infty$ (See Figure 2.8(a).)	$x \to +\infty, \quad P(x) \to -\infty$ $x \to -\infty, \quad P(x) \to -\infty$ (See Figure 2.8(b).)
Odd	$x \to +\infty, \quad P(x) \to +\infty$ $x \to -\infty, \quad P(x) \to -\infty$ (See Figure 2.8(c).)	$x \to +\infty, \quad P(x) \to -\infty$ $x \to -\infty, \quad P(x) \to +\infty$ (See Figure 2.8(d).)

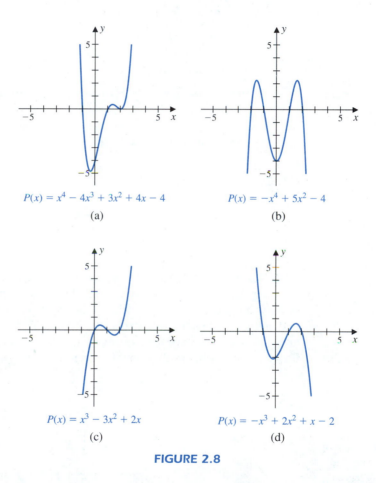

$P(x) = x^4 - 4x^3 + 3x^2 + 4x - 4$

(a)

$P(x) = -x^4 + 5x^2 - 4$

(b)

$P(x) = x^3 - 3x^2 + 2x$

(c)

$P(x) = -x^3 + 2x^2 + x - 2$

(d)

FIGURE 2.8

Here a graphing device is very useful in visualizing the end behavior of a polynomial, by allowing us to plot the graph of the function using very large horizontal and vertical views.

Of course, the end behavior only applies for x-values with large magnitude. Figure 2.9 shows two views of the graph of the third-degree polynomial $f(x) = x^3 - 3x^2 + 2x$. For x-values with large magnitude the polynomial behaves like $y = x^3$. However, as Figure 2.9(b) shows, for small values of x, the polynomial does not behave at all like $y = x^3$. In particular, for $0 \le x \le 2$, $f(x) = x^3 - 3x^2 + 2x$ crosses the x-axis three times, whereas $y = x^3$ crosses the x-axis only at 0.

A number c is called a **zero** of the function f if $f(c) = 0$. It is also common in this case to say that c is a **root** of the equation $f(x) = 0$. The graph of $y = f(x)$ crosses the x-axis at $x = c$ precisely when c is a zero of f.

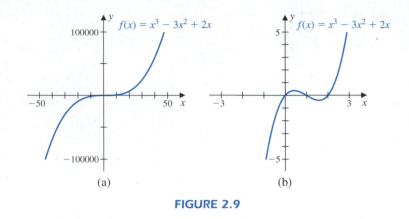

(a)   (b)

FIGURE 2.9

Calculus is primarily
concerned with
continuous functions.
The definition of
continuous requires a
precise definition of
the limit of a
function, which is
generally where the
first calculus course
begins.

Determining—or even approximating—the zeros of a polynomial can greatly aid in sketching its graph, since every polynomial function has a property called *continuity*. This essentially means that its graph has no breaks or interruptions. An important consequence of continuity is the *Intermediate Value Theorem*.

Intermediate Value Theorem

If f is continuous on $[a, b]$ and if K is a number between $f(a)$ and $f(b)$, then some number c in (a, b) exists with $f(c) = K$.

The Intermediate Value Theorem tells us that continuous functions do not skip over values in the range of the functions. In particular, a polynomial that is both positive and negative in some interval must take on the value zero somewhere between the positive and negative values. As a consequence, the graph of a polynomial between successive zeros is either always positive or always negative, and the graph will lie either entirely above or entirely below the x-axis between successive zeros. (See Figure 2.10.)

In calculus you will
frequently need to
know intervals on
which continuous
functions are positive
and negative, which
is easy to determine
if you know when the
functions are zero.
The Intermediate
Value Theorem is
implicitly used in
these instances.

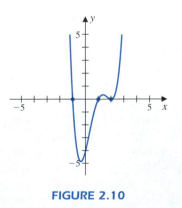

FIGURE 2.10

In the next example we show how knowing the zeros of a polynomial together with the end behavior can be used to produce the general shape of the graph.

EXAMPLE 5   Sketch the graph of $y = f(x) = x^3 - x^2 - 2x = x(x + 1)(x - 2)$.

Solution   In factored form it is easy to determine the points where the graph crosses the x-axis since

$$f(x) = x(x + 1)(x - 2) = 0$$

is satisfied precisely when one of the factors is 0. Hence the zeros of f are $x = -1$, $x = 0$, and $x = 2$, and the points in the plane where the graph crosses the x-axis are $(-1, 0)$, $(0, 0)$, and $(2, 0)$. To determine whether the graph is above or below the x-axis between two successive zeros, we can select an x-value between two zeros and determine the sign of $f(x)$, as shown in Figure 2.11.

```
        x              - - - - 0 + + + + + + + + + +
      (x + 1)          - - 0 + + + + + + + + + + + +
      (x - 2)          - - - - - - - - - 0 + + + + + +
f(x) = x(x + 1)(x - 2) - - 0 + 0 - - - 0 + + + + + +
                       +--+--+--+--+--+--+--+--+----►
                      -2 -1  0  1  2  3  4  5   x
```

FIGURE 2.11

To determine the end behavior of the graph, we consider $f(x)$ in its original form:

$$f(x) = x^3 - x^2 - 2x.$$

The polynomial is of odd degree, 3, and the leading coefficient is the positive number 1, so the polynomial behaves like x^3 for x large in magnitude (See Table 2.2 on page 115).

This information together with the results in the chart in Figure 2.11 permits us to sketch the graph shown in Figure 2.12.   ■

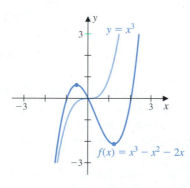

FIGURE 2.12

Many of the applications in a first calculus course involve finding the precise location of local extrema. Graphing devices can be used to estimate these values in many cases.

The graph in Figure 2.12 has a high point somewhere between $x = -1$ and $x = 0$ and has a low point between $x = 0$ and $x = 2$. These points are called a *local maximum* and *local minimum*, respectively (collectively they are sometimes called *local extrema*).

One of the most valuable applications of calculus is determining exact locations of these points. Without calculus these points cannot generally be determined exactly, but by zooming in with a graphing device we can approximate them.

EXAMPLE 6   Sketch the graph of $f(x) = x^4 - 3x^3 + 2x^2 = x^2(x-1)(x-2)$.

Solution   The zeros of f are $x = 0$, $x = 1$, and $x = 2$, and the sign chart in Figure 2.13 indicates there are two sign changes.

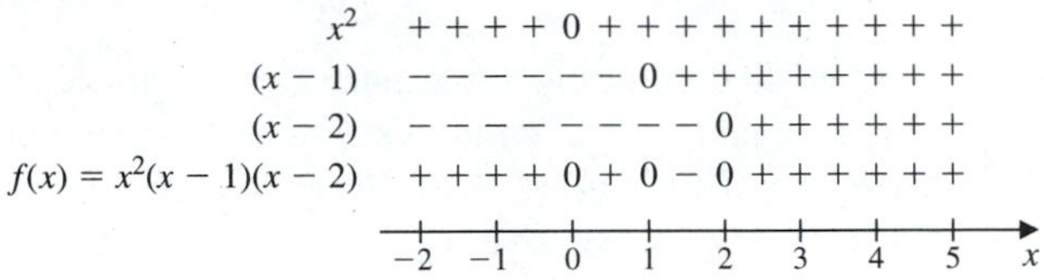

$$
\begin{array}{rl}
x^2 & +\,+\,+\,+\,0\,+\,+\,+\,+\,+\,+\,+\,+\,+\,+ \\
(x-1) & -\,-\,-\,-\,-\,-\,0\,+\,+\,+\,+\,+\,+\,+\,+ \\
(x-2) & -\,-\,-\,-\,-\,-\,-\,-\,-\,0\,+\,+\,+\,+\,+\,+ \\
f(x) = x^2(x-1)(x-2) & +\,+\,+\,+\,0\,+\,0\,-\,0\,+\,+\,+\,+\,+\,+
\end{array}
$$

$$-2 \quad -1 \quad 0 \quad 1 \quad 2 \quad 3 \quad 4 \quad 5 \quad x$$

FIGURE 2.13

The degree of the polynomial is 4, and the leading coefficient is positive, so the end behavior of the graph is the same as that of $y = x^4$. That is,

$$\text{as } x \to \infty, \ f(x) \to \infty \qquad \text{and as } x \to -\infty, \ f(x) \to \infty.$$

Plotting the points $(0,0)$, $(1,0)$, and $(2,0)$ helps us to infer that the graph is similar to that shown in Figure 2.14.

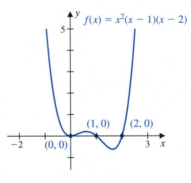

FIGURE 2.14

It seems clear from the graph that a local minimum occurs at $x = 0$, and by zooming in on $[0, 1]$ and $[1, 2]$, we could approximate the other local extrema. However, we would not be able to determine them precisely in this way since graphing devices only give rational numbers, and the extrema happen to occur at the irrational numbers $\frac{1}{8}(9 \pm \sqrt{17})$.

Notice also that the graph has both a zero and a local minimum at $x = 0$. This is due to the factor x^2, which caused $x = 0$ to be a *repeated* zero of *multiplicity* two.   ∎

If a polynomial $f(x)$ has a factor of the form $(x - c)^k$, where $k > 1$, then $x = c$ is a **repeated** zero of f of **multiplicity** k. If k is even, then the graph flattens and just touches the x-axis at $x = c$, as occurs at the origin in Figure 2.15(a). If k is odd, the graph flattens and then crosses the x-axis at $x = c$, as occurs at the origin in Figure 2.15(b).

Suppose we change the function in Example 6 slightly, to

$$g(x) = (x^2 + 0.25)(x-1)(x-2).$$

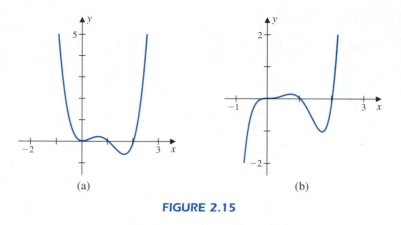

FIGURE 2.15

Then g will only have two zeros, at $x = 1$ and $x = 2$, and there will still be two sign changes. As shown in Figure 2.16(a), the graph of g now has only one local extreme point, a local minimum between $x = 1$ and $x = 2$.

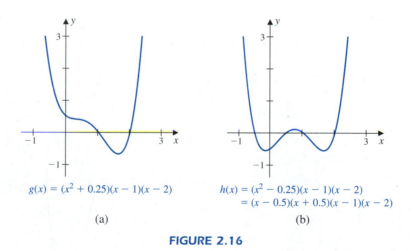

$g(x) = (x^2 + 0.25)(x - 1)(x - 2)$

$h(x) = (x^2 - 0.25)(x - 1)(x - 2)$
$= (x - 0.5)(x + 0.5)(x - 1)(x - 2)$

(a)   (b)

FIGURE 2.16

On the other hand, if we change the function to

$$h(x) = (x^2 - 0.25)(x - 1)(x - 2),$$

we can factor the quadratic term as

$$x^2 - 0.25 = (x - 0.5)(x + 0.5).$$

Then there are four zeros, at -0.5, 0.5, 1, and 2. The graph now appears as shown in Figure 2.16(b). Local minimums occur between -0.5 and 0.5 and between 1 and 2. A local maximum occurs between 0.5 and 1. The lesson in this is that small changes in the function can often result in a significant change in the graph.

EXAMPLE 7   Sketch the graph of $f(x) = x^3 - x^2 - 4x + 4$.

Solution   We begin by grouping the terms of f and factoring:

$$f(x) = x^3 - x^2 - 4x + 4 = x^2(x - 1) - 4(x - 1).$$

Since $(x - 1)$ is common to both terms, we have

$$f(x) = (x - 1)(x^2 - 4) = (x - 1)(x - 2)(x + 2).$$

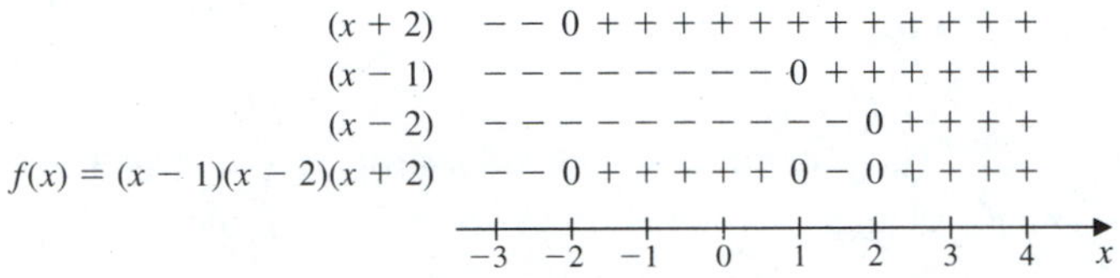

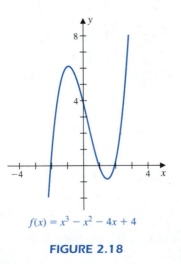

```
(x + 2)                   – – 0 + + + + + + + + + + + +
(x – 1)                   – – – – – – – – 0 + + + + + +
(x – 2)                   – – – – – – – – – – 0 + + + +
f(x) = (x – 1)(x – 2)(x + 2)   – – 0 + + + + + 0 – 0 + + + +
```

FIGURE 2.17

We see that f has three zeros and Figure 2.17 implies that there are three sign changes. The curve is shown in Figure 2.18. Notice that there are two local extreme points and that the graph has the same end behavior as $y = x^3$.   ■

$$f(x) = x^3 - x^2 - 4x + 4$$

FIGURE 2.18

Graphing higher-degree polynomials without a graphing device is generally difficult unless the zeros of the polynomial are easy to determine, in which case we can find intervals where the polynomial is positive and negative. We examine this topic in the next section.

EXERCISE SET 2.2

In Exercises 1–4, determine the lowest possible degree for the polynomial whose graph is shown.

1.

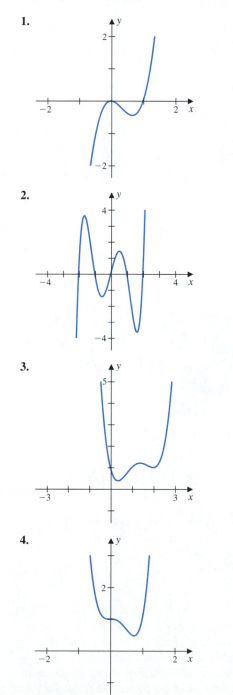

2.

3.

4.

In Exercises 5–12, sketch each of the following curves by shifting the graph in the accompanying figure, and determine the coordinates of the labeled points.

5. $y = f(x - 2)$ **6.** $y = f(x + 1)$

7. $y = f(x - 1) + 2$ **8.** $y = -f(x)$

9. $y = f(-x)$ **10.** $y = -f(-x - 2) - 3$

11. $y = |f(x)|$ **12.** $y = f(|x|)$

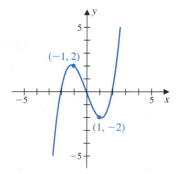

13. Sketch the graph of a polynomial $P(x)$ that has zeros of multiplicity one at $x = 1$, $x = -1$, and $x = 2$ and satisfies $P(x) \to \infty$ as $x \to \infty$, and $P(x) \to -\infty$ as $x \to -\infty$. What is the least possible degree of this polynomial?

14. Sketch the graph of a polynomial $P(x)$ that has zeros of multiplicity one at $x = 0$ and $x = 1$, has a zero of multiplicity three at $x = 3$, and satisfies $P(x) \to -\infty$ as $x \to \infty$ and $P(x) \to \infty$ as $x \to -\infty$. What is the least possible degree of this polynomial?

15. The cubic polynomial $P(x)$ has zeros at $x = 1$, $x = 2$, and $x = -2$ and y-intercept at 2. Determine $P(x)$ and sketch its graph.

16. The cubic polynomial $P(x)$ has a zero of multiplicity one at $x = 2$, a zero of multiplicity two at $x = 1$, and $P(-1) = 4$. Determine $P(x)$ and sketch its graph.

17. Determine the polynomial of degree four whose graph is shown in the figure.

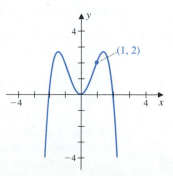

18. Use a graphing device to sketch the graphs of the polynomials $P(x) = x^4 + x^3 - x^2 - x$ and $Q(x) = 4 - x^2$ and determine all the values of x for which $P(x) < Q(x)$.

In Exercises 19–22, use a graphing device to sketch the graph of the given function and approximate

a. Intervals on which the function is increasing;
b. Local maximum and minimum values.

19. $f(x) = x^4 + x^3 - 2x^2$

20. $f(x) = x^4 - 2x^3$

21. $f(x) = \frac{1}{5}x^5 - \frac{5}{4}x^4 + \frac{5}{3}x^3 + \frac{5}{2}x^2 - 6x + 1$

22. $f(x) = x^5 - 2x^4 + 2x^2 - x$

In Exercises 23–26, use a graphing device to sketch the graph of the given function and then sketch the graphs of $y = -f(x)$, $y = f(-x)$, $y = |f(x)|$, and $y = f(x + a) + b$ for the given values of a and b.

23. $f(x) = x^3 - 2x^2 + x + 2;\ a = -3,\ b = 2$

24. $f(x) = -x^3 + 7x^2 - 36;\ a = 7,\ b = -30$

25. $f(x) = x^4 - 400x^2;\ a = -10,\ b = 30,000$

26. $f(x) = x^5 + x^4 - 13x^3 - x^2 + 48x - 36;\ a = 5,\ b = -10$

27. A box without a lid is constructed from a 20-in. × 20-in. piece of cardboard, as shown in the following figure. (a) Determine the volume of the box as a function of the variable x. (b) Use a graphing device to approximate the value of x that produces a volume of 500 in.^{3}.

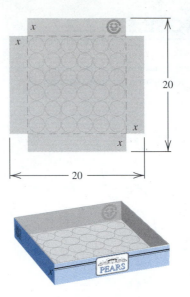

28. A fuel tank is in the shape of a right circular cylinder of length 10 m capped on either end with a hemisphere, as shown in the figure. Determine the radius r of the tank if the volume of the tank is 50 m^3.

2.3   FINDING FACTORS AND ZEROS OF POLYNOMIALS

We saw in Section 2.2 that the zeros of a polynomial give important graphing information. Finding the zero of a linear polynomial is easy, and the quadratic formula

$$x = \frac{-b \pm \sqrt{b^2 - 4ac}}{2a}$$

for finding the zeros of the quadratic polynomial $ax^2 + bx + c$ has been known for at least 2000 years. But finding the zeros of higher-degree polynomials is generally difficult.

There are formulas for finding the exact values of the zeros for polynomials of degrees three and four. These were discovered in the middle of the 16th century by mathematicians in the north of Italy. This area was the birthplace of the Renaissance and produced many of the artists, musicians, and scientists of the period. There was a great deal of scandal and intrigue surrounding the development of these formulae, but the third degree formula is now credited to Nicolo Fontana (known as Tartaglia "the stammerer") and the fourth-degree formula to his rival Ludovico Ferrari. Like the quadratic formula, these are *algebraic solutions*, formulas that require only common arithmetic operations together with the extraction of roots. But unlike the quadratic formula, they are quite difficult to apply.

In the 16th through the early 19th centuries, efforts were made by many of the outstanding mathematicians of the time to determine formulas for finding the zeros of higher-degree polynomials. However, in 1824 the 22-year-old Norwegian Niels Abel proved that there is no algebraic solution for finding the zeros of all fifth-degree polynomials. Certain fifth degree polynomials can be solved exactly; for example, $x^5 - x^4$ has the zeros 0 and 1, but there is no formula that will work for all fifth-degree polynomials. Just a few years later a brilliant teenage mathematician in France, Evariste Galois, developed results that showed precisely which polynomials of degree five and higher have an algebraic solution. Unfortunately, he was killed in a duel soon afterward, and his work was so complicated that its full significance was not appreciated for nearly 40 years.

In this section we will see how we can decide which polynomials have zeros that we can determine and how this information can be useful to us. A major role in this discussion is played by the technique of polynomial division, which is where we begin.

EXAMPLE 1   Find the zeros of $P(x) = x^3 + 2x^2 - x - 2$.

Solution   By inspection we see that $P(1) = 0$, so $x = 1$ is a root of the polynomial equation $P(x) = 0$. Let us now divide the linear factor $x - 1$ into the cubic $P(x)$. Division of polynomials is similar to division of numbers. To divide $x - 1$ (the divisor) into $x^3 + 2x^2 - x - 2$ (the dividend), we have

$$
\begin{array}{r}
x^2 + 3x + 2 \\
x - 1 \overline{\smash{\big)}\ x^3 + 2x^2 - x - 2} \\
\underline{x^3 - x^2} \\
3x^2 - x - 2 \\
\underline{3x^2 - 3x} \\
2x - 2 \\
\underline{2x - 2} \\
0
\end{array}
$$

The first step in determining the quotient $x^2 + 3x + 2$ is to divide the leading term, x, of the divisor $x - 1$ into the leading term, x^3, of the dividend $x^3 + 2x^2 - x - 2$. This gives the leading term in the quotient. This term, x^2, is then multiplied by the divisor $x - 1$ to get the term $x^3 - x^2$, which is shown in the second row of the calculations. We now subtract this term, just like in the division of numbers, from the dividend.

The process is repeated using the divisor and the reduced dividend at the bottom of the calculations until this line has degree less than the degree of the divisor. In our example the calculations stop when the bottom line is a constant since the degree of the divisor is 1.

Since the constant in the remainder of our example is 0, the linear factor $x - 1$ divides evenly into the polynomial $P(x)$, and we have factored $P(x)$ as

$$P(x) = x^3 + 2x^2 - x - 2 = (x - 1)(x^2 + 3x + 2).$$

The quadratic on the right of this equation factors as

$$x^2 + 3x + 2 = (x + 1)(x + 2),$$

so the complete factorization of $P(x)$ is

$$P(x) = (x - 1)(x^2 + 3x + 2) = (x - 1)(x + 1)(x + 2).$$

When a polynomial is written as a product of linear terms, the zeros are the numbers that make the linear factors zero. So the polynomial $P(x)$ has three distinct zeros, 1, -1, and -2. Using the fact that $P(0) = -2$ and that its end behavior is similar to that of $f(x) = x^3$ gives a graph similar to that shown in Figure 2.19.   ■

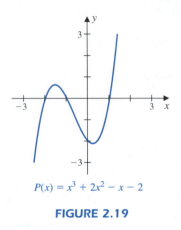

$$P(x) = x^3 + 2x^2 - x - 2$$

FIGURE 2.19

In the previous example we used the following result in the special case where the divisor was the linear term $D(x) = x - 1$.

The Division Algorithm

Suppose $D(x)$ and $P(x)$ are polynomials with $D(x) \neq 0$ and the degree of $D(x)$ is less than the degree of $P(x)$. Then there exists unique polynomials $Q(x)$ and $R(x)$, where $R(x)$ is either 0 or has degree less than the degree of $D(x)$, such that

$$P(x) = Q(x) \cdot D(x) + R(x) \quad \text{or, equivalently,} \quad \frac{P(x)}{D(x)} = Q(x) + \frac{R(x)}{D(x)}.$$

EXAMPLE 2   Find the quotient and remainder when the polynomial $2x^5 + x^4 - x^3 - x - 1$ is divided by the polynomial $x^2 - 2x + 1$.

Solution   Let $P(x) = 2x^5 + x^4 - x^3 - x - 1$ and $D(x) = x^2 - 2x + 1$. It is usually easier to perform the division if a position is reserved for every power of x starting with the highest-power term (in this case $2x^5$). So before performing the division we insert a $0x^2$ term into $P(x)$.

Since

$$
\begin{array}{r}
2x^3 + 5x^2 + 7x + 9 \\
x^2 - 2x + 1 \overline{\smash{\big)}\ 2x^5 + x^4 - x^3 + 0x^2 - x - 1} \\
\underline{2x^5 - 4x^4 + 2x^3} \\
5x^4 - 3x^3 + 0x^2 - x - 1 \\
\underline{5x^4 - 10x^3 + 5x^2} \\
7x^3 - 5x^2 - x - 1 \\
\underline{7x^3 - 14x^2 + 7x} \\
9x^2 - 8x - 1 \\
\underline{9x^2 - 18x + 9} \\
10x - 10
\end{array}
$$

we have

$$2x^5 + x^4 - x^3 - x - 1 = (2x^3 + 5x^2 + 7x + 9)(x^2 - 2x + 1) + (10x - 10),$$

or, equivalently,

$$\frac{2x^5 + x^4 - x^3 - x - 1}{x^2 - 2x + 1} = (2x^3 + 5x^2 + 7x + 9) + \frac{(10x - 10)}{(x^2 - 2x + 1)}. \qquad \blacksquare$$

If the polynomial $P(x)$ is divided by a linear factor $x - c$, then the remainder produced by the Division Algorithm must be a constant. Calling this constant R gives

$$P(x) = (x - c)Q(x) + R,$$

and setting $x = c$ gives

$$P(c) = (c - c)Q(c) + R = 0 + R = R.$$

Hence the remainder R in the division of $P(x)$ by $x - c$ is the value of the polynomial $P(x)$ at $x = c$. In particular, this implies that the constant R will be zero precisely when $x - c$ is a factor of $P(x)$.

Review the Division Algorithm and Factor Theorem carefully. They are frequently needed in calculus.

> **The Factor Theorem**
>
> If the polynomial $P(x)$ is divided by the linear factor $(x - c)$, then the remainder is $P(c)$. As a consequence, the linear term $(x - c)$ is a factor of the polynomial $P(x)$ if and only if $P(c) = 0$.

EXAMPLE 3   Factor $P(x) = x^4 + 5x^3 + 5x^2 - 5x - 6$, and sketch the graph of P.

Solution   First note that both $P(1) = 0$ and $P(-1) = 0$, so the Factor Theorem implies that $x - 1$ and $x - (-1) = x + 1$ are factors of $P(x)$. To find any remaining factors, divide $P(x)$ by the product $(x - 1)(x + 1) = x^2 - 1$ to produce

$$P(x) = (x^2 - 1)(x^2 + 5x + 6) = (x - 1)(x + 1)(x^2 + 5x + 6).$$

Finally, we factor the quadratic as

$$x^2 + 5x + 6 = (x + 2)(x + 3)$$

and write

$$P(x) = (x - 1)(x + 1)(x + 2)(x + 3).$$

The end behavior of $P(x)$ is the same as that of $f(x) = x^4$, so the graph of $y = P(x)$ is similar to that shown in Figure 2.20. ∎

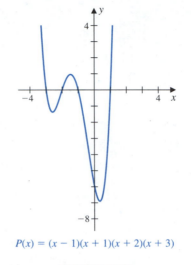

$$P(x) = (x - 1)(x + 1)(x + 2)(x + 3)$$

FIGURE 2.20

If the problem in Example 3 is changed slightly to $P(x) = x^4 + 5x^3 + 6x^2 - 5x - 7$, we still have $P(1) = 0$ and $P(-1) = 0$, so we still have the factors $x - 1$ and $x + 1$. But division by $x^2 - 1$ for this polynomial produces

$$P(x) = (x - 1)(x + 1)(x^2 + 5x + 7).$$

Applying the quadratic formula to the quadratic equation $x^2 + 5x + 7 = 0$, we find that

$$x = \frac{-5}{2(1)} \pm \frac{\sqrt{5^2 - 4(1)(7)}}{2(1)} = -\frac{5}{2} \pm \frac{\sqrt{-3}}{2}.$$

Since $\sqrt{-3}$ is not a real number, the quadratic has no real zeros, and the Factor Theorem implies that the quadratic has no real factors. Hence the factorization

$$P(x) = (x - 1)(x + 1)(x^2 + 5x + 7)$$

is complete. In Figure 2.21 we have shown the graph of $P(x) = x^4 + 5x^3 + 6x^2 - 5x - 7$. Notice that the graph crosses the x-axis at only two points, compared to the four x-intercepts for the graph in Figure 2.20.

Polynomials are basic functions that are useful for many applications. It is common, for example, to fit a polynomial to a collection of data points and then use the polynomial as a model for the data. For instance, the polynomial $P(x) = 1 + \frac{1}{2}(x - 1) - \frac{1}{8}(x - 1)^2 + \frac{1}{16}(x - 1)^3$ gives a good approximation to the function $f(x) = \sqrt{x}$ near $x = 1$, as shown in Figure 2.22.

In the next example we consider the special case of finding a polynomial that has specific zeros.

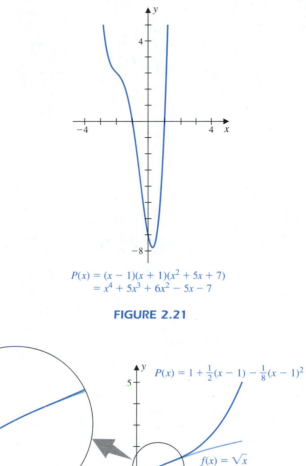

$$P(x) = (x - 1)(x + 1)(x^2 + 5x + 7)$$
$$= x^4 + 5x^3 + 6x^2 - 5x - 7$$

FIGURE 2.21

The polynomial $P(x)$ shown in Figure 2.22 is a *Taylor polynomial* for the function f at $x = 1$. In calculus you will see how to construct Taylor polynomials, which can be used to approximate functions.

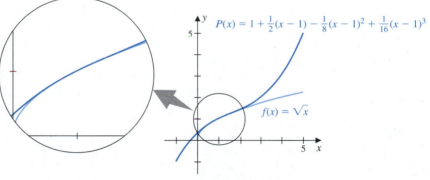

FIGURE 2.22

EXAMPLE 4   Find a fourth-degree polynomial that has zeros at $x = 1, -1, \frac{1}{2}$, and 2.

Solution   Since $1, -1, \frac{1}{2}$, and 2 are zeros of the polynomial, the Factor Theorem implies that the linear expressions $x - 1$, $x + 1$, $x - \frac{1}{2}$, and $x - 2$ are factors of the polynomial. So a fourth-degree polynomial that has the given zeros is

$$P(x) = (x - 1)(x + 1)\left(x - \frac{1}{2}\right)(x - 2) = x^4 - \frac{5}{2}x^3 + \frac{5}{2}x - 1.$$

If we wanted a polynomial with *integer* coefficients having the same zeros, we could multiply by 2 to produce the polynomial $\hat{P}(x) = 2x^4 - 5x^3 + 5x - 2$. The graphs of these two polynomials are shown in Figure 2.23.

The graph of $y = \hat{P}(x)$ is a vertical elongation, by a factor of 2, of the graph of $y = P(x)$.   ∎

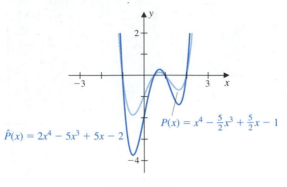

$\hat{P}(x) = 2x^4 - 5x^3 + 5x - 2$

$P(x) = x^4 - \frac{5}{2}x^3 + \frac{5}{2}x - 1$

FIGURE 2.23

The Rational Zero Test

In our examples to this point we have been able quite easily to determine the factors and zeros of the polynomials. When the factors are not so readily determined, there are some simple tests that we can apply to help isolate the zeros. The first test we consider is a method for finding rational zeros for polynomials that have rational coefficients.

The Rational Zero Test applies only to polynomials whose coefficients are rational numbers. (Recall that a rational number has the form p/q, where p and q are integers with $q \neq 0$.) So, for example, we cannot apply this test to $P(x) = \sqrt{2}x^2 - x + 1$ or to $P(x) = x^5 - \pi x^3 - x + 1$. In fact, we really need consider only polynomials with integer coefficients since if we multiply a polynomial with rational coefficients by the common denominator of the coefficients, we obtain a polynomial having the same zeros and all integer coefficients. For example, the equation

$$0 = \frac{1}{2}x^4 - \frac{2}{3}x^2 - x + 2$$

has the same zeros as the integer polynomial equation

$$0 = 6\left[\frac{1}{2}x^4 - \frac{2}{3}x^2 - x + 2\right] = 3x^4 - 4x^2 - 6x + 12.$$

As a consequence, we can concentrate on polynomials with only integer coefficients.

To develop the Rational Zero Test, let

$$P(x) = a_n x^n + a_{n-1}x^{n-1} + \cdots + a_1 x + a_0$$

be a polynomial where $a_0, a_1, \ldots, a_n$ are all integers and $a_n \neq 0$. Suppose $x = p/q$ is a rational zero of $P(x)$, where all possible cancellation has been done so that p and q have no common factors.

Consider the possibilities for p and q. Since p/q is a zero of $P(x)$, we have

$$a_n \frac{p^n}{q^n} + a_{n-1}\frac{p^{n-1}}{q^{n-1}} + \cdots + a_1\frac{p}{q} + a_0 = 0.$$

Multiplying both sides of this equation by q^n gives

$$a_n p^n + a_{n-1}p^{n-1}q + a_{n-2}p^{n-2}q^2 + \cdots + a_1 pq^{n-1} + a_0 q^n = 0$$

Although the Rational Zero Test is not essential for calculus, it can be useful for quickly visualizing the graph of a polynomial.

or

$$p\left(a_n p^{n-1} + a_{n-1} p^{n-2} q + a_{n-2} p^{n-3} q^2 + \cdots + a_1 q^{n-1}\right) = -a_0 q^n.$$

Since p divides the left side of this last equation, it must also divide the right side. But p has no factors in common with q, so p cannot divide q^n. Hence p must divide a_0.

In a similar manner, we can rewrite the equation as

$$\left(a_{n-1} p^{n-1} + a_{n-2} p^{n-2} q + \cdots + a_1 p q^{n-2} + a_0 q^{n-1}\right) q = -a_n p^n,$$

and deduce that q must divide a_n.

Combining these observations gives the *Rational Zero Test*.

The Rational Zero Test

Let $\dfrac{p}{q}$ be a rational zero of

$$P(x) = a_n x^n + a_{n-1} x^{n-1} + \cdots + a_1 x + a_0,$$

where $a_0, \ldots, a_n$ are integers and $a_n \neq 0$. Then p divides a_0 and q divides a_n.

EXAMPLE 5   Find all the rational numbers that are possibilities for zeros of the polynomial $P(x) = x^3 - \frac{13}{4}x^2 + \frac{11}{4}x - \frac{1}{2}$.

Solution   Solving the equation $P(x) = x^3 - \frac{13}{4}x^2 + \frac{11}{4}x - \frac{1}{2} = 0$ is equivalent to solving the equation $Q(x) = 4P(x) = 4x^3 - 13x^2 + 11x - 2 = 0$.

The only possible rational zeros of $Q(x)$ are the factors of -2 divided by the factors of 4. That is,

$$\pm\frac{1}{1}, \ \pm\frac{1}{2}, \ \pm\frac{1}{4}, \ \pm\frac{2}{1}, \ \pm\frac{2}{2}, \ \pm\frac{2}{4}.$$

Simplifying these fractions and eliminating duplication gives the possibilities

$$\pm 1, \ \pm\frac{1}{2}, \ \pm\frac{1}{4}, \ \pm 2.$$

The following table shows $Q(x)$ evaluated at each of the possible rational zeros.

x	1	-1	2	-2	$\frac{1}{4}$	$-\frac{1}{4}$	$\frac{1}{2}$	$-\frac{1}{2}$
$Q(x)$	0	-30	0	-108	0	$-\frac{45}{8}$	$\frac{3}{4}$	$-\frac{45}{4}$

The polynomial $Q(x)$, and hence also $P(x)$, has zeros $x = 1$, $x = 2$, and $x = \frac{1}{4}$. Since $P(x)$ has a leading coefficient of 1, it is completely factored as

$$P(x) = x^3 - \frac{13}{4}x^2 + \frac{11}{4}x - \frac{1}{2} = (x - 1)(x - 2)\left(x - \frac{1}{4}\right).$$

If we are interested in finding exact values for zeroes of a polynomial, the graph obtained using a graphing device can eliminate many possibilities.

A graphing device can be used to eliminate many of the possible rational numbers that could be zeros. Figure 2.24 shows a graph of $P(x) = x^3 - \frac{13}{4}x^2 + \frac{11}{4}x - \frac{1}{2}$. It is clear from this graph, for example, that no negative numbers are zeros. ∎

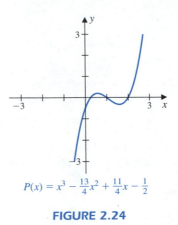

$$P(x) = x^3 - \frac{13}{4}x^2 + \frac{11}{4}x - \frac{1}{2}$$

FIGURE 2.24

EXAMPLE 6   Find all solutions to $P(x) = 2x^5 - x^4 - 12x^3 - 8x^2 - 5x + 6$.

Solution   If we can find three of the zeros of P, then $P(x)$ can be factored as the product of linear terms involving those zeros and a quadratic term. Since quadratics can be solved by the quadratic formula, we will be able to find all the zeros of P.

The only possible rational zeros are the quotients of the divisors of 6 divided by the divisors of 2. On eliminating duplicates we have

$$\pm 1, \; \pm 2, \; \pm 3, \; \pm 6, \; \pm \frac{1}{2}, \; \pm \frac{3}{2}$$

If we use a graphing device we can sketch the graph of $P(x)$, as shown in Figure 2.25. The graph indicates that it is likely that there are zeros at 3, at -2 and between 0 and 1. Many of the rational possibilities are clearly eliminated.

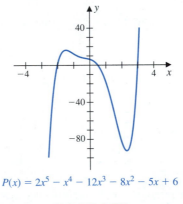

$$P(x) = 2x^5 - x^4 - 12x^3 - 8x^2 - 5x + 6$$

FIGURE 2.25

We can verify that $P(3) = 0$ and $P(-2) = 0$, and at the only rational possibility between 0 and 1 we have $P(\frac{1}{2}) = 0$. So $(x - 3)$, $(x + 2)$, and $(2x - 1)$ are all factors of $P(x)$. Dividing by these factors gives

$$P(x) = (x - 3)(x + 2)(2x - 1)(x^2 + x + 1).$$

The quadratic formula applied to $x^2 + x + 1 = 0$ gives

$$x = -\frac{1}{2} \pm \frac{\sqrt{(1)^2 - 4(1)(1)}}{2} = -\frac{1}{2} \pm \frac{\sqrt{-3}}{2},$$

which are not real numbers. So the only real zeros of $P(x)$ are $x = 3$, $x = -2$, and $x = \frac{1}{2}$. ∎

Descartes' Rule of Signs

The previous examples have shown that the list of possible rational zeros can be lengthy. The 17th century philosopher and mathematician René Descartes is credited with discovering a test that is helpful in eliminating some of these rational numbers from the list of possible zeros.

Suppose that $P(x)$ is a polynomial written in descending powers of x. We say that $P(x)$ has a *variation in sign* whenever adjacent coefficients are opposite in sign. For example, the polynomial

$$4x^7 + \underbrace{3x^5 - x^4}_{+ \text{ to } -} \quad \underbrace{-2x^3 + 5x^2}_{- \text{ to } +} \quad \underbrace{+2x - 7}_{+ \text{ to } -}$$

has three variations in sign. The number of sign changes gives an indication of the number of possible positive zeros.

Although you might not need Descartes' Rule of Signs in calculus, it is a simple result related to zeros of polynomials.

> **Descartes' Rule of Signs**
>
> Let $P(x)$ be a polynomial with real coefficients.
>
> (i) The number of positive zeros of $P(x)$ is either equal to the number of variations in sign of $P(x)$ or less than this by an even number.
> (ii) The number of negative real zeros is either equal to the number of variations in sign of $P(-x)$ or less than this by an even number.

Notice that Descartes' Rule of Signs gives information about the number of *real* zeros, not just about the number of *rational* zeros.

EXAMPLE 7   Use Descartes' Rule of Signs to determine the number of possible positive and negative real solutions of the equation

$$P(x) = 2x^7 + x^6 - 3x^4 - x^3 - 2 = 0.$$

Solution   Since $P(x)$ has only one variation in sign (between the term x^6 and $-3x^4$), the polynomial has exactly one positive real zero. To determine the number of possible negative real zeros, replace x with $-x$ in $P(x)$ to produce

$$P(-x) = 2(-x)^7 + (-x)^6 - 3(-x)^4 - (-x)^3 - 2 = -2x^7 + x^6 - 3x^4 + x^3 - 2.$$

There are four variations of sign in $P(-x)$, so there are either four, two, or zero negative real zeros. Hence the polynomial can have either one, three, or five real zeros. The

Rational Zero Test tells us that the only rational possibilities for zeros are

$$\frac{\text{divisors of } -2}{\text{divisors of } 2} = \pm 1, \pm 2, \pm \frac{1}{2}.$$

In fact, none of these possibilities satisfy the original equation, so any real solutions must be irrational numbers.   ■

It is good practice to create a summary list when you need to solve a complicated problem. It organizes your thoughts and makes sure that you fully understand the problem.

Listed next is an outline that summarizes a procedure for finding zeros and factors of a polynomial.

Finding Zeros and Factors of a Polynomial

(i)   List all possible rational zeros using the Rational Zero Test.

(ii) Apply Descartes' Rule of Signs to determine the number of positive and negative zeros the polynomial can have. At this step it might be determined that there are no positive or no negative real zeros.

(iii) Check the candidates for zeros substituting the values from the smallest in magnitude to the largest.

(iv) When a zero is found, factor the polynomial and repeat the process on the quotient. There is no need to check possible zeros of the quotient that have already been eliminated from the list of zeros at the previous stage, but check again any zeros that have been found at the previous stage, since one or more may have a multiplicity greater than one.

(v)   If the polynomial has been factored to linear terms times a quadratic term, factor the quadratic, using the quadratic formula if necessary.

EXERCISE SET 2.3

In Exercises 1–6, find the quotient $Q(x)$ and remainder $R(x)$ when the polynomial $P(x)$ is divided by the polynomial $D(x)$.

1. $P(x) = 3x^2 - 2x + 2$, $D(x) = x - 1$

2. $P(x) = 2x^2 - 3x + 4$, $D(x) = x + 2$

3. $P(x) = x^3 + x^2 - 2$, $D(x) = x - 1$

4. $P(x) = 3x^4 + 2x^3 - x + 2$, $D(x) = x^2 + 2x - 1$

5. $P(x) = 2x^4 - 2x^3 + 4x^2 + x - 2$, $D(x) = x^2 - x - 1$

6. $P(x) = 3x^5 - 2x^4 + x^3 + 5x^2 - x - 1$,
$\quad D(x) = x^4 - 2x^3 + x + 1$

In Exercises 7–10, use the Factor Theorem to show that $x - c$ is a factor of $P(x)$ for the given values of c, and factor $P(x)$ completely.

7. $P(x) = x^3 - 5x^2 + 8x - 4$, $c = 1$

8. $P(x) = 3x^4 + 5x^3 - 5x^2 - 5x + 2$, $c = 1$, $c = -2$

9. $P(x) = 2x^4 + 3x^3 - 12x^2 - 7x + 6$, $c = -1$, $c = -3$

10. $P(x) = 2x^4 - 2x^3 - 11x^2 - 4x + 3$, $c = -1$, $c = 3$

In Exercises 11–16, determine all the possibilities for rational roots.

11. $x^4 - 2x^3 - 3x^2 - 2x + 6 = 0$

12. $x^5 + 2x^4 - 8x^3 + 4x^2 + 12x - 16 = 0$

13. $10x^5 - 14x^3 + 18x^2 + 6x - 4 = 0$

14. $6x^5 - 23x^4 + 40x^3 - 47x^2 + 24x - 4 = 0$

15. $\frac{2}{3}x^3 - \frac{19}{3}x^2 + 18x - 15 = 0$

16. $\frac{3}{4}x^3 + \frac{1}{8}x^2 - \frac{9}{4}x + 1 = 0$

In Exercises 17–20, use Descartes' Rule of Signs to determine the maximum number of positive and negative zeros of the polynomial.

17. $6x^4 + 5x^3 - 14x^2 + x + 2$

18. $9x^4 - 9x^3 - 19x^2 + x + 2$

19. $x^5 + 2x^4 - x - 2$

20. $x^5 - 2x^4 - 9x^3 + 8x^2 - 22x + 24$

In Exercises 21–26, find all rational zeros of the polynomial, then determine any irrational zeros, and factor the polynomial completely.

21. $x^3 - 5x^2 + 2x + 8$ **22.** $3x^3 - 8x^2 + 3x + 2$

23. $x^3 - 3x^2 + x - 3$ **24.** $2x^4 - 3x^3 - 3x^2 + 2x$

25. $x^4 - 4x^3 + 3x^2 + 4x - 4$

26. $2x^5 + x^4 - 12x^3 + 10x^2 + 2x - 3$

27. Determine the polynomial $P(x)$ of least degree whose graph could be as shown in the figure.

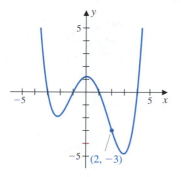

28. Find a third-degree polynomial $P(x)$ that has zeros at $x = -1$, $x = 1$, and $x = 2$ and whose x-term has coefficient 3.

29. Find a fifth-degree polynomial that has a zero of multiplicity 2 at $x = 1$, a zero at $x = 4$, and the factor $x^2 + x + 1$.

30. The polynomial $P(x) = x^4 - 4x^3 + 4x^2 - 1$ has a local maximum at $(1, 0)$ and local minima at $(0, -1)$ and $(2, -1)$. Factor the polynomial completely and sketch its graph. Then determine how many real zeros the polynomial $Q(x) = P(x) + c$ has for each constant c.

31. Sketch the graphs of $f(x) = x^3$ and $g(x) = 2x^2 + x - 2$, and determine all the points of intersection.

32. For each positive integer n, let $f_n(x) = x^n$. Compare the sizes of $f_n(x)$ and $f_{n+1}(x)$ on the intervals $0 < x < 1$ and $x > 1$.

2.4   RATIONAL FUNCTIONS

A rational number is the quotient of two integers, the most basic of numbers. In a similar manner, we define a *rational function* to be the quotient of two of the most basic functions, polynomials.

Rational Functions

A rational function has the form

$$f(x) = \frac{P(x)}{Q(x)},$$

where $P(x)$ and $Q(x)$ are polynomials. The domain of f is the set of all real numbers x with $Q(x) \neq 0$.

The functions described by

$$f(x) = \frac{x+2}{x^2-1}, \quad g(x) = \frac{1}{x}, \quad \text{and} \quad h(x) = \frac{x^2-2x+1}{x^4+2x^2+5}$$

are all rational. The domain of f is the set of all real numbers other than 1 and -1. The domain of g is the set of all nonzero real numbers. The domain of h is the set of all real numbers, $\mathbb{R}$, since the denominator is never zero.

If the numerator and denominator of a rational function have common factors, we can cancel these factors and write the fraction in simplest form. For example, if

$$f(x) = \frac{x^2-9}{x-3},$$

we can factor the numerator to obtain

$$f(x) = \frac{(x-3)(x+3)}{x-3} = x+3.$$

However, $x = 3$ is excluded from the domain of the function f, since f is undefined at $x = 3$ in the original form. The graph of f is shown in Figure 2.26. The open circle at the point $(3, 6)$ is used to indicate that $x = 3$ is not in the domain of f but that $f(x)$ approaches 6 as x approaches 3. This limiting behavior is so common in calculus that it is often abbreviated by writing $f(x) \to 6$ as $x \to 3$. You will not see this feature if you are using a graphing device to sketch the graph unless the zero in the denominator occurs at one of the x-values used by the device in plotting the graph.

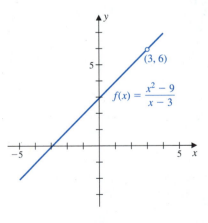

FIGURE 2.26

EXAMPLE 1   Find the domain, x-intercepts, and y-intercepts of the rational function f given by

$$f(x) = \frac{x^2+x-2}{x-2}.$$

Solution   The denominator of $f(x)$ is zero precisely when $x = 2$, so the domain of f consists of all real numbers except $x = 2$.

The y-intercept of the graph occurs at $(0, f(0)) = \left(0, \frac{-2}{-2}\right) = (0, 1)$.

To find the x-intercepts, we need to determine when the numerator of $f(x)$ is 0. Factoring gives

$$\frac{x^2 + x - 2}{x - 2} = \frac{(x + 2)(x - 1)}{x - 2},$$

and $f(x) = 0$ precisely when $x = -2$ or $x = 1$. The graph crosses the y- and x-axes at the points $(0, 1)$, $(-2, 0)$, and $(1, 0)$, as shown in Figure 2.27.

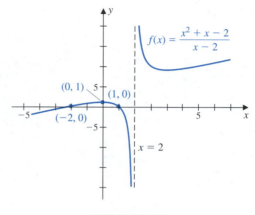

FIGURE 2.27

Figure 2.27 suggests that as x gets close to the value 2 from the left side (gets close to 2 but remains less than 2), the values of the function become negative and large in magnitude. As x gets close to 2 from the right side (gets close to 2 but remains greater than 2), the values of the function become positive and large. Table 2.3 shows $f(x)$ for values of x close to 2 from the left and from the right, and provides additional evidence of this.

TABLE 2.3

x	$f(x)$	x	$f(x)$
1.9	-35.1	2.1	45.1
1.99	-395.01	2.01	405.01
1.999	-3995.001	2.001	4005.001
1.9999	-39995.0001	2.0001	40005.0001

An **asymptote** of a graph is a line that the graph approaches. The vertical line $x = a$ is a *vertical asymptote* of the graph of f provided $f(x)$ approaches ∞ or $-\infty$ as x approaches a from the left side or from the right side. For example, Figure 2.27 shows the dashed vertical line at $x = 2$ as a vertical asymptote of the graph in Example 1.

In Figure 2.28 we see the various possibilities that can produce vertical asymptotes, together with some notation that is used to describe the situations. The plus sign as an exponent of a means that x approaches a from the right and the minus sign as an exponent on a means that x approaches a from the left.

Asymptotes are closely connected with the concept of *limit* that you will encounter in calculus. In calculus you will be asked to make accurate sketches of rational functions, including their asymptotes. When used properly, graphing devices will display the behavior of a function near a vertical asymptote.

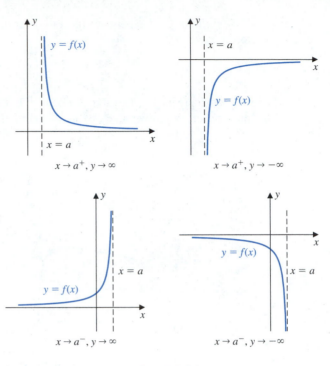

FIGURE 2.28

Vertical Asymptotes

The vertical line $x = a$ is a **vertical asymptote** for the graph of f provided that $f(x) \to \infty$ or $f(x) \to -\infty$ as x approaches a from the left, written $x \to a^-$, or as x approaches a from the right, written $x \to a^+$.

A rational function f can have a vertical asymptote at $x = a$ only if $x - a$ is a factor of its denominator—that is, if a is not in the domain of f. Such a number need not be a vertical asymptote, however. Consider the rational function defined by

$$f(x) = \frac{x^2 + 2x - 3}{x^2 - 1}.$$

Its domain is the set of all real numbers except -1 and 1, the zeros of the denominator. But factoring the numerator and denominator gives

$$f(x) = \frac{x^2 + 2x - 3}{x^2 - 1} = \frac{(x - 1)(x + 3)}{(x - 1)(x + 1)} = \frac{x + 3}{x + 1}.$$

A vertical asymptote occurs at $x = -1$ since the simplified denominator has -1 as a zero, which implies that $f(x)$ becomes large in magnitude as x approaches -1. But there is no vertical asymptote at $x = 1$. Instead, the graph has a hole at the point $(1, 2)$, as shown in Figure 2.29. The original expression for $f(x)$ tells us where vertical asymptotes *can* occur, but the reduced expression tells us where they *do* occur.

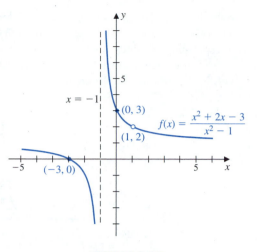

$f(x) = \dfrac{x^2 + 2x - 3}{x^2 - 1}$

FIGURE 2.29

EXAMPLE 2 Sketch the graph of

$$f(x) = \frac{x^2 - 1}{x^3 - 3x^2 + 2x} \qquad \text{for } x \text{ in } [-3, 3].$$

Solution First we factor the numerator and the denominator to obtain

$$f(x) = \frac{(x + 1)(x - 1)}{x(x - 1)(x - 2)} = \frac{x + 1}{x(x - 2)}, \quad \text{if } x \ne 1.$$

The domain of f excludes the numbers that make the original denominator zero, namely, 0, 1, and 2. Vertical asymptotes occur at $x = 0$ and at $x = 2$, but a "hole" occurs in the graph when $x = 1$. The y-coordinate of the hole is the value at $x = 1$ of the simplified fraction, that is,

$$y = \frac{1 + 1}{1(1 - 2)} = -2.$$

The vertical lines $x = 0$ and $x = 2$ are vertical asymptotes for the graph. To determine how the graph approaches these asymptotes, we need to determine whether the function is positive or negative near these values of x.

The function can change sign only when the function is zero or when passing over points that are not in the domain. Hence the sign of $f(x)$ must remain the same in intervals that are bracketed by these values. By determining the sign of $f(x)$ at a single

```
    x       − − − − − − 0 + + + + + + + + + +
 (x − 2)    − − − − − − − − − − − 0 + + + + + +
 (x + 1)    − − − − 0 + + + + + + + + + + + +
 (x + 1)
 ─────      − − − − 0 + ☐ − − − ☐ + + + + + +
 x(x − 2)
          ─┼──┼──┼──┼──┼──┼──┼──┼──┼──→
          −3  −2  −1   0   1   2   3   4   5    x
```

As $x \to 0^-, f(x) \to \infty$ As $x \to 2^-, f(x) \to -\infty$
As $x \to 0^+, f(x) \to -\infty$ As $x \to 2^+, f(x) \to \infty$

FIGURE 2.30

point in each of these intervals, we can determine the sign of $f(x)$ on its entire domain. The chart in Figure 2.30 shows the situation for our function. We have used the ☐ symbol to represent the values at which the function f is undefined.

Using the information on the chart and evaluating $f(x)$ at the endpoints -3 and 3 verifies the features on the computer-generated graph in Figure 2.31.   ■

We cannot analyze the graph of the rational function completely without calculus. But from what we know the computer-generated graph in Figure 2.31 appears reasonable and one we can accept with confidence.

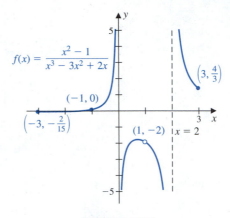

FIGURE 2.31

In Section 2.2 we discussed the end behavior of polynomials and showed that the leading coefficient of the polynomial (the term with highest exponent) determines the behavior of the polynomial as $x \to \infty$ and as $x \to -\infty$. In a similar manner, the leading terms of the polynomials in the numerator and denominator of a rational function determine the behavior of the rational function as $x \to \infty$ and as $x \to -\infty$.

Suppose that the rational function f has the form

$$f(x) = \frac{p(x)}{q(x)} = \frac{a_n x^n + a_{n-1} x^{n-1} + \cdots + a_1 x + a_0}{b_m x^m + b_{m-1} x^{m-1} + \cdots + b_1 x + b_0},$$

where a_n and b_m are nonzero. Then as $x \to \infty$ and $x \to -\infty$, $f(x)$ behaves in the same way as

$$\frac{a_n x^n}{b_m x^m} = \frac{a_n}{b_m} x^{n-m}.$$

That is,

(i) When $n = m$: $f(x) \to \frac{a_n}{b_m}$ as $x \to \infty$ and as $x \to -\infty$.

(ii) When $n < m$: $f(x) \to 0$ as $x \to \infty$ and as $x \to -\infty$.

(iii) When $n > m$: The behavior of $f(x)$ as $x \to \infty$ and as $x \to -\infty$ depends on whether $n - m$ is even or odd and also on the sign of the quotient $\frac{a_n}{b_m}$.

When $n = m$, the graph approaches the horizontal line $y = a_n/b_m$, which is a *horizontal asymptote* to the graph. (See Figure 2.32(a).) When $n < m$, the graph approaches the horizontal asymptote $y = 0$. (See Figure 2.32(b).)

Figure 2.33 illustrates some of the possibilities that can occur in the case $n > m$, when the graph has no horizontal asymptote.

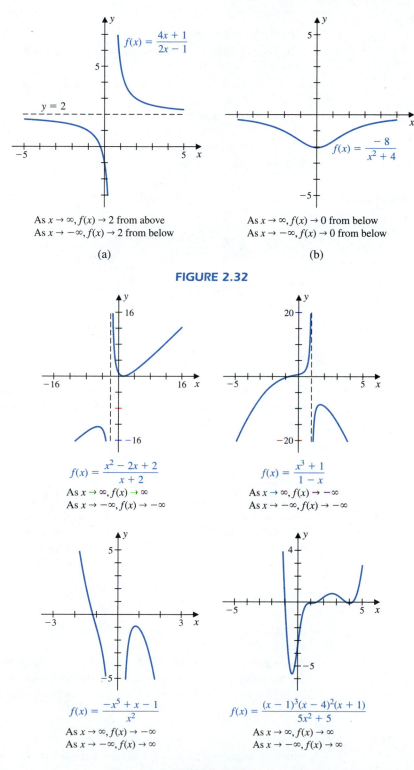

As $x \to \infty$, $f(x) \to 2$ from above
As $x \to -\infty$, $f(x) \to 2$ from below

(a)

As $x \to \infty$, $f(x) \to 0$ from below
As $x \to -\infty$, $f(x) \to 0$ from below

(b)

FIGURE 2.32

$f(x) = \dfrac{x^2 - 2x + 2}{x + 2}$
As $x \to \infty$, $f(x) \to \infty$
As $x \to -\infty$, $f(x) \to -\infty$

$f(x) = \dfrac{x^3 + 1}{1 - x}$
As $x \to \infty$, $f(x) \to -\infty$
As $x \to -\infty$, $f(x) \to -\infty$

$f(x) = \dfrac{-x^5 + x - 1}{x^2}$
As $x \to \infty$, $f(x) \to -\infty$
As $x \to -\infty$, $f(x) \to \infty$

$f(x) = \dfrac{(x - 1)^3(x - 4)^2(x + 1)}{5x^2 + 5}$
As $x \to \infty$, $f(x) \to \infty$
As $x \to -\infty$, $f(x) \to \infty$

FIGURE 2.33

> **Horizontal Asymptotes**
>
> The horizontal line $y = a$ is a **horizontal asymptote** to the graph of f if $f(x) \to a$ as $x \to \infty$ or as $x \to -\infty$.

One way to keep from confusing horizontal from vertical asymptotes is to remember that horizontal means parallel to the horizon.

A rational function can have at most one horizontal asymptote since if $f(x)$ approaches a finite value as x approaches ∞, it approaches that same value as x approaches $-\infty$. In the next section we will see examples of non-rational functions that have two distinct horizontal asymptotes, one corresponding as $x \to \infty$ and the other as $x \to -\infty$.

EXAMPLE 3   For each of the following rational functions, determine if the graph has a horizontal asymptote.

a. $f(x) = \dfrac{3x^4 - 200x + 10000}{-7x^4 + 50x^3 + 1000x^2 + 5000x + 10000}$

b. $f(x) = \dfrac{x^3 - 2x + 2}{x^4 - x^3 + 5x^2 - 3}$

c. $f(x) = \dfrac{2x^4 - 3x^3 + 4x - 1}{5x^3 + 7x + 2}$

d. $f(x) = \dfrac{2x^4 - 3x^3 + 4x - 1}{5x^2 + 7x + 2}$

Solution   a. Since the degree, 4, of the polynomial in the numerator agrees with the degree of the polynomial in the denominator, the graph, shown in Figure 2.34(a), has the horizontal asymptote

$$y = \frac{3}{-7} = -\frac{3}{7}.$$

b. The degree of the polynomial in the numerator, 3, is less than the degree, 4, of the polynomial in the denominator, so $f(x) \to 0$ as $x \to \infty$ and as $x \to -\infty$. As a consequence, $y = 0$ is a horizontal asymptote to the graph, as shown in Figure 2.34(b).

c. Since the degree of the numerator, 4, is greater than the degree, 3, of the denominator, there is no horizontal asymptote to the graph. In fact, for x large in magnitude, the graph is similar to the graph of

$$\frac{2x^4}{5x^3} = \frac{2}{5}x.$$

So $f(x) \to \infty$ as $x \to \infty$, and $f(x) \to -\infty$ as $x \to -\infty$ along the line $y = \frac{2}{5}x$, as shown in Figure 2.34(c).

d. As in part (c), the degree of the numerator, 4, is greater than the degree of the denominator, and there is no horizontal asymptote to the graph. Now, however, for x large in magnitude, the graph of f behaves as the graph of

$$\frac{2x^4}{5x^2} = \frac{2}{5}x^2,$$

which approaches ∞ as $x \to \infty$ and also approaches ∞ as $x \to -\infty$, as shown in Figure 2.34(d). ∎

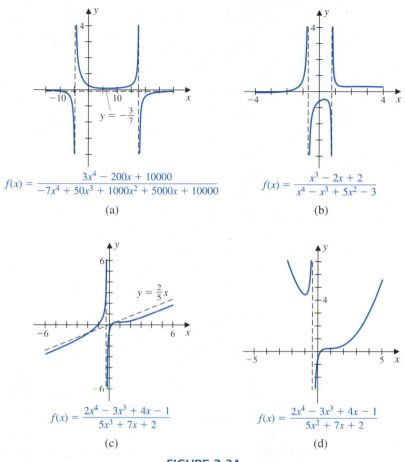

$$f(x) = \frac{3x^4 - 200x + 10000}{-7x^4 + 50x^3 + 1000x^2 + 5000x + 10000}$$

(a)

$$f(x) = \frac{x^3 - 2x + 2}{x^4 - x^3 + 5x^2 - 3}$$

(b)

$$f(x) = \frac{2x^4 - 3x^3 + 4x - 1}{5x^3 + 7x + 2}$$

(c)

$$f(x) = \frac{2x^4 - 3x^3 + 4x - 1}{5x^2 + 7x + 2}$$

(d)

FIGURE 2.34

EXAMPLE 4   Sketch the graph of

$$f(x) = \frac{x^2 - 1}{x^2 - 4}.$$

Solution   The numerator and denominator both have degree two and leading coefficient 1, so $y = 1$ is a horizontal asymptote to the graph.

Since $f(0) = \frac{1}{4}$, the y-intercept for the graph is at $(0, \frac{1}{4})$. To determine any x-intercepts and vertical asymptotes, we factor $f(x)$ as

$$f(x) = \frac{(x - 1)(x + 1)}{(x - 2)(x + 2)}.$$

Since the fraction has no common factors, the graph has x-intercepts at -1 and 1 and vertical asymptotes $x = -2$ and $x = 2$. The sign chart given in Figure 2.35 tells us how the graph approaches the vertical asymptotes.

Since

$$f(-x) = \frac{(-x)^2 - 1}{(-x)^2 - 4} = \frac{x^2 - 1}{x^2 - 4} = f(x),$$

the graph is symmetric with respect to the y-axis and likely appears as shown in Figure 2.36. ∎

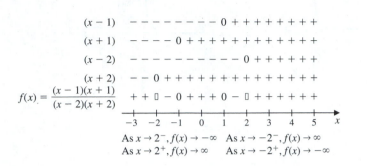

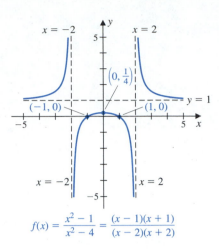

As $x \to 2^-, f(x) \to -\infty$ As $x \to -2^-, f(x) \to \infty$
As $x \to 2^+, f(x) \to \infty$ As $x \to -2^+, f(x) \to -\infty$

$$f(x) = \frac{x^2 - 1}{x^2 - 4} = \frac{(x - 1)(x + 1)}{(x - 2)(x + 2)}$$

FIGURE 2.35   **FIGURE 2.36**

EXAMPLE 5   Sketch the graph of the function defined by

$$f(x) = \frac{x^2 + 2x - 3}{2x^2 - 3x - 2}.$$

Solution   Since the degrees of the numerator and denominator are the same,

$$f(x) \to \frac{1}{2} \qquad \text{as } x \to \infty \text{ and as } x \to -\infty,$$

and $x = \frac{1}{2}$ is a horizontal asymptote.

The y-intercept to the graph of f is $(0, \frac{3}{2})$. To find the x-intercepts and vertical asymptotes, we first factor the numerator and denominator of $f(x)$ as

$$f(x) = \frac{x^2 + 2x - 3}{2x^2 - 3x - 2} = \frac{(x - 1)(x + 3)}{(x - 2)(2x + 1)}.$$

From this factored form we see that the x-intercepts are $(1, 0)$ and $(-3, 0)$ and that the graph has vertical asymptotes $x = 2$ and $x = -\frac{1}{2}$. The chart in Figure 2.37 shows the sign of $f(x)$ on the various regions of the domain and indicates how the graph approaches the vertical asymptotes.

```
         (x − 1)   − − − − − − − − − − 0 + + + + + +
         (x + 3)   − − 0 + + + + + + + + + + + + + +
         (x − 2)   − − − − − − − − − − − 0 + + + +
        (2x + 1)   − − − − − − − 0 + + + + + + + + +
f(x) = (x−1)(x+3)  + + 0 − − − − □ + + 0 − □ + + + +
       (x−2)(2x+1)
                   ┼──┼──┼──┼─┼┼┼─┼──┼──┼──┼──→
                  −4  −3  −2 −1−½ 0   1   2  3  4   x
```

As $x \to -\frac{1}{2}^-, f(x) \to -\infty$ As $x \to 2^-, f(x) \to -\infty$
As $x \to -\frac{1}{2}^+, f(x) \to \infty$ As $x \to 2^+, f(x) \to \infty$

FIGURE 2.37

This information permits us to sketch the graph, shown in Figure 2.38, with confidence.   ■

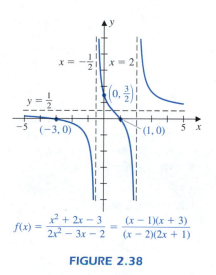

$$f(x) = \frac{x^2 + 2x - 3}{2x^2 - 3x - 2} = \frac{(x-1)(x+3)}{(x-2)(2x+1)}$$

FIGURE 2.38

We have seen that a rational function will not have a horizontal asymptote when the degree of the numerator is greater than the degree of the denominator. However, in the special case where the degree of the numerator is just 1 more than the degree of the denominator, the graph will approach a nonhorizontal line as $x \to \infty$ and as $x \to -\infty$. This line is called a **slant**, or *oblique*, **asymptote** to the graph. (See Figure 2.39.)

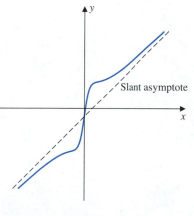

Slant asymptote

FIGURE 2.39

To determine the equation of the slant asymptote, suppose that

$$f(x) = \frac{P(x)}{Q(x)},$$

where the degree of the polynomial $P(x)$ is exactly 1 greater than the degree of the polynomial $Q(x)$. The Division Algorithm implies that there exists a linear quotient, say, $ax + b$, such that

$$P(x) = (ax + b)Q(x) + R(x).$$

So

$$f(x) = \frac{P(x)}{Q(x)} = (ax + b) + \frac{R(x)}{Q(x)},$$

where the degree of the polynomial $R(x)$ is less than the degree of $Q(x)$. Because of this, $R(x)/Q(x) \to 0$ as $x \to \infty$ and as $x \to -\infty$. As a consequence, the graph of f approaches the line $y = ax + b$ as $x \to \infty$ and as $x \to -\infty$. The line $y = ax + b$ is the slant asymptote for the graph.

EXAMPLE 6   Sketch the graph of

$$f(x) = \frac{x^2 - 2x - 3}{x + 2}.$$

Solution   Since $f(0) = -3/2$, the y-intercept for the graph occurs at $(0, -3/2)$.
Factoring the numerator of $f(x)$ gives

$$f(x) = \frac{(x + 1)(x - 3)}{x + 2},$$

which implies that the x-intercepts are at $(-1, 0)$ and $(3, 0)$, and a vertical asymptote occurs at $x = -2$. The chart in Figure 2.40 shows the sign of $f(x)$ in the various intervals of the domain of f and shows how the graph approaches the vertical asymptote.

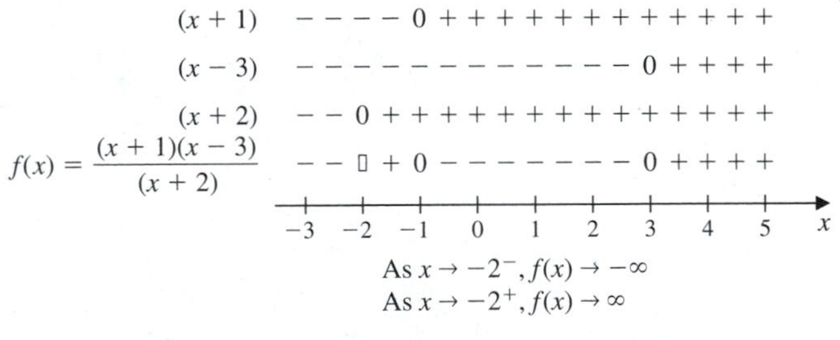

$$(x + 1) \qquad - - - - \ 0 + + + + + + + + + + + + +$$
$$(x - 3) \qquad - - - - - - - - - - - - - \ 0 + + + +$$
$$(x + 2) \qquad - - \ 0 + + + + + + + + + + + + + + +$$
$$f(x) = \frac{(x + 1)(x - 3)}{(x + 2)} \qquad - - \ \square + 0 - - - - - - - - 0 + + + +$$

As $x \to -2^-, f(x) \to -\infty$
As $x \to -2^+, f(x) \to \infty$

FIGURE 2.40

There are no horizontal asymptotes to the graph since the degree of the numerator is greater than the degree of the denominator. However, if we divide the denominator, $x + 2$, into the original numerator, $x^2 - 2x - 3$, we find that

$$f(x) = x - 4 + \frac{5}{x + 2}.$$

The line $y = x - 4$ is a slant asymptote to the graph, as shown in Figure 2.41.   ■

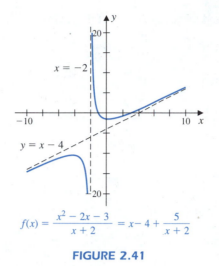

$$f(x) = \frac{x^2 - 2x - 3}{x + 2} = x - 4 + \frac{5}{x + 2}$$

FIGURE 2.41

EXAMPLE 7

This example is a typical optimization problem involving a rational function of the type you will encounter in calculus. To find the exact minimum value requires calculus. It is a good use of technology to use the graph to approximate this value. The graph not only allows us to estimate the minimum but shows why the minimum exists.

An aluminum can is to be constructed that will contain 2000 cm³ of liquid. Approximate the dimensions of the can that will minimize the amount of required material.

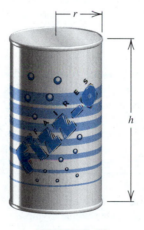

FIGURE 2.42

Solution   Let r represent the radius of the can and h represent the height, as shown in Figure 2.42. The total area A of the can is the sum of the areas of the top, bottom, and side. The areas of the top and bottom are both πr^2, the area of the circle with radius r. To determine the area of the side, cut the can vertically and roll out the side into a rectangle with length h and width the circumference of the top, $2\pi r$. This gives an area of $h(2\pi r)$. So

$$A = \pi r^2 + \pi r^2 + 2\pi rh = 2\pi r^2 + 2\pi rh.$$

We first determine A as a function only of the variable r by using the fact that the volume V of the can is 2000 cm³. Since the volume of the can is also $V = \pi r^2 h$, we have

$$2000 = \pi r^2 h, \quad \text{so} \quad h = \frac{2000}{\pi r^2},$$

and the area in terms of the radius is

$$A(r) = 2\pi r^2 + 2\pi r \left(\frac{2000}{\pi r^2}\right) = 2\pi r^2 + \frac{4000}{r}.$$

FIGURE 2.43

The graph of $A(r)$ in Figure 2.43 indicates that the minimum value of $A(r)$ occurs when r is approximately 7. As a consequence, the minimal amount of material needed to construct the can is approximately

$$A(7) = 2\pi(7)^2 + \frac{4000}{7} \approx 879 \text{ cm}^2.$$

Methods of calculus can be used to show that the true value of r that minimizes the area is $r = (1000/\pi)^{1/3} \approx 6.83$. This value gives an area, accurate to two decimal places, of 878.76, which does not significantly differ from our approximate value. ■

EXERCISE SET 2.4

In Exercises 1–6, the graph of a rational function is given. Specify the domain and the vertical and horizontal asymptotes of the function.

1.

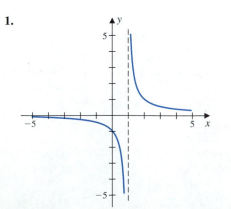

2.

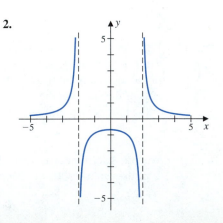

3.

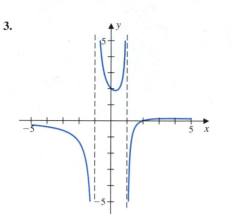

6.

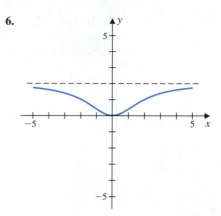

4.

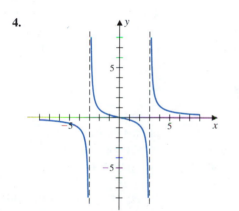

5.

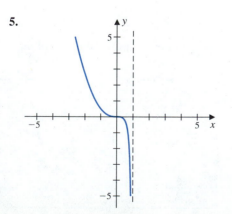

In Exercises 7–12, find the domain and any x- and y-intercepts of the rational function.

7. $f(x) = \dfrac{x - 3}{x + 1}$

8. $f(x) = \dfrac{(x - 1)(x + 2)}{(x + 3)(x - 4)}$

9. $f(x) = \dfrac{x^2 - x - 6}{x^2 - 4}$

10. $f(x) = \dfrac{x^2 + 3x - 10}{x^2 - 4x + 4}$

11. $f(x) = \dfrac{x^3 - 2x^2 - x + 2}{x^2}$

12. $f(x) = \dfrac{x^5 - 2x^4 - x + 2}{x^3 - 1}$

In Exercises 13–22, sketch the graph labeling any horizontal and vertical asymptotes and x- and y-intercepts.

13. $f(x) = \dfrac{2}{x - 1}$

14. $f(x) = \dfrac{4}{x + 3}$

15. $f(x) = \dfrac{x}{x - 3}$

16. $f(x) = \dfrac{2x}{3x - 1}$

17. $f(x) = \dfrac{2x - 3}{x^2 - x - 6}$

18. $f(x) = \dfrac{-2x}{x^2 - 9}$

19. $f(x) = \dfrac{x^2 - 1}{x^2 - 3x + 2}$

20. $f(x) = \dfrac{x^2 - x - 2}{x^2 - 2x - 3}$

21. $f(x) = \dfrac{x^2 - 9}{x^2 - 16}$

22. $f(x) = \dfrac{x^3 - x}{x + 1}$

In Exercises 23–26, the graph of the rational function crosses the horizontal asymptote one time. Sketch each graph and label the point of intersection of the curve with the horizontal asymptote.

23. $f(x) = \dfrac{x}{x^2 - 3x + 2}$

24. $f(x) = \dfrac{x^2 - 6x + 8}{x^2 - 4x + 3}$

25. $f(x) = \dfrac{3 - x^2}{2x^2 + x - 3}$

26. $f(x) = \dfrac{x^2}{x^2 - 5x + 6}$

In Exercises 27–30, each function has a graph with a slant asymptote. Use a graphing device to sketch the graph, and label any vertical and slant asymptotes and any x- and y-intercepts.

27. $f(x) = \dfrac{2x^2 + 3x - 1}{x + 2}$ **28.** $f(x) = \dfrac{x^2 - 2x - 1}{x - 1}$

29. $f(x) = \dfrac{x^3 - x^2 - 4x + 4}{x^2 - x + 6}$

30. $f(x) = \dfrac{x^3 - 5x^2 + 2x + 8}{x^2 + 2x - 3}$

31. Define a rational function that satisfies all the following conditions:

   i. Has a vertical asymptote $x = 2$;

  ii. Has a horizontal asymptote $y = 0$;

 iii. Never crosses the x-axis;

 iv. Has y-intercept -2.

32. Define a rational function that satisfies all the following conditions:

   i. Has the vertical asymptotes $x = 2$ and $x = -3$;

  ii. Has the horizontal asymptote $y = 1$;

 iii. Has x-intercepts at 3 and 4.

33. Define a rational function that satisfies all the following conditions:

   i. Has a vertical asymptote $x = 2$;

  ii. Has a slant asymptote $y = 2x + 1$;

 iii. Has y-intercept 2.

34. Rework the problem posed in Example 7 of this section with the change that the can to be constructed has a bottom but no top.

35. A rectangular box with a square base of length x and height h is to have a volume of 20 ft^3. The cost of the top and bottom of the box is 20 cents per square foot and the cost for the sides is 8 cents per square foot. Express the cost of the box in terms of (a) the variables x and h; (b) the variable x only; and (c) the variable h only; (d) Use a graphing device to approximate the dimensions of the box that will minimize the cost.

2.5   OTHER ALGEBRAIC FUNCTIONS

Polynomial and rational functions are examples of a larger class of functions called *algebraic*, which are obtained from polynomials by a finite combination of the operations of addition, subtraction, multiplication, division, raising to integral powers, and extracting roots.

The simplest algebraic functions that are not polynomials or rational functions are the *rational power functions*, which have the form

$$f(x) = x^{m/n},$$

where m/n is a rational number, n is an integer greater than 1, and the integers m and n have no common factors. In this case we can alternatively write $f(x)$ as

$$x^{m/n} = (\sqrt[n]{x})^m = \sqrt[n]{(x^m)}$$

The domain depends on whether the number n is odd or even, and on the sign of m.

The Domain of the Rational Power Function

The domain of $f(x) = x^{m/n}$ is

  (i) $[0, \infty)$ if m is positive and n is even;

 (ii) $(-\infty, \infty)$ if m is positive and n is odd;

(iii) $(0, \infty)$ if m is negative and n is even;

(iv) $(-\infty, 0) \cup (0, \infty)$ if m is negative and n is odd.

The graphs of some rational power functions for these various cases are shown in Figure 2.44. The graphs in parts (c) and (d) can be obtained by applying the reciprocal graphing techniques to the graphs in parts (a) and (b).

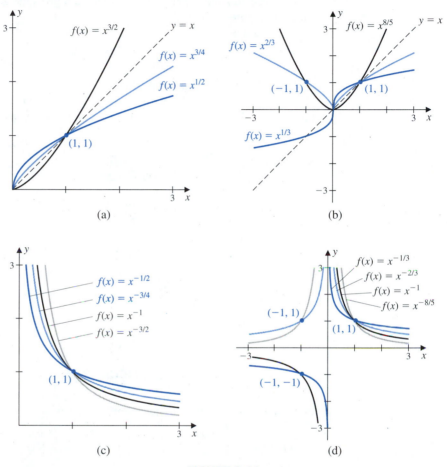

FIGURE 2.44

Notice that the rational power functions with odd denominators have symmetry. If the numerator is even, the symmetry is with respect to the y-axis. If the numerator is odd, the symmetry is with respect to the origin. There can be no symmetry when the denominator is even since no negative numbers are in the domain of the function.

In Section 1.9 we examined the graphs of functions that occur as translations and changes of scale of the square root function,

$$f(x) = \sqrt{x} = x^{1/2}.$$

The first example shows how this translation technique can be applied to any of the rational power functions.

EXAMPLE 1   Sketch the following graphs.

a. $f(x) = 2(\sqrt[3]{x + 2} - 1)$

b. $f(x) = (4x - 3)^{3/2}$

Solution a. As x becomes large in magnitude, the graph of $y = 2(\sqrt[3]{x+2} - 1)$ approaches the graph of $y = 2\sqrt[3]{x} = 2x^{1/3}$. So we first graph the rational power function $y = x^{1/3}$ and then stretch the vertical scale 2 units to give the graph of $y = 2\sqrt[3]{x}$.

Translating this graph two units to the left gives the graph of $y = 2\sqrt[3]{x+2}$, as shown in Figure 2.45(a).

The graph of the

$$f(x) = 2(\sqrt[3]{x+2} - 1) = 2\sqrt[3]{x+2} - 2$$

is obtained by moving this graph downward 2 units, as shown in Figure 2.45(b).

The graph of $y = 2\sqrt[3]{x}$ is symmetric with respect to the origin. Because of the translations, the new graph is symmetric with respect to the point $(-2, -2)$.

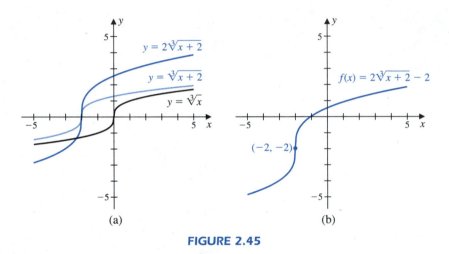

FIGURE 2.45

b. Factoring the 4 from within the parenthesis gives

$$f(x) = (4x - 3)^{3/2} = \left(4\left(x - \frac{3}{4}\right)\right)^{3/2} = 4^{3/2}\left(x - \frac{3}{4}\right)^{3/2} = 8\left(x - \frac{3}{4}\right)^{3/2}.$$

The graph of $y = x^{3/2}$ is shown in Figure 2.46(a), the graph of $y = 8x^{3/2}$ in Figure 2.46(b), and the final graph of $f(x) = (4x - 3)^{3/2}$ in Figure 2.46(c). ∎

In the next example we consider the effect of composing the square root function with a rational function.

EXAMPLE 2 Sketch the graph of

$$y = f(x) = \sqrt{\frac{x - 2}{x + 2}}$$

Solution The domain consists of all real numbers that make the expression under the radical positive, that is, those values of x such that

$$\frac{x - 2}{x + 2} \geq 0.$$

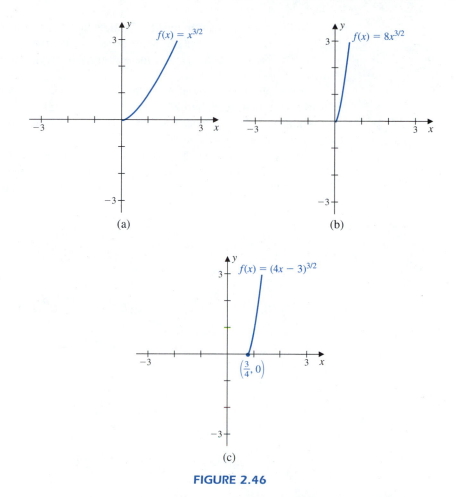

FIGURE 2.46

This quotient equals 0 when the numerator is 0, and the quotient is greater than 0 when the numerator and denominator are both positive or both negative. The sign graph in Figure 2.47 shows that the only zero of the quotient is at $x = 2$. The ☐ in the chart is used to indicate that the denominator is zero at $x = -2$.

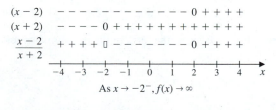

FIGURE 2.47

To determine the sign of the quotient $\frac{x-2}{x+2}$ in the intervals $(-\infty, -2)$, $(-2, 2)$, and $(2, \infty)$, we select a test value in each interval. Since, the quotient is positive at $x = -3$, it is positive on the entire interval $(-\infty, -2)$. Similarly, the quotient is negative on $(-2, 2)$ since it is negative at $x = 0$, and it is positive on $(2, \infty)$ since it is positive at $x = 3$.

Putting this information together implies that $\frac{x-2}{x+2}$ is nonnegative precisely when x is in $(-\infty, -2) \cup [2, \infty)$, and this is the domain of the function.

The x-intercept is $(2, 0)$, and there is no y-intercept since $x = 0$ is not in the domain.

The graph has the vertical asymptote $x = -2$ since the denominator of the function goes to zero at this value, but the numerator does not. Since the square root function is always positive, $f(x) \to \infty$ as $x \to -2^-$. There is no limit from the right at -2 since these values are not in the domain.

Finally, as $x \to \infty$ and $x \to -\infty$ the quotient under the radical approaches 1, so $f(x)$, the square root of this quotient, also approaches 1, and $y = 1$ is a horizontal asymptote. This analysis implies that the graph is similar to that shown in Figure 2.48. ■

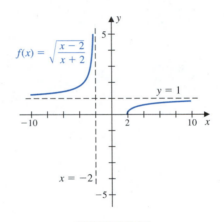

$$f(x) = \sqrt{\frac{x-2}{x+2}}$$

FIGURE 2.48

The final example illustrates a function with two distinct horizontal asymptotes.

EXAMPLE 3   Sketch the graph of

$$f(x) = \frac{x-2}{\sqrt{x^2 + x - 2}}.$$

Solution   First we factor the term in the denominator and rewrite $f(x)$ as

$$f(x) = \frac{x-2}{\sqrt{(x-1)(x+2)}}.$$

In this form we see that for x to be in the domain of f, we must have

$$(x-1)(x+2) > 0.$$

The sign graph in Figure 2.49 implies that the domain is $(-\infty, -2) \cup (1, \infty)$. The numbers $x = -2$ and $x = 1$ are eliminated from the domain since the denominator of the fraction cannot be zero.

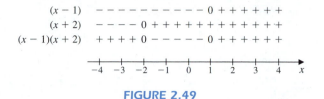

FIGURE 2.49

We could also use a graphing device to get a quick sketch of the parabola $y = x^2 + x - 2$ and use the sketch to help us find the domain of the function.

We could also find the domain of f by graphing the equation $y = (x - 1)(x + 2)$, as shown in Figure 2.50. From this graph we can see that $(x - 1)(x + 2) > 0$ precisely when x is in $(-\infty, -2) \cup (1, \infty)$.

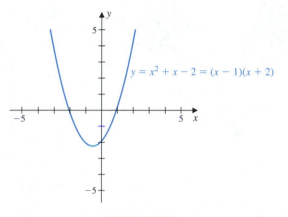

FIGURE 2.50

The x-intercept is $(2, 0)$, and since 0 is not in the domain there is no y-intercept. Vertical asymptotes occur at both $x = -2$ and $x = 1$. To be in the domain of the function, x can approach -2 only from the left side and can approach 1 only from the right side. Since the denominator of $f(x)$ is positive for all x in the domain, $f(x) > 0$ for $x > 2$ and $f(x) < 0$ when x is in $(-\infty, -2) \cup (1, 2)$.

So,

$$f(x) \to -\infty \quad \text{as } x \to -2^-$$

and

$$f(x) \to -\infty \quad \text{as } x \to 1^+.$$

When x is large in magnitude—that is, for $|x|$ large—the numerator of

$$f(x) = \frac{x - 2}{\sqrt{x^2 + x - 2}}$$

is approximately x, and the denominator is approximately $\sqrt{x^2} = |x|$. As a consequence, when $|x|$ is large

$$f(x) = \frac{x - 2}{\sqrt{x^2 + x - 2}} \approx \frac{x}{\sqrt{x^2}} = \frac{x}{|x|} = \begin{cases} 1, & \text{if } x > 0, \\ -1, & \text{if } x < 0. \end{cases}$$

So

$$f(x) \rightarrow -1 \qquad \text{as } x \rightarrow -\infty$$

and

$$f(x) \rightarrow 1 \qquad \text{as } x \rightarrow \infty.$$

The graph has two distinct horizontal asymptotes: $y = -1$, which is approached as $x \rightarrow -\infty$, and $y = 1$, which is approached as $x \rightarrow \infty$, as shown in Figure 2.51. ■

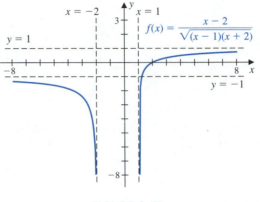

FIGURE 2.51

EXERCISE SET 2.5

In Exercises 1–6, a function f is described. Determine the domain of the function $g(x) = \sqrt{f(x)}$; that is, determine the values in the domain of f for which $f(x) \geq 0$.

1. $f(x) = x^2 + 4x - 12$

2. $f(x) = (x - 1)(x - 4)(x + 3)$

3. $f(x) = \dfrac{x - 1}{x + 3}$

4. $f(x) = \dfrac{7 - 2x}{x - 5}$

5. $f(x) = \dfrac{x + 1}{x^2 + 2x - 3}$

6. $f(x) = \dfrac{x^2 - 3x - 28}{x^2 - 4}$

In Exercises 7–18, sketch the graph and label any axis intercepts and asymptotes.

7. $f(x) = (x - 1)^{1/3} + 1$

8. $f(x) = (x + 1)^{1/4} - 1$

9. $f(x) = -2(x + 1)^{2/3} - 2$

10. $f(x) = 3(x - 2)^{3/4} + 1$

11. $f(x) = (1 - x)^{3/4} - 1$

12. $f(x) = (4 - x)^{3/2} + 1$

13. $f(x) = \sqrt{\dfrac{x + 2}{x - 1}}$

14. $f(x) = \sqrt{\dfrac{x - 1}{x + 2}}$

15. $f(x) = \sqrt{\dfrac{2 - x}{x + 2}}$

16. $f(x) = \sqrt{\dfrac{1 - x}{x + 3}}$

17. $f(x) = \dfrac{x - 1}{\sqrt{(x + 1)(x - 2)}}$

18. $f(x) = \dfrac{x}{\sqrt{4 - x^2}}$

19. A family of functions is defined by

$$g_n(x) = \frac{1}{x^n},$$

where $x \neq 0$ and n is a positive integer. On one set of axes sketch the graph of $y = g_n(x)$ for $n = 1$, 3, and 5. On another set of axes sketch the graph of $y = g_n(x)$ for $n = 2$, 4, and 6. Describe the general shape of the curves when n is odd and when n is even.

2.6   COMPLEX ROOTS OF POLYNOMIALS

We have seen throughout this chapter that finding the zero of a function is equivalent to determining where the graph of the function crosses the x-axis. What we have somewhat avoided to this point is the fact that some of the most elementary of functions have graphs that fail to cross the x-axis. Consider, for example, the graphs of $f(x) = x^2 - 1$, $g(x) = x^2$, and $h(x) = x^2 + 1$ shown in Figure 2.52. The graph of f has two distinct zeros, one at $x = -1$ and the other at $x = 1$. The graph of g has a single zero of multiplicity two at $x = 0$. The graph of h does not cross the x-axis, so there is no real number that makes $h(x) = 0$.

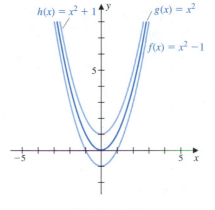

FIGURE 2.52

To examine the zeros of functions like h, we must extend our number system once more to the set of *complex numbers*. To define the set of complex numbers we introduce a number i whose square is -1.

The Square Root of -1

We define

$$i = \sqrt{-1} \quad \text{so that} \quad i^2 = -1.$$

The number i can not be a real number since there are no real numbers x that satisfy $h(x) = x^2 + 1 = 0$, and by definition we have $i^2 + 1 = -1 + 1 = 0$.

Complex Numbers

The set of **complex numbers** consists of all expressions of the form

$$z = a + bi,$$

where a and b are real numbers. The set of complex numbers is often denoted $\mathbb{C}$.

The number a is called the **real part** of the complex number z, and the number b is called the **imaginary part** of z.

When $b = 0$, the complex number is simply a real number, so all real numbers are also complex. To simplify the language in this section, however, when we speak of complex numbers we will be assuming that the imaginary part is nonzero.

Two complex numbers are equal precisely when their real and imaginary parts are the same, so

$$a + bi = c + di \qquad \text{precisely when} \qquad a = c \quad \text{and} \quad b = d.$$

Addition and subtraction of complex numbers is defined in a natural way, if we keep in mind that the real and imaginary part of a complex number must be kept separate:

$$(a+bi)+(c+di) = (a+c)+(b+d)i, \quad \text{and} \quad (a+bi)-(c+di) = (a-c)+(b-d)i.$$

Multiplication is also straightforward; we simply use the standard rules of algebra and the fact that $i^2 = -1$. So

$$(a + bi)(c + di) = ac + bci + adi + bd\, i^2 = ac + bci + adi - bd$$

$$= (ac - bd) + (ad + bc)i.$$

For example,

$$(2 + 3i) + (5 - 7i) = (2 + 5) + (3 + (-7))i = 7 - 4i$$

and

$$(2 + 3i) \cdot (5 - 7i) = (2 \cdot 5 - 3(-7)) + (2(-7) + 3 \cdot 5)i = 31 + i.$$

The quadratic formula tells us that the solutions to the quadratic equation $ax^2 + bx + c = 0$ are

$$x = \frac{-b}{2a} \pm \frac{\sqrt{b^2 - 4ac}}{2a}.$$

When $b^2 - 4ac > 0$, there are two real solutions to this equation,

$$x_1 = \frac{-b}{2a} - \frac{\sqrt{b^2 - 4ac}}{2a} \qquad \text{and} \qquad x_2 = \frac{-b}{2a} + \frac{\sqrt{b^2 - 4ac}}{2a}.$$

When $b^2 - 4ac = 0$, there is a single solution of multiplicity 2 at

$$x = \frac{-b}{2a}.$$

However, when $b^2 - 4ac < 0$, there are two complex solutions to the equation. To recover these from the quadratic formula we first observe that since $b^2 - 4ac < 0$, we have $4ac - b^2 > 0$, so $\sqrt{4ac - b^2}$ is a real number. We can then write

$$\sqrt{b^2 - 4ac} = \sqrt{(4ac - b^2)(-1)} = \sqrt{4ac - b^2}\sqrt{-1} = \sqrt{4ac - b^2}\, i.$$

The two complex solutions to the equation $ax^2 + bx + c = 0$ when $b^2 - 4ac < 0$ are

$$x_1 = \frac{-b}{2a} - \frac{\sqrt{4ac - b^2}}{2a}\, i \qquad \text{and} \qquad x_2 = \frac{-b}{2a} + \frac{\sqrt{4ac - b^2}}{2a}\, i.$$

EXAMPLE 1   Determine all the solutions to the equation $x^2 - 4x + 13 = 0$.

Solution   The quadratic formula applied to this equation gives

$$x = \frac{-(-4)}{2(1)} \pm \frac{\sqrt{(-4)^2 - 4(1)(13)}}{2(1)} = 2 \pm \frac{1}{2}\sqrt{-36} = 2 \pm \frac{1}{2}6\,i,$$

so the two solutions are

$$x_1 = 2 + 3\,i \qquad \text{and} \qquad x_2 = 2 - 3\,i,$$

and

$$x^2 - 4x + 13 = (x - (2 + 3i)) \cdot (x - (2 - 3i)).$$ ∎

The two complex solutions to the quadratic equation in Example 1 have the same real part, and imaginary parts which differ only in sign. These two numbers are called *complex conjugates*.

Complex Conjugate

The **complex conjugate** of the complex number $z = a + b\,i$ is the complex number $\bar{z} = a - b\,i$.

For numbers z and $\bar{z}$ we have

$$z + \bar{z} = (a + b\,i) + (a - b\,i) = 2a$$

and

$$z \cdot \bar{z} = (a + b\,i) \cdot (a - b\,i) = (a^2 - b^2\,i^2) + (ba - ab)\,i = a^2 + b^2,$$

which are both real numbers.

EXAMPLE 2   Determine the complex conjugates of $z = 4 + 3\,i$ and $w = 2 - 5\,i$, and then find $\bar{z} + \bar{w}$, $\overline{z + w}$, $\bar{z} \cdot \bar{w}$, and $\overline{z \cdot w}$.

Solution   The conjugates of z and w are

$$\bar{z} = 4 - 3\,i \qquad \text{and} \qquad \bar{w} = 2 + 5\,i,$$

so

$$\bar{z} + \bar{w} = 6 + 2\,i \qquad \text{and} \qquad \bar{z} \cdot \bar{w} = (8 + 15) + (20 - 6)\,i = 23 + 14\,i.$$

Since

$$z + w = (4 + 3\,i) + (2 - 5\,i) = 6 - 2\,i$$

and

$$z \cdot w = (4 + 3\,i) \cdot (2 - 5\,i) = (8 + 15) + (6 - 20)\,i = 23 - 14\,i,$$

we also have

$$\overline{z + w} = 6 + 2\,i \quad \text{and} \quad \overline{z \cdot w} = 23 + 14\,i.$$ ∎

Notice in the previous example that we have

$$\overline{z + w} = 6 + 2\,i = \overline{z} + \overline{w} \quad \text{and} \quad \overline{z \cdot w} = 23 + 14\,i = \overline{z} \cdot \overline{w}.$$

Results of this type are true in general.

Conjugate Results

For any pair of complex numbers z and w and for any integer n we have

$$\overline{z} + \overline{w} = \overline{z + w}, \quad \overline{z} \cdot \overline{w} = \overline{z \cdot w}, \quad \text{and} \quad \overline{z}^n = \overline{z^n}$$

We have seen from the quadratic formula that complex zeros of quadratic equations with real coefficients come in conjugate pairs. This result holds in a much broader situation. If z is a complex zero of a polynomial with real coefficients,

$$P(x) = a_n x^n + a_{n-1} x^{n-1} + \cdots + a_1 x + a_0,$$

then

$$a_n z^n + a_{n-1} z^{n-1} + \cdots + a_1 z + a_0 = 0.$$

Since $0, a_n, a_{n-1}, \ldots, a_1$, and a_0 are all real numbers, their complex conjugates are just themselves. The conjugate results imply that

$$
\begin{aligned}
0 = \overline{0} &= \overline{a_n z^n + a_{n-1} z^{n-1} + \cdots + a_1 z + a_0} \\
&= \overline{a_n z^n} + \overline{a_{n-1} z^{n-1}} + \cdots + \overline{a_1 z} + \overline{a_0} \\
&= a_n \overline{z^n} + a_{n-1} \overline{z^{n-1}} + \cdots + a_1 \overline{z} + a_0 \\
&= a_n \overline{z}^n + a_{n-1} \overline{z}^{n-1} + \cdots + a_1 \overline{z} + a_0,
\end{aligned}
$$

so $\overline{z}$ is also a solution to $P(x) = 0$.

EXAMPLE 3   One solution to the equation

$$x^4 - 6x^3 + 14x^2 - 22x + 5 = 0$$

is the complex number $1 + 2\,i$. Determine all the solutions to this equation.

Solution   Since $1 + 2\,i$ is a solution, its complex conjugate, $1 - 2\,i$, is also a solution. So the quadratic term

$$
\begin{aligned}
(x - (1 + 2\,i))\,(x - (1 - 2\,i)) &= x^2 - (1 + 2\,i + 1 - 2\,i)x + (1 + 2\,i)(1 - 2\,i) \\
&= x^2 - 2x + 5
\end{aligned}
$$

is a factor of the polynomial $x^4 - 6x^3 + 14x^2 - 22x + 5$, and polynomial division gives

$$x^4 - 6x^3 + 14x^2 - 22x + 5 = \left(x^2 - 2x + 5\right)\left(x^2 - 4x + 1\right).$$

We can now apply the quadratic formula to find the solutions to $x^2 - 4x + 1 = 0$, and these are the remaining solutions to the original equation.

If

$$x^2 - 4x + 1 = 0$$

then

$$x = \frac{-(-4)}{2(1)} \pm \frac{\sqrt{(-4)^2 - 4(1)(1)}}{2(1)} = 2 \pm \sqrt{3}.$$

As a consequence, the four solutions to the equation

$$x^4 - 6x^3 + 14x^2 - 22x + 5 = 0$$

are $1 + 2i$, $1 - 2i$, $2 + \sqrt{3}$, and $2 - \sqrt{3}$, and the polynomial is factored completely as

$$x^4 - 6x^3 + 14x^2 - 22x + 5 = (x - (2 + \sqrt{3}))(x - (2 - \sqrt{3}))(x - (1 + 2i))(x - (1 - 2i)). \quad \blacksquare$$

We have seen numerous results in this chapter concerning the zeros of polynomials but have not discussed the fundamental question in this regard, namely, which polynomials have zeros, and how many do they have? The answer to this question is that all polynomials have zeros, and, in a sense, as many zeros as the degree of the polynomial. The first result in this direction is the Fundamental Theorem of Algebra.

The Fundamental Theorem of Algebra

If

$$P(x) = a_n x^n + a_{n-1} x^{n-1} + \cdots + a_1 x + a_0$$

is a polynomial of degree $n > 0$ with real or complex coefficients, then P has at least one real or complex zero.

In fact, P has precisely n zeros, provided that a zero of multiplicity m is counted m times.

In addition, we have the following result.

Conjugate Pairs of Zeros

If the coefficients of

$$P(x) = a_n x^n + a_{n-1} x^{n-1} + \cdots + a_1 x + a_0$$

are all real numbers and z is a complex zero of P of multiplicity m, then its complex conjugate $\bar{z}$ is also a zero of multiplicity m.

The Fundamental Theorem of Algebra is first known to have been stated by the French mathematician Albert Girard in 1629, but with no accompanying demonstration of truth. Most of the greatest mathematicians in the 17th and 18th century, including the foremost scientist of all time, Isaac Newton, and the most prolific mathematician of record, Leonhard Euler, tried to prove this result. But it was Carl Friedrich Gauss, considered by many to be the greatest mathematician of all time, who proved the Fundamental Theorem in 1799, when he was only 20, for his doctoral dissertation. The proof is much more difficult than the statement of the theorem would imply, and all four of the proofs that Gauss eventually constructed introduced new mathematical ideas.

Because complex zeros must occur in conjugate pairs, a polynomial of even degree might have only complex zeros, but one of odd degree must have at least one zero that is real. If we add the information from Descartes' Rule of Signs and the Intermediate Value Theorem, we can often determine the character of the zeros, if not their precise location, as shown in the following example.

EXAMPLE 4   Determine possibilities for the zeros of the polynomials.

a. $P(x) = x^3 - 5x^2 + 2x + 1$
b. $Q(x) = 2x^5 - 4x^4 + 3x^3 - x + 3$

Solution   a. There are two sign changes for $P(x)$, so Descartes' Rule of Signs implies that there are either 2 or 0 positive zeros. But $P(0) > 0$, $P(1) < 0$, and $P(5) > 0$, so there must be real zeros lying in the intervals $(0, 1)$ and $(1, 5)$. Since $P(x)$ has only one additional zero, it cannot be complex, since complex zeros come in conjugate pairs. So the remaining zero must be negative. This is also implied from Descartes' Rule of Signs and the fact that

$$P(-x) = -x^3 - 5x^2 - 2x + 1$$

has only one change in sign. Since $P(-1) < 0$ and $P(0) > 0$, the negative root lies in the interval $(-1, 0)$. The computer-generated graph of $y = P(x)$ is shown in Figure 2.53.

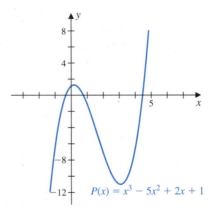

FIGURE 2.53

b. The degree of the polynomial $Q(x)$ is odd, so it must have at least one real zero. Since complex zeros occur in conjugate pairs, the number of real zeros must either be 1, 3, or 5. There are four changes in sign, so Descartes' Rule of Signs implies that there are either 4, 2 or 0 positive real zeros. Since $Q(x) > 0$ for $x = 1, 2,$ and 3, we might suspect that there are no positive zeros, but this is inconclusive.

The polynomial

$$Q(-x) = -2x^5 - 4x^4 - 3x^3 + x + 3$$

has only one sign change, so Descartes' Rule of Signs implies that there is precisely one negative real zero of Q. In fact, this zero lies in the interval $(-1, 0)$, since $Q(-1) < 0$ and $Q(0) > 0$.

From this analysis we can conclude that there are three possibilities for the zeros of $Q(x)$:

1 negative real zero	4 positive real zeros	0 complex zeros
1 negative real zero	2 positive real zeros	2 complex zeros
1 negative real zero	0 positive real zeros	4 complex zeros

We can also conclude that the last possibility is the most likely. The computer-generated graph of $y = Q(x)$ shown in Figure 2.54 verifies that this is, in fact, the case. ■

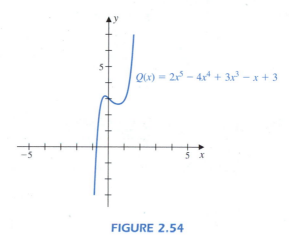

$Q(x) = 2x^5 - 4x^4 + 3x^3 - x + 3$

FIGURE 2.54

EXERCISE SET 2.6

In Exercises 1–16, write the complex number in standard complex form $a + bi$.

1. $(2 + i) + (-3 - 2i)$ **2.** $(3 - i) + (2 + 4i)$

3. $(-3 + 5i) - (2 - 3i)$ **4.** $(5 - 7i) - (2 - 4i)$

5. $2i \cdot (3 + i)$ **6.** $i \cdot (-2 - i)$

7. $(2 - i) \cdot (3 + i)$ **8.** $(5 - 2i) \cdot (4 - i)$

9. $(6 + 5i) \cdot (-3 - 2i)$ **10.** $(7 - 6i) \cdot (2 - 3i)$

11. $(3 - 8i) \cdot \overline{(2 + i)}$ **12.** $(4 - i) \cdot \overline{(4 - i)}$

13. i^5 **14.** i^6

15. i^{100} **16.** i^{101}

In Exercises 17–22, find the zeros of the quadratic function and write the function in factored form.

17. $f(x) = x^2 + 4$ **18.** $f(x) = 2x^2 + 18$

19. $f(x) = x^2 - 2x + 2$ **20.** $f(x) = x^2 - x + 1$

21. $f(x) = 2x^2 - x + 2$ **22.** $f(x) = 3x^2 + 2x + 1$

In Exercises 23–26, show that the given value of x is a solution of the equation, and then find all solutions.

23. $x^3 - 2x^2 + 9x - 18 = 0$; $x = 3i$

24. $x^3 + 3x^2 + 16x - 20 = 0$; $x = -2 + 4i$

25. $x^4 - 2x^3 - 2x^2 - 2x - 3 = 0$; $x = i$

26. $x^5 - 2x^4 + x^3 + 2x^2 - 2x = 0$; $x = 1 - i$

In Exercises 27–30, find all zeros and factor completely the given function.

27. $f(x) = x^3 - 3x^2 + 9x - 27$

28. $f(x) = x^4 - 1$

29. $f(x) = x^4 - x^2 - 2x + 2$

30. $f(x) = x^5 - 2x^4 - 4x^3 + 4x^2 - 5x + 6$

In Exercises 31–34, find a polynomial with integer coefficients that satisfies the given conditions.

31. Degree three and zeros 2 and $2i$

32. Degree four, a zero of multiplicity 2 at 1, and a zero at $2 + i$

33. Degree four, zeros $\sqrt{3}i$ and $3i$, and x^2 term 24

34. Degree five, zeros i and $3 - i$, and passing through the origin

35. Complex conjugation can be used to place the quotient of two complex numbers in standard complex form. Show that when $c + di \neq 0$, we can multiply the numerator and denominator of $\dfrac{a + bi}{c + di}$ by $c - di$ to produce

$$\frac{a + bi}{c + di} = \frac{ac + bd}{c^2 + d^2} + \frac{bc - ad}{c^2 + d^2}i.$$

36. Use the technique in Exercise 35 to write the following expressions in standard complex form $a + bi$.

a. $\dfrac{1}{i}$ **b.** $\dfrac{1}{1 - i}$

c. $\dfrac{1}{2 + 3i}$ **d.** $\dfrac{1}{3 - 4i}$

e. $\dfrac{2 - 3i}{2 + 3i}$ **f.** $\dfrac{5 + 6i}{5 - 6i}$

REVIEW EXERCISES FOR CHAPTER 2

In Exercises 1–4, sketch the graph of the given function by transforming a curve of the form $y = x^n$.

1. $f(x) = -2(x - 1)^2 + 2$ **2.** $f(x) = -2(x - 2)^3 - 1$

3. $f(x) = -x^4 - 3$ **4.** $f(x) = (x + 1)^4 + 2$

In Exercises 5–10, give a reasonable sketch of the graph of the function by plotting the x- and y-intercepts and using the end-behavior.

5. $f(x) = (x + 1)(x + 2)(x - 3)$

6. $f(x) = x^2(x - 1)$

7. $f(x) = \dfrac{1}{2}(x - 1)^3(x + 2)$

8. $f(x) = -\dfrac{1}{16}(x - 2)^3(x + 1)(x + 2)$

9. $f(x) = x^3 - \dfrac{1}{2}x^2 - \dfrac{1}{2}x$

10. $f(x) = x^5 + 2x^4 + 4x + 8$

In Exercises 11–14, the graph of a polynomial is given. What is the lowest possible degree for the polynomial and the sign of the leading coefficient?

11.

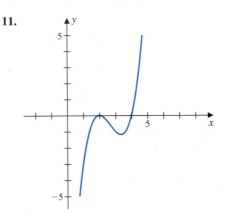

12.

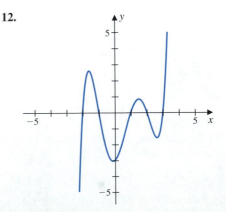

13.

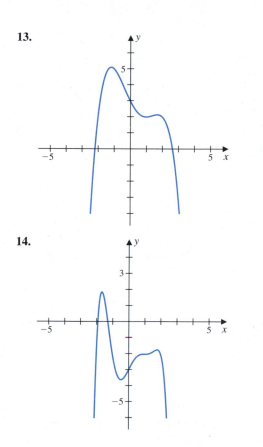

14.

In Exercises 15–18, find the quotient $Q(x)$ and remainder $R(x)$ when $P(x)$ is divided by $D(x)$.

15. $P(x) = 4x^2 + 2x - 1, D(x) = x - 2$

16. $P(x) = 3x^2 - 4x + 7, D(x) = -2x + 1$

17. $P(x) = 3x^3 + 2x^2 - 2x + 1, D(x) = x + 2$

18. $P(x) = x^5 + x^4 - 5x^2 - 2x - 3, D(x) = x^4 - 2x^3 + x + 1$

In Exercises 19–22, (a) list all the possibilities for rational roots, (b) use Descartes' Rule of Signs to determine the maximum number of positive and negative roots, (c) use the Factor Theorem to show that $x - c$ is a factor of the polynomial $P(x)$ for the given value of c, and (d) factor $P(x)$ completely in terms of real factors.

19. $P(x) = 3x^4 - 9x^3 - 2x^2 + 5x + 3, c = 3$

20. $P(x) = x^4 + 4x^3 + 6x^2 + 5x + 2, c = -2$

21. $P(x) = x^5 - 3x^4 - 5x^3 + 27x^2 - 32x + 12, c = -3$

22. $P(x) = x^6 - 5x^5 + 5x^4 + 9x^3 - 14x^2 - 4x + 8, c = 1$

In Exercises 23–28, find the domain, any x- and y-intercepts, and any vertical and horizontal intercepts of the rational functions.

23. $f(x) = \dfrac{x - 4}{x - 1}$

24. $f(x) = \dfrac{(x - 2)(x + 2)}{(x - 3)(x + 1)}$

25. $f(x) = \dfrac{x^2 - 2x + 1}{2x^2 - 18}$

26. $f(x) = \dfrac{x^2 + 4x - 12}{x^2 - x - 6}$

27. $f(x) = \dfrac{x^3 + 2x^2 - x - 2}{x^3}$

28. $f(x) = \dfrac{x^4 - 2x^3 + x^2}{x^3 - 1}$

In Exercises 29–36, sketch the graph showing any horizontal and vertical asymptotes and any axis intercepts.

29. $f(x) = \dfrac{3}{x - 2}$

30. $f(x) = \dfrac{-4}{x + 2}$

31. $f(x) = \dfrac{2}{(x - 1)(x - 2)}$

32. $f(x) = \dfrac{-3}{(x + 3)(x - 4)}$

33. $f(x) = \dfrac{4}{x^2 - 4}$

34. $f(x) = \dfrac{x - 5}{x^2 - 2x - 3}$

35. $f(x) = \dfrac{x^2 - 1}{x^2 + 2x}$

36. $f(x) = \dfrac{4 - x^2}{x^2 - 9}$

In Exercises 37 and 38, each graph has a slant asymptote. Sketch the graph showing any vertical and slant asymptotes and the x- and y-intercepts.

37. $f(x) = \dfrac{x^2 - 2x + 1}{x + 1}$

38. $f(x) = \dfrac{x^3 - 2x^2 + 4x - 3}{x^2 - 3x + 2}$

In Exercises 39–44, draw a rough sketch of the graph showing any axis intercepts and asymptotes, and then use a graphing device to check your work.

39. $f(x) = \sqrt{\dfrac{x - 2}{x + 1}}$

40. $f(x) = \sqrt{\dfrac{x - 3}{x + 3}}$

41. $f(x) = \sqrt{\dfrac{4 - x}{x + 4}}$

42. $f(x) = \dfrac{x - 3}{\sqrt{(x - 1)(x + 2)}}$

43. $f(x) = \dfrac{x^2}{\sqrt{4 - x^2}}$

44. $f(x) = \dfrac{\sqrt{x^2 - 9}}{x - 2}$

In Exercises 45–48, use a graphing device to sketch the graph of the given function and estimate

a. Intervals where the function is increasing and where the function is decreasing;

b. Local maximum and local minimum points.

45. $f(x) = x^3 - 2x^2 - x + 2$

46. $f(x) = x^4 - 2x^3$

47. $f(x) = \dfrac{2}{3}x^3 + \dfrac{7}{2}x^2 - 12x$

48. $f(x) = x^5 - 2x^4 + 2x^2 - x$

In Exercises 49–58, write the complex number in standard form $a + bi$.

49. $\dfrac{2 + i\sqrt{2}}{4}$

50. $\dfrac{-5 - \sqrt{-4}}{6}$

51. $(3 - i) - (2 - 3i)$

52. $(-2 + 6i) + (-3 + i)$

53. $(2 - i) \cdot \overline{(2 + i)}$

54. $(4 - 6i) \cdot \overline{(3 - 2i)}$

55. i^{20}

56. i^{21}

57. $\dfrac{2 + 3i}{4 - 7i}$

58. $\dfrac{-5 + 3i}{2 - 3i}$

59. Sketch a graph of a polynomial $P(x)$ that has zeros of multiplicity one at $x = 2$, $x = -2$, and $x = 3$ and satisfies $P(x) \to \infty$ as $x \to \infty$, and $P(x) \to -\infty$ as $x \to -\infty$.

60. Sketch a graph of a polynomial $P(x)$ that has a zero of multiplicity one at $x = 0$, a zero of multiplicity two at $x = -2$, a zero of multiplicity three at $x = 3$, and satisfies $P(x) \to \infty$ as $x \to \infty$, and $P(x) \to \infty$ as $x \to -\infty$.

61. The cubic polynomial $P(x)$ has zeros at $x = 1$, $x = 3$, and $x = -1$, and $P(0) = 1$. Find $P(x)$ and sketch its graph.

62. The fourth degree polynomial $P(x)$ has zeros of multiplicity one at $x = 1$ and $x = -1$, a zero of multiplicity two at $x = 2$, and $P(-2) = 2$. Find $P(x)$ and sketch its graph.

63. Sketch the graph of a rational function that satisfies all the following conditions:

  i. $f(x) \to \infty$ $x \to 2^+$

  ii. $f(x) \to -\infty$ $x \to 2^-$

  iii. $f(x) \to \infty$ $x \to 0^+$

  iv. $f(x) \to -\infty$ $x \to 0^-$

  v. Has a horizontal asymptote $y = 0$

  vi. $f(1) = 0$

64. Sketch the graph of a rational function that satisfies all the following conditions:

  i. $f(x) \to \infty$ $x \to 1^+$

  ii. $f(x) \to -\infty$ $x \to 1^-$

  iii. $f(x) \to -\infty$ $x \to -2^+$

  iv. $f(x) \to \infty$ $x \to -2^-$

  v. Has a horizontal asymptote $y = 0$

  vi. Never crosses the x-axes

  vii. Has a local maximum at $(-1, -2)$

65. Find the points of intersection of the curves $f(x) = x^3$ and $g(x) = -2x^2 + 9x + 18$. Sketch both curves on the same set of axes and label the points of intersection.

66. Repeat Exercise 65 with $f(x) = x^3 - x$ and $g(x) = -x^2 + 1$.

67. Find a polynomial of degree three that has integer coefficients and zeros at 1 and $-i$.

68. Find a polynomial of degree four that has integer coefficients, $3 - i$ as a zero, and -2 as a zero of multiplicity two.

69. Find a polynomial of degree four that has integer coefficients, zeros at $\sqrt{2}i$ and $2i$, and constant term 8.

70. Find a polynomial of degree five that has integer coefficients, zeros at i and $2 + i$, and a graph that passes through $(0, 5)$.

CHAPTER 2: CALCULUS PREVIEW EXERCISES

1. Sketch the graphs of $y = x - 1$ and $y = x - 2$ on the same set of axes and observe where the y-values on the graphs are zero, positive, and negative. Use this knowledge, together with the end-behavior, to sketch an approximate graph of each of the following.

  a. $f(x) = (x - 1)(x - 2)$

  b. $g(x) = \dfrac{x - 1}{x - 2}$

2. Sketch the graphs of $y = x^2 - 1$ and $y = x - 2$ on the same set of axes and observe where the y-values

on the graphs are zero, positive, and negative. Use this knowledge, together with the end-behavior, to sketch an approximate graph of each of the following.

  a. $f(x) = (x^2 - 1)(x - 2)$

  b. $g(x) = \dfrac{x^2 - 1}{x - 2}$

3. Factor the polynomial $P(x) = 2x^3 - 3x^2$ completely and sketch its graph, using the fact that a relative minimum occurs at $(1, -1)$ and a relative maximum occurs at (0, 0). For each value of the constant c, determine the number of real zeros of the polynomial $Q(x) = P(x) + c$.

4. Factor the polynomial $P(x) = x^4 - 4x^3 + 4x^2 - 1$ completely and sketch its graph, using the fact that relative minima occur at $(0, -1)$ and $(2, -1)$ and a relative maximum occurs at $(1, 0)$. For each value of the constant c, determine the number of real zeros of the polynomial $Q(x) = P(x) + c$.

5. Show that if $P(x)$ is a linear polynomial, then the composition $Q(x) = (P \circ P)(x)$ is also a linear polynomial and that $y = Q(x)$ has positive slope.

6. Suppose that $P(x)$ is a quadratic polynomial. Determine the degree of the polynomial $Q(x) = (P \circ P)(x)$ and how its coefficients depend on the coefficients of $P(x)$.

7. Define a rational function that satisfies the following conditions and sketch its graph.

   a. Has a vertical asymptote $x = 2$

   b. Has a horizontal asymptote $y = 0$

   c. Never crosses the x-axis

   d. Has y-intercept 2

8. Define a rational function that satisfies the following conditions and sketch its graph.

   a. Has vertical asymptotes $x = 2$ and $x = -3$

   b. Has a horizontal asymptote $y = 1$

   c. Has x-intercepts 3 and 4

9. Define a rational function that satisfies the following conditions and sketch its graph.

   a. As $x \to 1^+$, $f(x) \to \infty$

   b. As $x \to 1^-$, $f(x) \to -\infty$

   c. As $x \to -2^+$, $f(x) \to -\infty$

   d. As $x \to -2^-$, $f(x) \to \infty$

   e. Has a horizontal asymptote $y = 2$

   f. Has its only zero at the origin

10. The graph of $y = f(x)$ is shown in the figure.

   a. What is the domain of the function f?

   b. What is the range of the function f?

   c. On which interval or intervals is the function increasing?

   d. On which interval or intervals is the function decreasing?

   e. Specify any horizontal asymptotes for the graph.

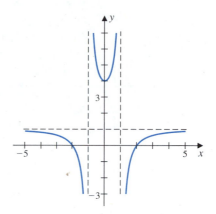

f. Specify any vertical asymptotes for the graph.

g. Does the function have an inverse on the interval $(-1, 1)$?

h. Does the function have an inverse on the interval $(1, \infty)$?

11. Match each equation with its graph, and explain your choices.

   a. $y = x^6$

   b. $y = (x + 1)^7$

   c. $y = (x - 1)(x + 2)(x - 3)$

   d. $y = (x - 1)^2(x + 2)(x - 3)$

i.

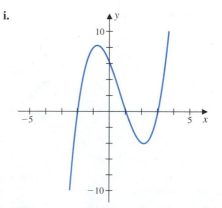

ii.

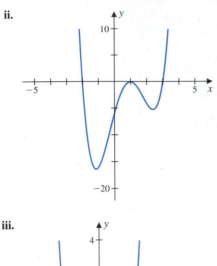

iii.

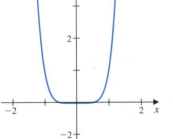

iv.

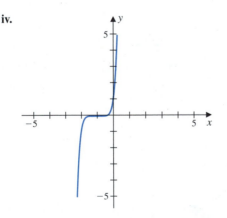

12. For which rational numbers k does the equation

$$x^3 + x^2 + kx - 3 = 0$$

have at least one rational root?

13. Show that $P(x) = 2x^4 - 4x^2 + 1$ has four real zeros. How many of these zeros are rational?

14. Show that $P(x) = x^4 - 8x^3 + 19x^2 - 11x - 4$ has exactly one rational zero and four real zeros.

15. a. Show that the linear polynomial defined by

$$P(x) = \frac{x - x_2}{x_1 - x_2} y_1 + \frac{x - x_1}{x_2 - x_1} y_2$$

passes through the points (x_1, y_1) and (x_2, y_2).

b. Use this fact to determine an equation of the line that passes through $(-1, 6)$ and $(1, -2)$.

16. a. Expand the technique in Exercise 15 to determine a formula for the quadratic polynomial that passes through the points (x_1, y_1), (x_2, y_2), and (x_3, y_3).

b. Use this formula to determine an equation of the quadratic polynomial that passes through the points $(-1, 6)$, $(1, -2)$, and $(2, 3)$.

17. Sketch the graph of $y = 1/x^2$, and use this graph to determine the graph of $y = a/x^2$ for various constants a.

18. Use the results of Exercise 17 to determine the possibilities for the graph of $f(x) = (a + bx^2)/x^2$, where a and b are constants.

19. A family of functions is defined by

$$g_n(x) = x^{1/n},$$

where n is a positive integer. On one set of axes sketch the graph of g_n for $n = 1, 3$, and 5. On another set of axes sketch the graph of g_n for $n = 2, 4$, and 6. Determine from these some properties of the graphs of $y = g_n(x)$ when n is odd and when n is even.

20. If $P(x) = x^2 + bx + c$ is a quadratic polynomial with roots r_1 and r_2, then we can write

$$x^2 + bx + c = (x - r_1)(x - r_2),$$

where $b = -(r_1 + r_2)$ and $c = r_1 r_2$. Determine a similar relationship between the zeros r_1, r_2, r_3 and the coefficients of the cubic polynomial $P(x) = x^3 + bx^2 + cx + d$.

21. For each positive integer n, the function $f_n(x) = (x - 1)^n (x + 1)^n$ has roots at $x = 1$ and at $x = -1$. Use a graphing device to determine how the multiplicity of the roots effects the shape of the graph.

22. In Section 1.8 we found that every quadratic equation could be obtained from the basic parabola $y = x^2$ using a series of transformations. This exercise details how the graph of any cubic equation can be obtained from a cubic curve of the form $y = x^3 + kx$. We start with a general cubic function

$$f(x) = ax^3 + bx^2 + cx + d, \quad \text{where } a \neq 0.$$

a. Rewrite $f(x)$ as $af_1(x)$, where

$$f_1(x) = x^3 + b_1 x^2 + c_1 x + d_1,$$

and determine b_1, c_1, and d_1 in terms of a, b, and c. Describe how the graph of f is related to the graph of f_1.

b. Let $f_2(x) = f_1\left(x - \frac{b_1}{3}\right)$. Show that

$$f_2(x) = x^3 + c_2 x + d_2,$$

and determine c_2 and d_2 in terms of b_1, c_1, and d_1. Describe how the graph of f_1 is related to the graph of f_2?

c. Let $f_3(x) = f_2(x) - d_2$, so that

$$f_3(x) = x^3 + c_2 x.$$

How is the graph of f_2 related to the graph of f_3?

d. Let $k = c_2$, and write

$$g(x) = x^3 + kx.$$

Use the results of parts (a), (b), and (c) to explain how the graph of f can be obtained from the graph of g.

e. Use a graphing device to sketch $y = g(x)$ for various values of k, describe the possible shapes of the graph, and how the shape depends on the value of k.

23. A pizza box with a lid is constructed from a 20-in. × 50-in. piece of cardboard, as shown in the following figure. Determine the volume of the box as a function of the variable x.

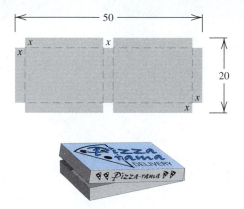

24. A rectangle is inscribed within a triangular region formed by the positive x-axis, the positive y-axis, and the graph of the line $x + y = 50$, as shown in the figure.

a. Determine the area as a function of the variable x.

b. Given the knowledge that there is a single value of x that produces a rectangle with maximum area, show that this rectangle must be a square.

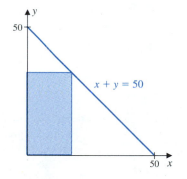

TRIGONOMETRIC FUNCTIONS 3

CALCULUS CONNECTIONS

Many of the common phenomena in our real world have an oscillatory, or periodic, behavior. We can observe this when riding in a boat on a rough weather day, when listening to an overloaded washer that shakes and rattles as it starts, and when listening to the beat of our own heart. You might not know that electricity flowing through the wires in your house, the motion of your car after it hits a hole in the road, the music that you hear on your CD player, and your watch, if it is digital, also exhibit this behavior. All this behavior can be modeled using equations based on the familiar sine and cosine functions, equations of the form

$$y = A + B\sin(Cx + D),$$

where A, B, C, and D are constants and sin represents the trigonometric sine function.

In this chapter we will see how to apply and construct functions of this type that will permit us to model behavior like that given in the following figure.

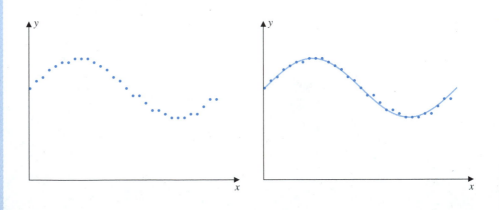

You have probably been exposed to sines and cosines in your previous mathematics courses, but you may not have thought of them as functions. Instead, your study in these courses might have emphasized their relationships to the sides of a right triangle. To study problems having periodic behavior we need calculus and differential equations, and these subjects require that trigonometry be given a functional approach. Since this is the approach you will need in your future work, it is the one we adopt. We will not discuss the connection with triangles until quite late in the chapter. Be assured, though, that no matter how we start, we all end up in the same place.

3.1 INTRODUCTION

Trigonometric functions have many applications. They are used to describe the behavior of such diverse topics as sound waves, vibrations, the motion of an automobile on a bumpy road, and the orbits of the planets. In fact, trigonometric functions are commonly required to describe the motion of any object that behaves in a circular, oscillating, or periodic manner.

Although the trigonometric functions are defined as functions whose domains are sets of real numbers, the applications in trigonometry often involve triangles and the angles of their vertices. Our first step, then, is to develop a connection between angles and the set of real numbers.

An angle consists of two *rays*, or half-lines, that originate at a common point, called the *vertex*, which we will denote as O. One of the rays is called the *initial* side of the angle, and the other ray is called the *terminal* side. The angle can be thought of as being generated by revolving the initial side about the point O until the terminal side is reached. We choose two points A and B on the two rays and denote the angle as $\angle AOB$, as shown in Figure 3.1.

FIGURE 3.1

The amount of rotation from the initial side to the terminal side is critical in the definition of an angle, but the particular points chosen to represent it are not, nor, in general, is the position of the angle in the plane. To better describe the rotation we superimpose an xy-coordinate system on the angle with the origin at the point O and the initial side along the positive x-axis, as shown in Figure 3.2(a). We can then rotate the angle with its coordinate system to the more familiar position illustrated in Figure 3.2(b).

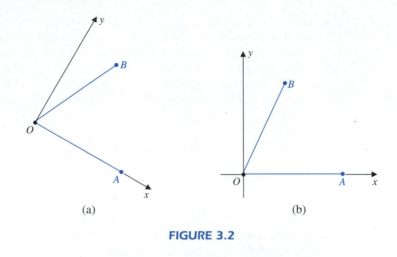

(a)   (b)

FIGURE 3.2

To provide a measure for an angle, we need to know how the angle was generated. The angle in Figure 3.3 could have been generated by any of the three rotations shown, or by an infinite number of other such rotations, one for each full revolution about the point O in the clockwise direction and another in the counterclockwise direction.

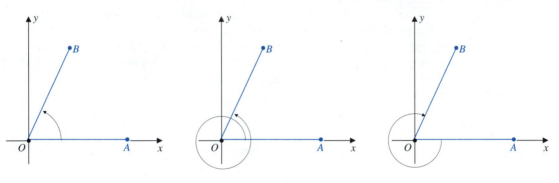

FIGURE 3.3

To define the measure of an angle, we first add to the coordinate system the unit circle U centered at the origin—that is, the circle with equation

$$x^2 + y^2 = 1,$$

as shown in Figure 3.4(a). Then we define a function P that sends each point on the real line $\mathbb{R}$ onto the unit circle U in the following manner:

1. If $t \geq 0$, then $P(t)$ is the point on the unit circle for which the length of the arc of the circle from $(1, 0)$ to $P(t)$ is t units, measured in the counterclockwise direction from $(1, 0)$. (See Figure 3.4(b).)
2. If $t < 0$, then $P(t)$ is the point on the unit circle for which the length of the arc of the circle from $(1, 0)$ to $P(t)$ is $-t > 0$ units, measured in the clockwise direction from $(1, 0)$. (See Figure 3.4(c).)

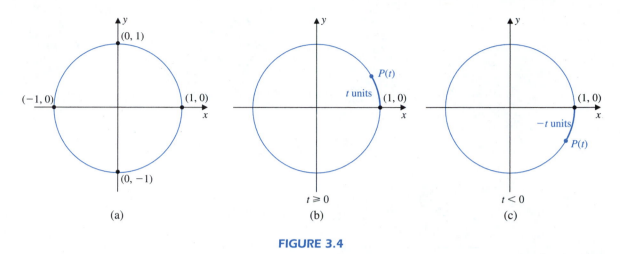

FIGURE 3.4

The mapping of t into $P(t)$ can be described by positioning a copy of the real line vertically with $t = 0$ coinciding with the point $(1, 0)$ on the unit circle (see Figure 3.5.). For any real number t the point $P(t)$ is obtained by "wrapping" the line around the circle and marking as $P(t)$ the point on the circle that corresponds to the position of t.

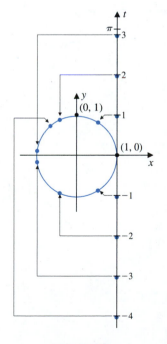

FIGURE 3.5

This wrapping corresponds to rotating the initial side of an angle around the point O until we reach the terminal side. The real number t that gives the point $P(t)$ on the terminal side of the angle is called the *radian measure* of the angle. (See Figure 3.6.)

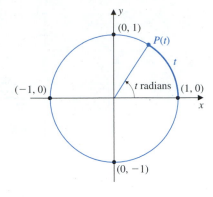

FIGURE 3.6

Radian measure of angles is the most useful system for the precalculus and calculus analysis we will be performing, although we will see a need for degree measure later in this chapter.

Radian Measure

For any real number t, the angle that is generated by rotating from the positive x-axis to the point $P(t)$ on the unit circle is said to have **radian measure** t.

The smallest positive real number that is mapped onto $(-1, 0)$ is π, an irrational number whose value is approximately 3.14159. Because of the symmetry of the circle, the number $\pi/2$ is associated with the point $(0, 1)$, the number $3\pi/2$ with the point $(0, -1)$, the number 2π with the point $(1, 0)$, and so on, as illustrated in Figure 3.7.

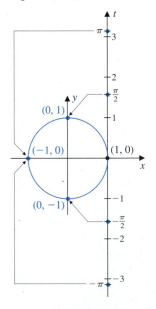

FIGURE 3.7

Corresponding to each real number t is a pair of xy-coordinates we denote $(x(t), y(t))$ that describes the point $P(t)$ on the unit circle. The basic trigonometric functions are defined in terms of these coordinates.

3.2   THE SINE AND COSINE FUNCTIONS

In the previous section we saw that corresponding to each real number t, and hence to any angle, there is a pair $(x(t), y(t))$ of xy-coordinates describing the point $P(t)$ on the unit circle U. These coordinates provide us with the two most basic trigonometric functions. The y-coordinate of $P(t)$ is the *sine* of t, and the x-coordinate of $P(t)$ is the *cosine* of t, as shown in Figure 3.8.

The Sine and Cosine Functions

Suppose that the coordinates of a point $P(t)$ on the unit circle are $(x(t), y(t))$. Then the **sine** of t, written $\sin t$, and the **cosine** of t, written $\cos t$, are defined by

$$\sin t = y(t) \qquad \text{and} \qquad \cos t = x(t).$$

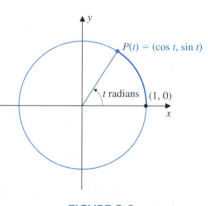

FIGURE 3.8

These definitions are also used for the sine and cosine of an angle with radian measure t. So the trigonometric functions serve two purposes, directly as functions with domain the set of real numbers and indirectly as functions with domain the set of all possible angles, where the angles are given in radian measure.

There are some results that follow quickly from these definitions of the sine and cosine. The first, and probably most important, identity in trigonometry follows from the fact that for any number t the point $P(t) = (\cos t, \sin t)$ lies on the circle with equation $x^2 + y^2 = 1$.

In your study of trigonometry you were probably exposed to a large number of identities. We will concentrate on just a few basic identities, those that are frequently used in mathematics.

The Pythagorean Identity

For all real numbers t,

$$(\sin t)^2 + (\cos t)^2 = 1.$$

In addition, we have the following.

Bounds on the Sine and Cosine

For all real numbers t,

$$-1 \le \sin t \le 1 \qquad \text{and} \qquad -1 \le \cos t \le 1.$$

The signs of the sine and cosine functions are also easily determined once it is known in which quadrant of the plane $P(t)$ lies, as shown in Figure 3.9.

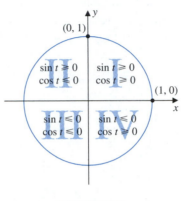

FIGURE 3.9

The points $P(t) = (\cos t, \sin t)$ and $P(-t) = (\cos(-t), \sin(-t))$ are obtained in the same manner, except that in the first instance the rotation is counterclockwise from $(1, 0)$ and in the second the rotation is clockwise from $(1, 0)$. A typical situation when $t > 0$ is illustrated in Figure 3.10.

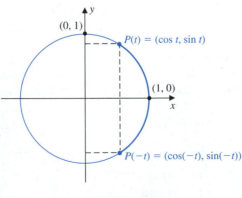

FIGURE 3.10

As a consequence, the x-coordinates of $P(t)$ and $P(-t)$ are always the same, and the y-coordinates of these points differ in sign but not in magnitude.

The Cosine Function is Even   The Sine Function is Odd

For all real numbers t,

$$\cos(-t) = \cos t \quad \text{and} \quad \sin(-t) = -\sin t.$$

We will now find some specific values for the sine and the cosine functions which are determined by the geometry of the unit circle. The circumference of a circle with radius r is $C = 2\pi r$, so the unit circle U has circumference 2π. The axis intercepts for the unit circle shown in Figure 3.11 give the values of the sine and cosine for the multiples of $\pi/2$ listed in Table 3.1.

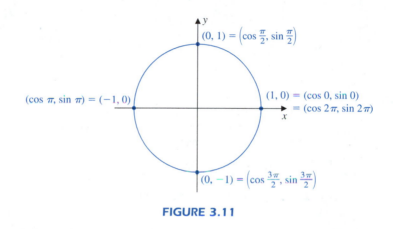

FIGURE 3.11

TABLE 3.1

t	0	$\pi/2$	π	$3\pi/2$	2π
$\cos t$	1	0	-1	0	1
$\sin t$	0	1	0	-1	0

Since we arrive at the same point on the unit circle after 2π, the sine at 0 and at 2π agree, as do the cosine. If we extended the table to consider the values of the sine and cosine at the next multiple of $\pi/2$, which is $5\pi/2$, we would have $\sin 5\pi/2 = \sin \pi/2$ and $\cos 5\pi/2 = \cos \pi/2$. Similarly, if we extended it backward to -2π, we have the values shown in Table 3.2.

TABLE 3.2

t	-2π	$-3\pi/2$	$-\pi$	$-\pi/2$	0	$\pi/2$	π	$3\pi/2$	2π
$\cos t$	1	0	-1	0	1	0	-1	0	1
$\sin t$	0	1	0	-1	0	1	0	-1	0

This behavior follows from the fact the sine and cosine are *periodic* functions.

Periodic Functions

A nonconstant function f is said to be **periodic** if a positive number T exists with

$$f(t + T) = f(t) \qquad \text{for all } t \text{ in the domain of } f.$$

The smallest positive number T for which this equation holds is called the **period** of f. Figure 3.12 shows some examples of periodic functions.

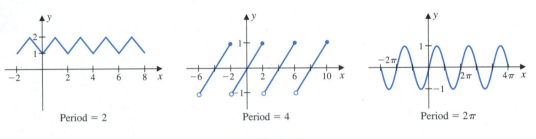

Period = 2   Period = 4   Period = 2π

FIGURE 3.12

Since the circumference of the unit circle is 2π, the coordinates of $P(t)$ and $P(t + 2\pi)$ are the same.

Period of the Sine and Cosine

The sine and cosine functions are periodic with **period** 2π. For every real number t,

$$\sin(t + 2\pi) = \sin t \qquad \text{and} \qquad \cos(t + 2\pi) = \cos t.$$

To determine values of the sine and cosine functions for all values of t, then, we need to determine only the values of these functions for $0 \le t < 2\pi$. The other values can be determined by finding a corresponding value in the interval $[0, 2\pi)$. For example, to find the sine and cosine of 17π, we note that the coordinates of $P(17\pi)$ agree with those of $P(15\pi), P(13\pi), \ldots$, and, finally, $P(\pi)$. So

$$\sin 17\pi = \sin \pi = 0 \qquad \text{and} \qquad \cos 17\pi = \cos \pi = -1.$$

In a similar manner, the coordinates of $P(-5\pi/2)$ are the same as those of $P(-\pi/2)$ and $P(3\pi/2)$. So the sine and cosine of $-5\pi/2$ agree with those of $3\pi/2$, which are given in Table 3.1 as -1 and 0, respectively.

This periodic information will not be of much use, of course, unless we can determine more values of $\sin t$ and $\cos t$ for t in the interval $[0, 2\pi)$. So our next step is to extend our knowledge to include other values of t that lie in the interval $[0, 2\pi)$. We will first determine the coordinates for $P(\pi/4)$, $P(\pi/3)$, and $P(\pi/6)$, and consequently the sines and cosines of $\pi/4$, $\pi/3$, and $\pi/6$. The symmetry of the unit circle will then permit us to determine the values of the sine and cosine functions for any integer multiple of these angles.

In calculus you will be expected to know the values of the sine and cosine at $\pi/6$, $\pi/4$, and $\pi/3$.

Determining $\sin\left(\frac{\pi}{4}\right)$ and $\cos\left(\frac{\pi}{4}\right)$

The point $P(\pi/4)$ lies on the unit circle midway between $P(0) = (1, 0)$ and $P(\pi/2) = (0, 1)$, as shown in Figure 3.13.

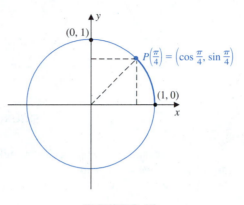

FIGURE 3.13

As a consequence, the x- and y-coordinates of $P(\pi/4)$ agree, and

$$\sin\frac{\pi}{4} = \cos\frac{\pi}{4}.$$

By this and the Pythagorean Identity we have

$$1 = \left(\sin\frac{\pi}{4}\right)^2 + \left(\cos\frac{\pi}{4}\right)^2 = 2\left(\cos\frac{\pi}{4}\right)^2,$$

so

$$\left(\cos\frac{\pi}{4}\right)^2 = \frac{1}{2} \qquad \text{and} \qquad \sin\frac{\pi}{4} = \cos\frac{\pi}{4} = \pm\frac{\sqrt{2}}{2}.$$

Since $P(\pi/4) = (\cos(\pi/4), \sin(\pi/4))$ lies in the first quadrant of the plane, both of its coordinates are positive, which implies the following.

The Sine and Cosine of $\dfrac{\pi}{4}$

$$\sin\frac{\pi}{4} = \frac{\sqrt{2}}{2} \qquad \text{and} \qquad \cos\frac{\pi}{4} = \frac{\sqrt{2}}{2}.$$

Determining $\sin\left(\frac{\pi}{3}\right)$ and $\cos\left(\frac{\pi}{3}\right)$

Consider the isosceles triangle AOB shown in Figure 3.14(a), where O is at the origin $(0, 0)$ of the plane, A is at $P(0) = (1, 0)$, and B is at $P(\pi/3) = (\cos\pi/3, \sin\pi/3)$.

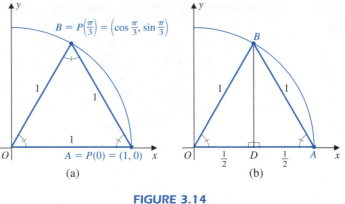

FIGURE 3.14

The base angles of an isosceles triangle are equal, so $\angle OAB = \angle OBA$. Since the sum of the angles in a triangle is π and $\angle AOB = \pi/3$, we must also have $\angle OAB = \angle OBA = \pi/3$. So $\triangle AOB$ is equilateral, with all its sides of length 1.

Drop a perpendicular from B to the x-axis meeting the axis at D, as shown in Figure 3.14(b). Then $\triangle DBO$ is congruent to $\triangle DBA$, and

$$OD = DA = \frac{1}{2} = \cos \frac{\pi}{3}.$$

The Pythagorean Theorem implies that the altitude of the triangle is

$$BD = \sqrt{1^2 - \left(\frac{1}{2}\right)^2} = \sqrt{\frac{3}{4}} = \frac{\sqrt{3}}{2} = \sin \frac{\pi}{3}.$$

Hence $P(\pi/3) = (1/2, \sqrt{3}/2)$, and we have the following.

The Sine and Cosine of $\dfrac{\pi}{3}$

$$\sin \frac{\pi}{3} = \frac{\sqrt{3}}{2} \quad \text{and} \quad \cos \frac{\pi}{3} = \frac{1}{2}.$$

Determining $\sin\left(\frac{\pi}{6}\right)$ and $\cos\left(\frac{\pi}{6}\right)$

In Figure 3.15, the triangles $\triangle AOE$ and $\triangle COD$ are congruent since they are both right triangles with hypotenuse of length 1 and

$$\angle AOE = \frac{\pi}{2} - \frac{\pi}{3} = \frac{\pi}{6} = \angle COD.$$

But we have just found that $P(\pi/3) = (1/2, \sqrt{3}/2)$, so

$$OD = OE = \frac{\sqrt{3}}{2} = \cos \frac{\pi}{6} \quad \text{and} \quad DC = EA = \frac{1}{2} = \sin \frac{\pi}{6}.$$

Hence, we have the following.

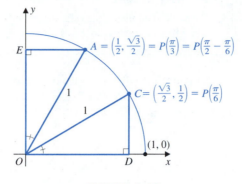

FIGURE 3.15

The Sine and Cosine of $\dfrac{\pi}{6}$

$$\sin \frac{\pi}{6} = \frac{1}{2} \quad \text{and} \quad \cos \frac{\pi}{6} = \frac{\sqrt{3}}{2}.$$

We now have the values of the sine and cosine functions for five values of t in the interval $[0, \pi/2]$. These are shown in Figure 3.16 and in Table 3.3 and form the basis for our knowledge of the values of these trigonometric functions.

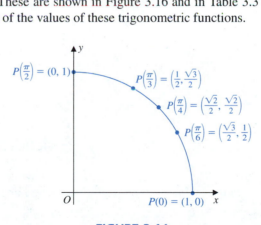

FIGURE 3.16

TABLE 3.3

t	0	$\pi/6$	$\pi/4$	$\pi/3$	$\pi/2$
$\cos t$	1	$\sqrt{3}/2$	$\sqrt{2}/2$	$1/2$	0
$\sin t$	0	$1/2$	$\sqrt{2}/2$	$\sqrt{3}/2$	1

Table 3.3 provides us with the basic values needed for the sine and cosine functions since the values of any multiple of $\pi/6$ or $\pi/4$ can be found using the entries in this table. To do this we introduce the notion of the *reference number*.

Reference Number

For any real number t, the **reference number** r associated with t is the shortest distance along the unit circle from t to the x-axis. For any t, the reference number is in $[0, \pi/2]$.

Figure 3.17 shows values of t in each of the four quadrants together with their reference number r. Notice that in each case the coordinates of $P(t)$ are closely related to the coordinates of $P(r)$.

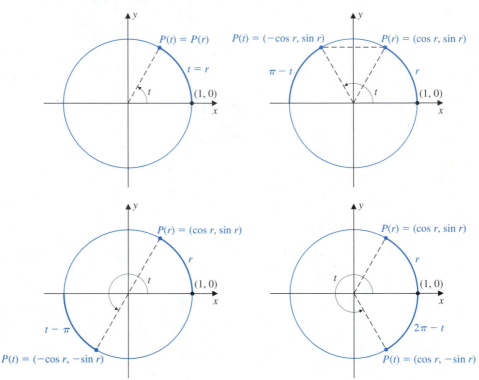

FIGURE 3.17

The following example shows how the reference numbers can be used with the values in Table 3.3 on page 179 to determine the sine and cosine of a multiple of $\pi/6$ or $\pi/4$.

EXAMPLE 1   Find the values of $\sin t$ and $\cos t$ when

a. $t = \dfrac{5\pi}{6}$ b. $t = -\dfrac{3\pi}{4}$ c. $t = \dfrac{17\pi}{3}$.

Solution   For each value of t we first determine the reference number in $[0, \pi/2]$ that corresponds to t. The coordinates of $P(t)$ are the same as those of the reference number, except for an adjustment of signs to reflect the quadrant in which $P(t)$ lies.

a. The reference number for $5\pi/6$ is $\pi/6$, so the point $P(5\pi/6)$, shown in Figure 3.18, has the same y-coordinate as the point $P(\pi/6)$. The x-coordinates of these two points have the same magnitude but differ in sign. So $P(5\pi/6) = (-\sqrt{3}/2, 1/2)$, which implies that

$$\sin \frac{5\pi}{6} = \frac{1}{2} \quad \text{and} \quad \cos \frac{5\pi}{6} = -\frac{\sqrt{3}}{2}.$$

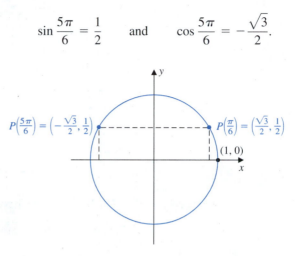

FIGURE 3.18

b. The point $P(-3\pi/4)$, shown in Figure 3.19, lies in the third quadrant with reference number $\pi/4$. Its x- and y-coordinates have the same magnitude as those of $P(\pi/4)$ but differ in sign. Hence,

$$\sin \left(-\frac{3\pi}{4}\right) = -\frac{\sqrt{2}}{2} \quad \text{and} \quad \cos \left(-\frac{3\pi}{4}\right) = -\frac{\sqrt{2}}{2}.$$

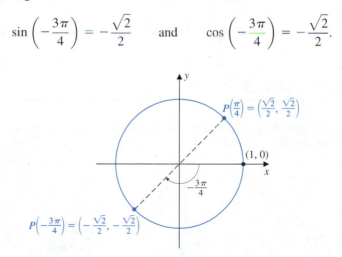

FIGURE 3.19

c. Since $17\pi/3$ exceeds 2π, it is a little more difficult to determine where $P(17\pi/3)$ lies on the unit circle. Notice, however that

$$\frac{17\pi}{3} = \frac{18\pi}{3} - \frac{\pi}{3} = 3(2\pi) - \frac{\pi}{3}.$$

This implies that $17\pi/3$ is just $\pi/3$ units short of being three counterclockwise revolutions of the circle, beginning at $(1, 0)$. So $P(17\pi/3)$ lies in the fourth quadrant and $17\pi/3$ has reference number $\pi/3$, as shown in Figure 3.20.

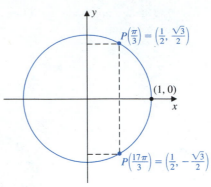

FIGURE 3.20

Since $P(\pi/3) = (1/2, \sqrt{3}/2)$, we have

$$P(17\pi/3) = P(5\pi/3) = (1/2, -\sqrt{3}/2).$$

This gives

$$\sin \frac{17\pi}{3} = -\frac{\sqrt{3}}{2} \quad \text{and} \quad \cos \frac{17\pi}{3} = \frac{1}{2}.$$ ■

When using a calculator to approximate values of the trigonometric functions, make sure the calculator is set to the correct mode. For our purposes you should be in radian mode, not in degree mode.

We have described in this section how we can obtain the sine and cosine of the most frequently used values of t. Any scientific calculator will give approximations of the sine and cosine functions for arbitrary values of t.

EXERCISE SET 3.2

In Exercises 1–8, show the approximate location on the unit circle of $P(t)$ for the given value of t.

1. $t = \dfrac{\pi}{8}$ **2.** $t = -\dfrac{\pi}{4}$

3. $t = \dfrac{7\pi}{6}$ **4.** $t = -\dfrac{\pi}{2}$

5. $t = -\dfrac{11\pi}{4}$ **6.** $t = -\dfrac{4\pi}{3}$

7. $t = \dfrac{315\pi}{4}$ **8.** $t = -\dfrac{35\pi}{2}$

In Exercises 9–14, find (a) the reference number r for the given value of t and (b) show $P(t)$ and $P(r)$ on the unit circle.

9. $t = \dfrac{2\pi}{3}$ **10.** $t = \dfrac{3\pi}{4}$

11. $t = -\dfrac{5\pi}{6}$ **12.** $t = -\dfrac{2\pi}{3}$

13. $t = -\dfrac{\pi}{4}$ **14.** $t = \dfrac{5\pi}{3}$

In Exercises 15–22, find $\sin t$ and $\cos t$ for the given value of t.

15. $t = \dfrac{\pi}{6}$ **16.** $t = \dfrac{\pi}{4}$

17. $t = \dfrac{13\pi}{4}$ **18.** $t = \dfrac{7\pi}{3}$

19. $t = -\dfrac{\pi}{3}$ **20.** $t = \dfrac{7\pi}{6}$

21. $t = -\dfrac{5\pi}{6}$ **22.** $t = -\dfrac{5\pi}{4}$

In Exercises 23–26, find all values of t in the interval $[0, 2\pi]$ that satisfy the given equation.

23. $\cos t = 1$ **24.** $\sin t = -\dfrac{\sqrt{3}}{2}$

25. $\cos \dfrac{t}{2} = \dfrac{1}{2}$

26. $\sin 3t = -\dfrac{\sqrt{2}}{2}$

In Exercises 27–30, determine whether the function is even, odd, or neither.

27. $f(x) = (\cos x)^2$

28. $f(x) = x^3 \sin x$

29. $f(x) = |x| \sin x$

30. $f(x) = \sin(\cos x)$

31. If $P(t)$ has coordinates $\left(\dfrac{3}{5}, \dfrac{4}{5}\right)$, find the coordinates of each of the following.

  a. $P(t + \pi)$ **b.** $P(-t)$

  c. $P(t - \pi)$ **d.** $P(-t - \pi)$

32. If $\sin t = -\dfrac{2\sqrt{2}}{3}$ and $\cos t = \dfrac{1}{3}$, find the sine and cosine of each of the following.

  a. $t + \pi$ **b.** $-t$

  c. $t + \dfrac{\pi}{2}$ **d.** $-t + \dfrac{\pi}{2}$

33. If $\sin t = \dfrac{3}{5}$ and $P(t)$ is in quadrant II, find $\cos t$.

34. If $\sin t = \dfrac{4}{5}$ and $0 < t < \pi/2$, find $\cos t$.

35. If $\cos t = \dfrac{2}{3}$ and $3\pi/2 < t < 2\pi$, find $\sin t$.

36. If $\cos t = -\dfrac{\sqrt{2}}{5}$ and $P(t)$ is in quadrant III, find $\sin t$.

37. Find all t in the interval $[0, 2\pi]$ satisfying $(\cos t)^2 + \cos t - 2 = 0$.

38. Find all t in the interval $[0, 2\pi]$ satisfying $\sin t + \cos t = 1$.

39. Find all t in the interval $[0, 2\pi]$ satisfying $\sin t \cos t - \sin t - \cos t + 1 = 0$.

40. Suppose the function f is even and periodic with period 2, that $f(-1) = 0$ and $f(0) = 1$, and that f is linear on the interval $[-1, 0]$ and on the interval $[0, 1]$.

  a. Make a sketch of the graph of $y = f(x)$.

  b. Determine the values of x for which $f(x) = 0$ and $f(x) = 1$ and the range of the function.

3.3 GRAPHS OF THE SINE AND COSINE FUNCTIONS

The information gained in Section 3.2 can be used to sketch reasonable graphs of the sine and cosine functions. If a more accurate graph is needed, it can be obtained by using a graphing device. In Section 3.2 the variable t was used to describe the domain of these functions so that an xy-coordinate system could be used to position the point $P(t)$. The coordinates of $P(t)$ simultaneously produced the sine and cosine of t. Now that we can determine values of these functions for representative real numbers, we can revert to the usual situation where the variable x represents the numbers in the domain of a function and the variable y represents the resulting values in the range.

We first consider the graph of the sine function, $f(x) = \sin x$. For x in the interval $[0, 2\pi]$ we have the values given in Table 3.4.

TABLE 3.4

x	0	$\pi/6$	$\pi/4$	$\pi/3$	$\pi/2$	$2\pi/3$	$3\pi/4$	$5\pi/6$
$\sin x$	0	$1/2$	$\sqrt{2}/2$	$\sqrt{3}/2$	1	$\sqrt{3}/2$	$\sqrt{2}/2$	$1/2$

x	π	$7\pi/6$	$5\pi/4$	$4\pi/3$	$3\pi/2$	$5\pi/3$	$7\pi/4$	$11\pi/6$	2π
$\sin x$	0	$-1/2$	$-\sqrt{2}/2$	$-\sqrt{3}/2$	-1	$-\sqrt{3}/2$	$-\sqrt{2}/2$	$-1/2$	0

Graphing devices will give very accurate graphs of the trigonometric functions. Remember to be certain that radian measure is being used.

Assuming a simple and smooth behavior based on these values gives the graph in Figure 3.21(a) for $x \in [0, 2\pi]$. The graph has x-intercepts at $x = 0$, at $x = \pi$, and at $x = 2\pi$. The sine function is increasing on $[0, \pi/2]$ and on $[3\pi/2, 2\pi]$ and is decreasing on $[\pi/2, 3\pi/2]$. A maximum of 1 occurs at $x = \pi/2$ and a minimum of -1 at $x = 3\pi/2$.

Since for every real number x we have $\sin(x + 2\pi) = \sin x$, the graph of the sine function extends indefinitely to the right and to the left, as indicated in Figure 3.21(b). Notice that the horizontal scale has been compressed to show more of the graph. This is often done when graphing trigonometric functions.

Radian measure is preferred in precalculus and calculus. However, to see one cycle of $y = \sin x$ on a graphing calculator set on degree mode, you will need to use a viewing rectangle $[0, 360] \times [-1, 1]$.

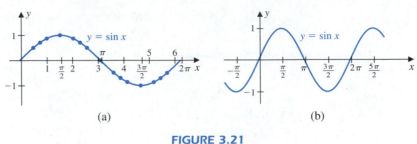

(a)   (b)

FIGURE 3.21

In a similar manner, the values of the cosine function, $f(x) = \cos x$, for x in $[0, 2\pi]$ given in Table 3.5 produce the graph shown in Figure 3.22(a).

TABLE 3.5

x	0	$\pi/6$	$\pi/4$	$\pi/3$	$\pi/2$	$2\pi/3$	$3\pi/4$	$5\pi/6$	
$\cos x$	1	$\sqrt{3}/2$	$\sqrt{2}/2$	$1/2$	0	$-1/2$	$-\sqrt{2}/2$	$-\sqrt{3}/2$	

x	π	$7\pi/6$	$5\pi/4$	$4\pi/3$	$3\pi/2$	$5\pi/3$	$7\pi/4$	$11\pi/6$	2π
$\cos x$	-1	$-\sqrt{3}/2$	$-\sqrt{2}/2$	$-1/2$	0	$1/2$	$\sqrt{2}/2$	$\sqrt{3}/2$	1

Recognizing the general shape and behavior of the graphs of the sine and cosine functions will make your work with these functions in calculus much easier.

The graph has x-intercepts at $x = \pi/2$ and at $x = 3\pi/2$. The cosine function is decreasing on $[0, \pi]$ and is increasing on $[\pi, 2\pi]$. A maximum of 1 occurs at $x = 0$ and at $x = 2\pi$, and a minimum of -1 occurs at $x = \pi$.

Since for every real number x we have $\cos(x + 2\pi) = \cos x$, the graph extends indefinitely to the right and to the left, as indicated in Figure 3.22(b).

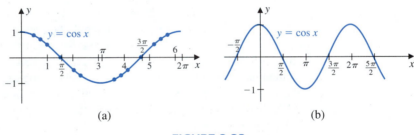

(a)   (b)

FIGURE 3.22

We found in Section 3.2 that the sine function is odd and the cosine is even. This implies that the graph of $y = \sin x$ is symmetric with respect to the origin and the graph of $y = \cos x$ is symmetric with respect to the y-axis.

This relationship between the sine and cosine is given a great deal of emphasis in calculus.

Notice also that the zeros of the sine function occur at integer multiples of π, that is, at $x = n\pi$ for all integers n, which is precisely where the maximum and minimum values of the cosine function occur. Moreover, the zeros of the cosine function occur when $x = n\pi + \pi/2$ for all integers n and coincide with the values producing the maximum and minimum values of the sine function.

EXAMPLE 1   Use the graphs of $y = \sin x$ and $y = \cos x$ to sketch the graphs of

$$y = \sin\left(x - \frac{\pi}{2}\right), \qquad y = \sin\left(x + \frac{\pi}{2}\right),$$

$$y = \cos\left(x - \frac{\pi}{2}\right), \qquad y = \cos\left(x + \frac{\pi}{2}\right).$$

Solution   The graph of $y = \sin(x - \pi/2)$ is a horizontal shift of the graph of $y = \sin x$ by $\pi/2$ units to the right, and the graph of $y = \sin(x + \pi/2)$ is a horizontal shift of the graph of $y = \sin x$ by $\pi/2$ units to the left. The graphs of these functions are shown in Figure 3.23.

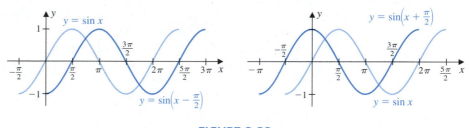

FIGURE 3.23

The graphs of $y = \cos(x - \pi/2)$ and $y = \cos(x + \pi/2)$ are obtained from the graph of $y = \cos x$ in a similar manner, as shown in Figure 3.24.   ∎

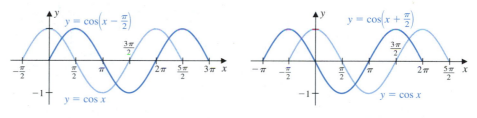

FIGURE 3.24

From the graphs in Figures 3.23 and 3.24 we can deduce the following relationships between the sine and cosine functions.

$$\sin\left(x + \frac{\pi}{2}\right) = \cos x, \qquad\qquad \sin\left(x - \frac{\pi}{2}\right) = -\cos x,$$

$$\cos\left(x - \frac{\pi}{2}\right) = \sin x, \qquad\text{and}\qquad \cos\left(x + \frac{\pi}{2}\right) = -\sin x,$$

By shifting by π units instead of $\pi/2$ we can also deduce that

$$\sin(x - \pi) = \sin(x + \pi) = -\sin x \qquad\text{and}\qquad \cos(x - \pi) = \cos(x + \pi) = -\cos x,$$

as shown in Figure 3.25. These results are special cases of some general identities we will consider in Section 3.5.

Identities involving the trigonometric functions can often be visualized using graphs, as in Figures 3.23, 3.24, and 3.25. Graphing devices become very useful in this respect.

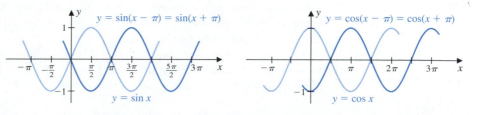

FIGURE 3.25

In the remainder of this section we consider some other shifts and extensions of the sine and cosine functions.

EXAMPLE 2   Sketch the graph of $y = -3 \sin (x - \pi/4)$.

Solution   As in previous cases, we build the graph of this function by using the graphs of more elementary functions. We start with $y = \sin x$ and stretch this graph vertically by a factor of 3 to produce the graph of $y = 3 \sin x$ shown in Figure 3.26(a). Then we shift this graph $\pi/4$ units to the right to produce the graph of $y = 3 \sin(x - \pi/4)$ shown in Figure 3.26(b). The graph of $y = -3 \sin(x - \pi/4)$, also shown in Figure 3.26(b), is the reflection about the x-axis of the graph of $y = 3 \sin(x - \pi/4)$.   ■

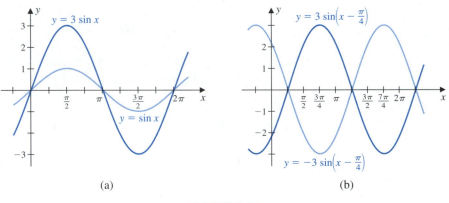

(a)   (b)

FIGURE 3.26

Examples 1 and 2 provide illustrations of the techniques needed to graph a member of the class of functions of the form

$$f(x) = A \sin(Bx + C) \qquad \text{and} \qquad f(x) = A \cos(Bx + C),$$

where $A \neq 0$ and $B > 0$. These functions are frequently seen in physics and engineering since they are involved in a variety of applications.

The graph of any function of this type is the compression or elongation and translation of a sine or cosine graph. The case of the sine and the cosine functions is similar, so let us consider the graph of $f(x) = A \sin(Bx + C)$.

The first step is to factor the constant B from the terms in $Bx + C$ to produce

$$f(x) = A \sin B \left(x + \frac{C}{B} \right).$$

First we sketch the graph of $y = A \sin x$. If $A > 0$, it simply involves a vertical compression or elongation, as shown in Figure 3.27(a). If $A < 0$, a reflection about the x-axis is also needed, as shown in Figure 3.27(b). The number $|A|$ is called the *amplitude* of f and determines the height of the "wave."

The transition from $y = A \sin x$ to $y = A \sin Bx$ requires a change in the period of the function. The sine function has period 2π, as does the function described by $y = A \sin x$. To find the period of $y = A \sin Bx$, we set

$$Bx = 2\pi, \qquad \text{which implies that} \qquad x = \frac{2\pi}{B},$$

It would be difficult to overemphasize the importance of being able to apply the basic graphing techniques we introduced in Chapter 1 to each new function you encounter.

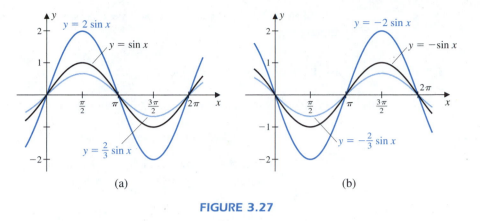

FIGURE 3.27

so the period is $2\pi/B$. Figure 3.28(a) shows examples of graphs of the form $y = A \sin Bx$ when $A > 0$. Figure 3.28(b) shows examples when $A < 0$.

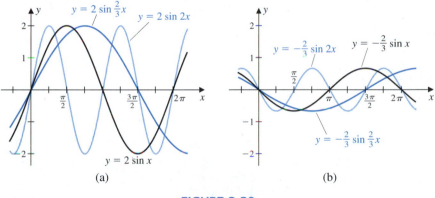

FIGURE 3.28

The final graph for $f(x) = A \sin(Bx + C) = A \sin B(x + C/B)$ requires a horizontal shift by the amount $|C/B|$. As usual, this *phase shift* is to the left if C/B is positive and to the right if C/B is negative. Some sample situations are shown in Figure 3.29.

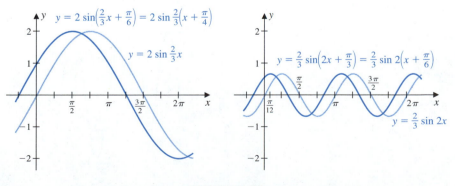

FIGURE 3.29

Sketching the graph of $y = A \cos(Bx + C)$ is similar, as illustrated in the following example.

EXAMPLE 3   Sketch the graph of $f(x) = 2 \cos \left(3x - \dfrac{\pi}{2} \right)$.

Solution   The first step is to factor the constant 3 from the terms in $3x - \pi/2$ to produce

$$y = 2 \cos \left(3x - \frac{\pi}{2} \right) = 2 \cos 3 \left(x - \frac{\pi}{6} \right).$$

The graph of $y = 2 \cos x$, shown in Figure 3.30, is a vertical stretch or elongation by a factor of 2 of the graph of $y = \cos x$.

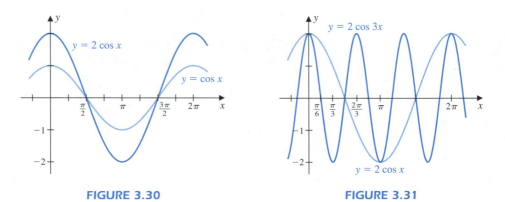

FIGURE 3.30   FIGURE 3.31

Since the period of $y = 2 \cos x$ is 2π, the period of $y = 2 \cos 3x$ is $2\pi/3$. As a consequence, the graph of $y = 2 \cos 3x$ is a horizontal compression of the graph of $y = 2 \cos x$, as shown in Figure 3.31.

Finally, the graph of

$$y = 2 \cos \left(3x - \frac{\pi}{2} \right) = 2 \cos 3 \left(x - \frac{\pi}{6} \right)$$

involves a translation of the graph of $y = 2 \cos 3x$ to the right $\pi/6$ units, as shown in Figure 3.32.   ■

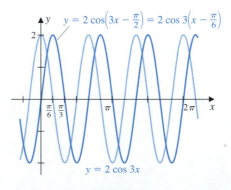

FIGURE 3.32

EXAMPLE 4   Use a graphing device to sketch the graph of $f(x) = \cos(75x)$.

Solution   Figures 3.33(a) and 3.33(b) show the graph of f using the viewing rectangles $[-10, 10] \times [-1.5, 1.5]$ and $[-8, 8] \times [-1.5, 1.5]$. About the only similarity between these graphs and the graph of the cosine function seems to be that both have range $[-1, 1]$. The problem the graphing device faces is that there is too much oscillation for its point–plotting technique to work effectively. To get a better representation for the graph, we first need to determine the period of the function so that we can prescribe a reasonable viewing rectangle. The period of $y = \cos x$ is 2π, so the period of f is

Although graphing devices are very useful in sketching the graphs of the trigonometric functions, understanding the general behavior of these functions is also needed, for example, to select a proper viewing window.

$$\frac{2\pi}{75} \approx 0.08.$$

In Figure 3.33(c), the graph is shown in the viewing rectangle $[-0.2, 0.2] \times [-1.5, 1.5]$, which includes

$$\frac{0.2 - (-0.2)}{0.08} = 5$$

periods of the function.

$f(x) = \cos(75x)$

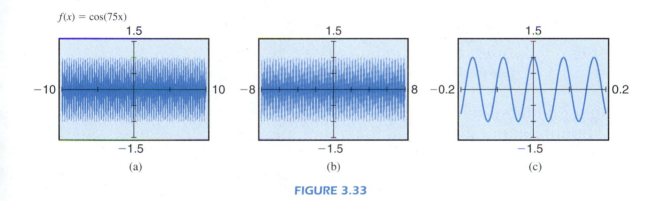

     (a)   (b)   (c)

FIGURE 3.33

EXAMPLE 5   Use a graphing device to sketch a representation of the graph of

$$f(x) = \frac{\sin x}{x},$$

and describe the important features of the graph.

Solution   In the previous example we needed a very narrow x-interval to obtain a good view of the graph. As Figure 3.34(a) and Figure 3.34(b) show, more information is found for this graph if a wide interval is used.

The function $f(x) = \sin x/x$ is important in calculus, where it is shown that

$$\text{as } x \to 0, \quad \text{we have} \quad \frac{\sin x}{x} \to 1.$$

This is why the computer-generated graph appears to fill in the point $(0, 1)$, even though 0 is not in the domain.

Another interesting feature of the graph of $f(x) = \sin x/x$, shown more clearly in Figure 3.34(b), is that $y = f(x)$ has a horizontal asymptote $y = 0$, since, as $x \to -\infty$ or $x \to \infty$, we have $f(x) \to 0$. Moreover, the graph crosses its horizontal asymptote

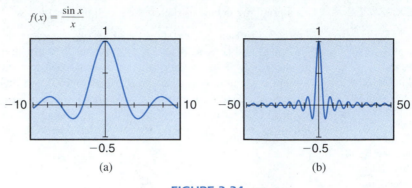

FIGURE 3.34

infinitely often, at every integer multiple of π. Finally, notice that both the numerator and denominator of f describe odd functions. Hence f itself is even since

$$f(-x) = \frac{\sin(-x)}{(-x)} = \frac{-\sin x}{-x} = \frac{\sin x}{x} = f(x)$$ ■

EXAMPLE 6   Fit a sine wave of the form

$$y = A \sin(Bx + C) + D$$

to the data points shown in Figure 3.35.

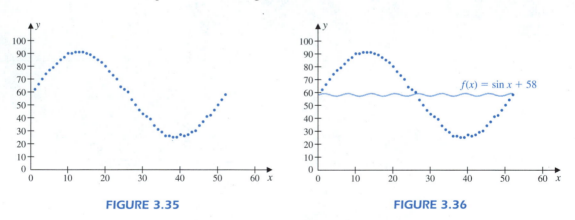

FIGURE 3.35   **FIGURE 3.36**

Solution   The data points displayed in Figure 3.35 exhibit an approximate sine wave pattern centered vertically on the value 58. Since the graph of $y = \sin x$ passes through $(0, 0)$ our first attempt at fitting the data might be to try

When trying to fit a curve to a collection of data points, a graphing device can often simplify the process.

$$y = \sin x + 58.$$

Figure 3.36 indicates the curve $y = \sin x + 58$ is far too flat.

The maximum and minimum data values appear to be 91 and 25, respectively, so that twice the height of the wave should be $91 - 25 = 66$. As a consequence, the amplitude is

$$\frac{1}{2}(66) = 33.$$

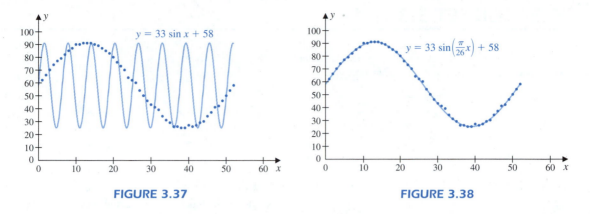

FIGURE 3.37   FIGURE 3.38

Figure 3.37 shows the original data points, along with

$$y = 33 \sin x + 58.$$

We now need to stretch the curve $y = 33 \sin x + 58$ horizontally to increase the period of the sine wave. The original data appear to have a period of 52, so we set

$$\frac{2\pi}{B} = 52, \qquad \text{which gives} \qquad B = \frac{2\pi}{52} = \frac{\pi}{26}.$$

Figure 3.38 shows the original data with the curve

$$y = 33 \sin \left(\frac{\pi}{26} x \right) + 58,$$

which appears to fit the data quite well.   ■

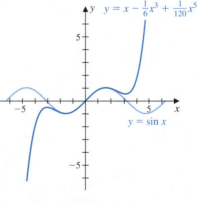

FIGURE 3.39

EXAMPLE 7   Use a graphing device to plot graphs of $y = \sin x$ and $y = x - \frac{1}{6}x^3 + \frac{1}{120}x^5$.

Solution   Figure 3.39 shows the computer-generated graphs in the viewing rectangle $[-2\pi, 2\pi] \times$ $[-5, 5]$. We can see that near the origin the polynomial is a good approximation to the sine curve. In particular, the approximate value for $\sin(1)$ given from a calculator is 0.84147098, and the polynomial evaluated at 1 is approximately 0.84166666.   ■

In calculus you will see how to determine a polynomial used to approximate $\sin x$.

EXERCISE SET 3.3

In Exercises 1–14, use the graphs of the sine and cosine to sketch one period of the graph of the function.

1. $y = \cos 2x$

2. $y = 2 \cos x$

3. $y = 4 \sin 3x$

4. $y = 2 \cos x - 1$

5. $y = 2 \sin \dfrac{x}{2}$

6. $y = \cos \left(x - \dfrac{\pi}{2} \right)$

7. $y = 2 + \sin \pi x$

8. $y = \sin(3x + \pi)$

9. $y = 2 - \cos(x - 1)$

10. $y = -2 \sin(x - 1) + 3$

11. $y = -2 + 3 \sin \left(3x - \dfrac{\pi}{2} \right)$

12. $y = \dfrac{1}{2} \cos \left(\dfrac{\pi}{2} - 2x \right)$

13. $y = |\cos x|$

14. $y = |\sin x|$

In Exercises 15–18, find a sine function or a cosine function whose graph matches the given curve.

15.

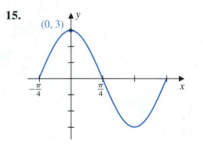

16.

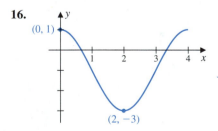

17.

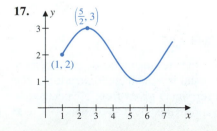

18.

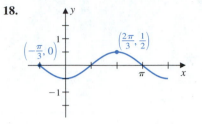

In Exercises 19–22, determine an appropriate viewing rectangle for the function and use it to sketch the graph.

19. $f(x) = \cos(100x)$

20. $f(x) = \sin \left(\dfrac{x}{50} \right)$

21. $f(x) = -5 \sin(20x)$

22. $f(x) = x^2 + 5 \sin(10x)$

In Exercises 23–26, use a graphing device to approximate the solutions to the equation.

23. $\cos x = x$

24. $\cos x = x^2$

25. $\sin x + \cos x = x$

26. $\sin x + x = x^3$

27. Find the smallest positive value of t for which

$$f(t) = -3 \cos(2t + \pi/6)$$

    a. Is equal to 0;

    b. Attains a maximum value;

    c. Attains a minimum value.

28. Write each of the following as the composition of two functions $h(x) = f(g(x))$.

    a. $h(x) = \sqrt{\sin x}$ **b.** $h(x) = 3 \cos(4x - 2)$

29. Fit a function of the form $f(x) = a + b \sin cx$ to approximate the given data.

x	0	0.5	1	1.5	2	−0.5	−1	−1.5	−2
y	1	2.4	3	2.4	1	−0.4	−1	−0.4	1

30. Let a, b, and c be positive constants. Describe the effect on the graph of $y = a \cos b(x + c)$ if

    a. b and c are fixed and a is doubled;

    b. a and c are fixed and b is doubled;

    c. a and b are fixed and c is doubled.

31. Use a graphing device to sketch each of the following graphs.

   a. $y = \sin x^2$ **b.** $y = (\sin x)^2$

   c. $y = \sin\left(\dfrac{1}{x}\right), x \neq 0$ **d.** $y = x \sin\left(\dfrac{1}{x}\right), x \neq 0$

   e. $y = \dfrac{(\sin x)^2}{x}, x \neq 0$ **f.** $y = \sqrt{|\sin x|}$

32. Use a graphing device to sketch the graphs of $y = f(x)$, $y = g(x)$, and $y = f(x) + g(x)$ on the same set of axes. Discuss the effect on the graph of $y = f(x)$ caused by adding the graph of $y = g(x)$.

   a. $f(x) = \sin x, g(x) = x$

   b. $f(x) = \sin x, g(x) = -x$

   c. $f(x) = \cos x, g(x) = x$

   d. $f(x) = \cos x, g(x) = -x$

3.4   OTHER TRIGONOMETRIC FUNCTIONS

In Section 3.2 we defined the sine and the cosine functions, the two most basic trigonometric functions. There are four additional and important trigonometric functions. These involve the quotients and reciprocals of the sine and cosine functions.

The Tangent, Cotangent, Secant, and Cosecant Functions

The **tangent, cotangent, secant,** and **cosecant** functions, written respectively as $\tan x$, $\cot x$, $\sec x$, and $\csc x$, are defined by the quotients

$$\tan x = \frac{\sin x}{\cos x}, \qquad \cot x = \frac{\cos x}{\sin x},$$

$$\sec x = \frac{1}{\cos x}, \qquad \csc x = \frac{1}{\sin x}.$$

The tangent and secant functions are defined whenever $x \neq \pi/2 + n\pi$ for an integer n.

    The cotangent and cosecant functions are defined whenever $x \neq n\pi$ for an integer n.

EXAMPLE 1   Suppose that $\sin x = 2/3$ and that $\pi/2 < x < \pi$. Determine the values of the other trigonometric functions.

Solution   We first find the value of the cosine function using the Pythagorean Identity $(\sin x)^2 + (\cos x)^2 = 1$. It gives

$$\cos x = \pm\sqrt{1 - (\sin x)^2} = \pm\sqrt{1 - \left(\frac{2}{3}\right)^2} = \pm\frac{\sqrt{5}}{3}.$$

Since x is in the interval $(\pi/2, \pi)$, the cosine is negative, and $\cos x = -\sqrt{5}/3$. The remaining trigonometric functions are easily determined now that we have the values of both the sine and the cosine. They are

$$\tan x = \frac{\sin x}{\cos x} = -\frac{2/3}{\sqrt{5}/3} = -\frac{2\sqrt{5}}{5} \qquad\qquad \cot x = \frac{\cos x}{\sin x} = -\frac{\sqrt{5}/3}{2/3} = -\frac{\sqrt{5}}{2}$$

$$\sec x = \frac{1}{\cos x} = -\frac{3}{\sqrt{5}} = -\frac{3\sqrt{5}}{5}, \qquad \text{and} \quad \csc x = \frac{1}{\sin x} = \frac{3}{2}. \qquad\blacksquare$$

Table 3.6 shows the signs of the trigonometric functions in the various quadrants of the plane. Notice that this information agrees with the results we obtained in Example 1.

TABLE 3.6

Quadrant	I	II	III	IV
$\sin x$	+	+	−	−
$\cos x$	+	−	−	+
$\tan x$	+	−	+	−
$\cot x$	+	−	+	−
$\sec x$	+	−	−	+
$\csc x$	+	+	−	−

Because they involve quotients, the domains of the tangent, cotangent, secant, and cosecant functions are restricted to those numbers for which the denominator is nonzero. The denominator in the case of the tangent and the secant is $\cos x$, which is zero at odd multiplies of $\pi/2$. In the case of the cotangent and cosecant the denominator is $\sin x$, which is zero at multiples of π. The values of the sine and cosine functions that we found in the previous sections give the entries in Table 3.7. A dash, —, occurs in the table to indicate a number that is not in the domain of the function.

TABLE 3.7

x	0	$\frac{\pi}{6}$	$\frac{\pi}{4}$	$\frac{\pi}{3}$	$\frac{\pi}{2}$	$\frac{2\pi}{3}$	$\frac{3\pi}{4}$	$\frac{5\pi}{6}$	π
$\sin x$	0	$\frac{1}{2}$	$\frac{\sqrt{2}}{2}$	$\frac{\sqrt{3}}{2}$	1	$\frac{\sqrt{3}}{2}$	$\frac{\sqrt{2}}{2}$	$\frac{1}{2}$	0
$\cos x$	1	$\frac{\sqrt{3}}{2}$	$\frac{\sqrt{2}}{2}$	$\frac{1}{2}$	0	$-\frac{1}{2}$	$-\frac{\sqrt{2}}{2}$	$-\frac{\sqrt{3}}{2}$	-1
$\tan x$	0	$\frac{\sqrt{3}}{3}$	1	$\sqrt{3}$	—	$-\sqrt{3}$	-1	$-\frac{\sqrt{3}}{3}$	0
$\cot x$	—	$\sqrt{3}$	1	$\frac{\sqrt{3}}{3}$	0	$-\frac{\sqrt{3}}{3}$	-1	$-\sqrt{3}$	—
$\sec x$	1	$\frac{2\sqrt{3}}{3}$	$\sqrt{2}$	2	—	-2	$-\sqrt{2}$	$-\frac{2\sqrt{3}}{3}$	-1
$\csc x$	—	2	$\sqrt{2}$	$\frac{2\sqrt{3}}{3}$	1	$\frac{2\sqrt{3}}{3}$	$\sqrt{2}$	2	—

The Graph of the Tangent Function

The tangent function is zero precisely when the sine function is zero and is undefined when the cosine function is zero. It is positive on $(0, \pi/2)$ since the sine and cosine are both positive there, but it is negative on $(\pi/2, \pi)$ since on this interval the sine is positive and the cosine is negative. The portion of the graph of the tangent function for values in its domain that lie in the interval $[0, \pi]$ is shown in Figure 3.40. Notice the vertical asymptote at $x = \pi/2$.

The sine and cosine functions have period 2π, so the tangent function must also repeat itself every 2π units. Its period, however, is smaller than 2π. In Section 3.3 we showed that

$$\sin(x + \pi) = -\sin x \quad \text{and} \quad \cos(x + \pi) = -\cos x.$$

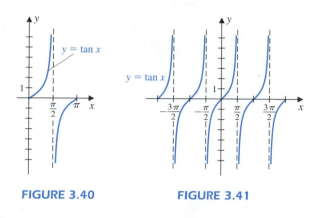

FIGURE 3.40   **FIGURE 3.41**

These results imply that

$$\tan(x + \pi) = \frac{\sin(x + \pi)}{\cos(x + \pi)} = \frac{-\sin x}{-\cos x} = \tan x,$$

so the tangent function repeats itself every π units. It is clear from the graph in Figure 3.40 that it repeats itself on no smaller interval, so the period of the tangent function is π. This means that we can sketch the graph of the entire tangent function from the portion shown in Figure 3.40. The graph given in Figure 3.41 shows that the tangent function is increasing on every interval that is completely contained in its domain and that the graph has vertical asymptotes at the points that are not in its domain.

EXAMPLE 2   Sketch each of the graphs.

a. $y = \tan 2x,$ b. $y = \tan\left(2x - \frac{\pi}{3}\right),$ c. $y = -\tan\left(2x - \frac{\pi}{3}\right) + 1.$

Solution   a. The graph of $y = \tan 2x$ is a horizontal compression, by a factor of 2, of the graph of the tangent function, so the resulting graph is as shown in Figure 3.42(a). Since the graph has been horizontally compressed by a factor of 2, the period of the function has been reduced by the same factor, and $y = \tan 2x$ has period $\pi/2$. Notice that the first positive vertical asymptote is $x = \pi/4$, and the first positive zero is $x = \pi/2$.

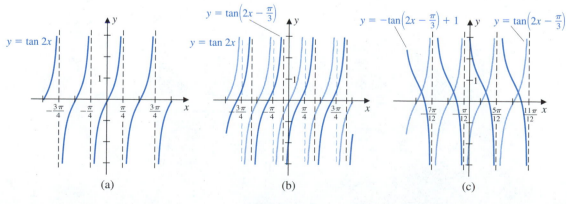

(a)   (b)   (c)

FIGURE 3.42

b. To sketch the graph of $y = \tan(2x - \pi/3)$ we first factor the 2 from the terms within the parentheses to produce

$$y = \tan\left(2x - \frac{\pi}{3}\right) = \tan 2\left(x - \frac{\pi}{6}\right).$$

This graph has the same shape as the graph of $y = \tan 2x$, but it is shifted $\pi/6$ units to the right, as shown in Figure 3.42(b). The first positive vertical asymptote is now at $x = \pi/4 + \pi/6 = 5\pi/12$, and the first two positive zeros are at $x = \pi/6$ and $x = \pi/2 + \pi/6 = 2\pi/3$.

c. The graph of $y = -\tan(2x - \pi/3)$ is the reflection about the x-axis of the graph of $y = \tan(2x - \pi/3)$. To obtain the graph of $y = -\tan(2x - \pi/3) + 1$, we shift the graph of $y = -\tan(2x - \pi/3)$ upward 1 unit, as shown in Figure 3.42(c). Notice that the vertical asymptotes for this graph are the same as for the graph in part (b).

The location of the zeros of $y = -\tan(2x - \pi/3) + 1$ is not obvious from the graph. To find them, first note that

$$0 = -\tan\left(2x - \frac{\pi}{3}\right) + 1 \qquad \text{implies that} \qquad \tan\left(2x - \frac{\pi}{3}\right) = 1.$$

But $y = \tan x$ is 1 precisely when $x = \pi/4 + n\pi$ for some integer n. So

$$\tan\left(2x - \frac{\pi}{3}\right) = 1 \qquad \text{precisely when} \qquad 2x - \frac{\pi}{3} = \frac{\pi}{4} + n\pi,$$

that is, when

$$x = \frac{1}{2}\left(\frac{\pi}{4} + \frac{\pi}{3} + n\pi\right) = \frac{7\pi}{24} + \frac{n\pi}{2}$$

for some integer n. The first positive zero is $7\pi/24$, and the second is $7\pi/24 + \pi/2 = 19\pi/24$. ∎

Now that we have the graphs of the sine, cosine, and tangent functions, we can sketch the graphs of the remaining trigonometric functions by using the reciprocal graphing technique. Figure 3.43 shows the graph of $y = \csc x$, the reciprocal of the sine function. Notice that $y = \csc x$ decreases as $y = \sin x$ increases, and $y = \csc x$ increases as $y = \sin x$ decreases. The local maxima and local minima of $y = \csc x$ occur at the same points as $y = \sin x$, but with the roles reversed. The graph of $y = \csc x$ has vertical asymptotes at those values of x for which $\sin x = 0$, that is, at integer multiples of π.

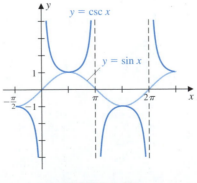

FIGURE 3.43

The graphs of $y = \sec x$ and $y = \cot x$ are obtained in a similar manner and are shown in Figure 3.44. Notice that the period of the cotangent function is π and the period of the secant function is 2π.

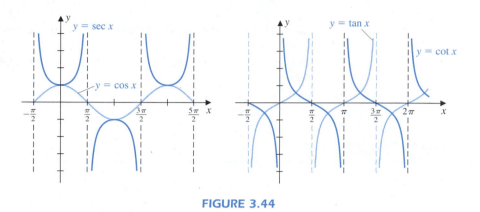

FIGURE 3.44

In Example 2, the graph in part (c) was systematically constructed from the graphs of $y = \tan 2x$ and $y = \tan(2x - \pi/3)$ needed in parts (a) and (b). In the next example no preliminary graphs are requested, only a single complicated graph. Try to determine the strategy for sketching the graph before reading the solution.

EXAMPLE 3   Sketch the graph of $y = \dfrac{1}{2} \csc\left(3x + \dfrac{\pi}{2}\right)$.

Solution   The strategy we have chosen is to start with the graph of $y = \csc x$, which was given in Figure 3.43. Then we modify this graph to produce the graphs of $y = \csc 3x$ and $y = \frac{1}{2}\csc 3x$. The graph that we want as our final result will be a horizontal shift of the graph of $y = \frac{1}{2}\csc 3x$.

     The graph of $y = \csc 3x$ is a horizontal compression of the graph of $y = \csc x$ by a factor of 3, as shown in Figure 3.45(a). Notice that the period has been reduced to $2\pi/3$.

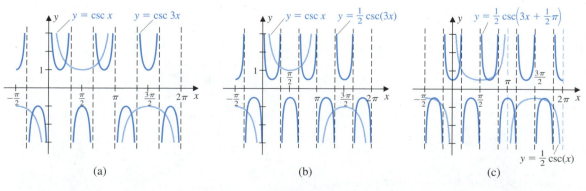

(a)   (b)   (c)

FIGURE 3.45

The graph of $y = \frac{1}{2} \csc 3x$, shown in Figure 3.45(b), is found by reducing by a factor of 2 all the y-values on the graph of $y = \csc 3x$. We now need to determine and perform the final horizontal shift on the graph of $y = \frac{1}{2} \csc 3x$. To do this we write

$$y = \frac{1}{2} \csc \left(3x + \frac{\pi}{2} \right) = \frac{1}{2} \csc 3 \left(x + \frac{\pi}{6} \right)$$

and find that the shift is $\pi/6$ units to the left, as shown in Figure 3.45(c).   ■

The graph in Figure 3.45(c) looks suspiciously like the graph of a secant function that has been compressed horizontally and vertically. Try to determine what you think the equation of this equivalent modified secant function would be. In the next section you will have a chance to verify your conjecture.

EXERCISE SET 3.4

In Exercises 1–6, find the quadrant in which $P(t)$ lies if the given conditions are satisfied.

1. $\sin t < 0$ and $\cos t > 0$ **2.** $\tan t < 0$ and $\cos t < 0$

3. $\sin t < 0$ and $\cot t > 0$ **4.** $\csc t > 0$ and $\tan t < 0$

5. $\sin t > 0$ and $\tan t > 0$ **6.** $\sin t < 0$ and $\sec t < 0$

In Exercises 7–12, find the values of all the trigonometric functions from the given information.

7. $\sin t = 4/5$, t is in quadrant I

8. $\cos t = -1/2$, $\pi \le t \le 2\pi$

9. $\sin t = 1/3$, t is in quadrant II

10. $\tan t = 2$, $0 < t < \pi/2$

11. $\csc t = -2$, $3\pi/2 < t < 2\pi$

12. $\cot t = 3$, t is in quadrant III

In Exercises 13–24, sketch one period of the given curve.

13. $y = 2 \tan x$ **14.** $y = -2 \tan x$

15. $y = \frac{1}{2} \cot x$ **16.** $y = -\frac{1}{2} \cot x$

17. $y = \tan \left(x + \frac{\pi}{4} \right)$ **18.** $y = \tan \left(x - \frac{\pi}{2} \right)$

19. $y = \sec \left(x + \frac{\pi}{4} \right)$ **20.** $y = \csc \left(x - \frac{\pi}{2} \right)$

21. $y = \tan \pi x$ **22.** $y = \cot \left(\frac{\pi x}{2} \right)$

23. $y = \tan \left(2x - \frac{\pi}{2} \right)$ **24.** $y = \sec \left(2x + \pi \right)$

In Exercises 25–28, find all values of t in the interval $[0, 2\pi]$ satisfying the given equation.

25. $|\tan t| = 1$ **26.** $(\cot t)^2 = 3$

27. $2 \sin 2t - \sqrt{2} \tan 2t = 0$ **28.** $\tan t - 3 \cot t = 0$

In Exercises 29–34, determine an appropriate viewing rectangle for the function and sketch the graph.

29. $f(x) = \tan(5x)$ **30.** $f(x) = \tan(8x - 10)$

31. $f(x) = \csc(100x)$ **32.** $f(x) = \sec \left(\frac{x}{50} \right)$

33. $f(x) = \tan \left(\frac{x}{100} \right)$

34. $f(x) = \tan(25x) - \csc(25x)$

35. Determine the values of the trigonometric functions of t if $P(t)$ lies in the fourth quadrant and on the line $y = -2x$.

36. Use the figure to show the following reduction formulas hold.

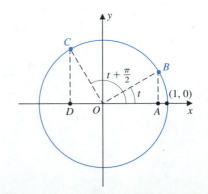

(a) $\sin\left(t + \dfrac{\pi}{2}\right) = \cos t$

(b) $\cos\left(t + \dfrac{\pi}{2}\right) = -\sin t$

(c) $\tan\left(t + \dfrac{\pi}{2}\right) = -\cot t$

[*Hint:* First show that triangles OAB and ODC are congruent.]

3.5   TRIGONOMETRIC IDENTITIES

In the first few sections of this chapter we have considered the basic properties and graphs of the trigonometric functions. We now look at some of the algebraic identities that simplify the work involving these functions and dramatically increase their usefulness in applications. But first a word of warning if you have been exposed to this subject in the past. Our emphasis is on developing a small core of identities that are continually needed and can be used to determine a much larger collection. These core identities must be known intimately if you are to succeed when you study the calculus of trigonometry. Fortunately, the number is quite small.

In calculus you will need to know the core identities *and be able to derive the others from this core.*

The first and most basic identity, and one we have already encountered, comes from the fact that for any real number x, the point $(\cos x, \sin x)$ lies on the unit circle centered at the origin. This gives the Pythagorean Identity we discovered at the beginning of Section 3.2:

$$(\sin x)^2 + (\cos x)^2 = 1.$$

By dividing this Pythagorean Identity by $(\cos x)^2$, assuming it is nonzero, we obtain

$$\frac{(\sin x)^2}{(\cos x)^2} + 1 = \frac{1}{(\cos x)^2},$$

or

$$(\tan x)^2 + 1 = (\sec x)^2.$$

In a similar manner, dividing the Pythagorean Identity by $(\sin x)^2$ produces

$$(\cot x)^2 + 1 = (\csc x)^2.$$

Since these two identities are just modified forms of the same basic identity, we call the collection the Pythagorean Identities.

The Pythagorean Identities

For each real number x

$$(\sin x)^2 + (\cos x)^2 = 1, \qquad (\tan x)^2 + 1 = (\sec x)^2, \qquad (\cot x)^2 + 1 = (\csc x)^2.$$

We recommend that you commit the original Pythagorean Identity to memory and derive the other two as needed.

The next identities we need are not so easy to derive. They involve the sine and cosine of the sum and difference of two numbers. This pair of identities should also be committed to memory since they are frequently needed.

The Sum and Difference Formulas for the Sine and Cosine

For every pair of real numbers x_1 and x_2 we have

$$\sin(x_1 \pm x_2) = \sin x_1 \cos x_2 \pm \cos x_1 \sin x_2$$

and

$$\cos(x_1 \pm x_2) = \cos x_1 \cos x_2 \mp \sin x_1 \sin x_2.$$

We will first use the definition of the sine and cosine functions to show that the identity

$$\cos(x_1 - x_2) = \cos x_1 \cos x_2 + \sin x_1 \sin x_2$$

holds for the case when $0 < x_1 - x_2 < \pi/2$. This restriction on x_1 and x_2 is simply for convenience of illustration; the proof holds in other cases as well. Once the identity for the cosine of the difference has been established, we can use it to verify the other identities.

Figure 3.46 gives a typical illustration of the situation that we are considering.

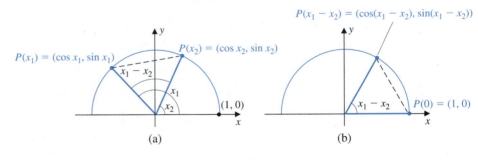

FIGURE 3.46

The triangle in Figure 3.46(b) is simply a clockwise rotation by x_1 radians of the triangle shown in Figure 3.46(a). So the length of the arc between the points $P(x_1 - x_2)$ and $P(0)$ in Figure 3.46(b) is $x_1 - x_2$, the same as the length of the arc between the points $P(x_1)$ and $P(x_2)$ shown in Figure 3.46 (a). Since the arc lengths between these pairs of points are the same, so are the straight-line distances. This implies that

$$(\cos(x_1 - x_2) - 1)^2 + (\sin(x_1 - x_2) - 0)^2 = (\cos x_1 - \cos x_2)^2 + (\sin x_1 - \sin x_2)^2.$$

Expanding this equation gives

$$(\cos(x_1 - x_2))^2 - 2\cos(x_1 - x_2) + 1 + (\sin(x_1 - x_2))^2$$
$$= (\cos x_1)^2 - 2\cos x_1 \cos x_2 + (\cos x_2)^2 + (\sin x_1)^2 - 2\sin x_1 \sin x_2 + (\sin x_2)^2.$$

But the Pythagorean Identity implies that

$$1 = (\sin(x_1 - x_2))^2 + (\cos(x_1 - x_2))^2 = (\sin x_1)^2 + (\cos x_1)^2 = (\sin x_2)^2 + (\cos x_2)^2,$$

so all the squared terms are eliminated, and we have

$$2 - 2\cos(x_1 - x_2) = 2 - 2\cos x_1 \cos x_2 - 2\sin x_1 \sin x_2.$$

Subtracting 2 from each side and then dividing by -2 gives the final identity,

$$\cos(x_1 - x_2) = \cos x_1 \cos x_2 + \sin x_1 \sin x_2.$$

Admittedly, the derivation of this identity was somewhat involved, but, as so often happens in mathematics, the other identities now follow rather easily. They can be converted into a form that uses the cosine of a difference.

To show the identity involving the cosine of the sum, we use the fact that the cosine function is even and the sine function is odd, that is,

$$\cos(-x) = \cos x \qquad \text{and} \qquad \sin(-x) = -\sin x.$$

Converting the cosine of a sum into a cosine of a difference gives

$$\cos(x_1 + x_2) = \cos(x_1 - (-x_2))$$

$$= \cos x_1 \cos(-x_2) + \sin x_1 \sin(-x_2)$$

$$= \cos x_1 \cos x_2 - \sin x_1 \sin x_2.$$

Combining this with the difference identity gives

$$\cos(x_1 \pm x_2) = \cos x_1 \cos x_2 \mp \sin x_1 \sin x_2$$

To prove the identities that involve the sine function, we use a result that we discovered in Example 1 of Section 3.3, that for any number x we have

$$\sin x = \cos\left(x - \frac{\pi}{2}\right) \qquad \text{and} \qquad \cos x = -\sin\left(x - \frac{\pi}{2}\right).$$

Thus,

$$\sin(x_1 \pm x_2) = \cos\left((x_1 \pm x_2) - \frac{\pi}{2}\right)$$

$$= \cos\left(x_1 \pm \left(x_2 - \frac{\pi}{2}\right)\right)$$

$$= \cos x_1 \cos\left(x_2 - \frac{\pi}{2}\right) \mp \sin x_1 \sin\left(x_2 - \frac{\pi}{2}\right)$$

$$= \cos x_1 \sin x_2 \mp \sin x_1(-\cos x_2),$$

which implies that

$$\sin(x_1 \pm x_2) = \sin x_1 \cos x_2 \pm \cos x_1 \sin x_2.$$

The next example shows how the sum and difference formulas for the sine and cosine can be used to extend our knowledge of these functions by allowing us to determine exact values for sine and cosine at some additional angles.

EXAMPLE 1   Determine (a) $\sin(\pi/12)$ and (b) $\cos(7\pi/12)$.

Solution   a. First we need to determine numbers whose sines and cosines are known and whose sum or difference produces the values we need. In the case of $\pi/12$ we have

$$\frac{\pi}{12} = \frac{\pi}{3} - \frac{\pi}{4},$$

so

$$\sin \frac{\pi}{12} = \sin \left(\frac{\pi}{3} - \frac{\pi}{4} \right)$$

$$= \sin \left(\frac{\pi}{3} \right) \cos \left(\frac{\pi}{4} \right) - \cos \left(\frac{\pi}{3} \right) \sin \left(\frac{\pi}{4} \right)$$

$$= \frac{\sqrt{3}}{2} \frac{\sqrt{2}}{2} - \frac{1}{2} \frac{\sqrt{2}}{2} = \frac{\sqrt{2}}{4} (\sqrt{3} - 1).$$

b. To find the cosine of $7\pi/12$ we could write

$$\frac{7\pi}{12} = \frac{\pi}{2} + \frac{\pi}{12},$$

but this would require finding the cosine of $\pi/12$ as well as the sine that we determined in part (a). Instead, notice that we can also write

$$\frac{7\pi}{12} = \frac{\pi}{3} + \frac{\pi}{4},$$

so

$$\cos \frac{7\pi}{12} = \cos \left(\frac{\pi}{3} + \frac{\pi}{4} \right) = \cos \left(\frac{\pi}{3} \right) \cos \left(\frac{\pi}{4} \right) - \sin \left(\frac{\pi}{3} \right) \sin \left(\frac{\pi}{4} \right)$$

$$= \frac{1}{2} \frac{\sqrt{2}}{2} - \frac{\sqrt{3}}{2} \frac{\sqrt{2}}{2} = \frac{\sqrt{2}}{4} (1 - \sqrt{3}). \qquad \blacksquare$$

We now have the three core identities for trigonometry. All the others we will need can be obtained from the Pythagorean Identities or the formulas for the sum and difference of the sine and cosine. For example, consider the formula for the tangent of the sum and difference of two values. The definition of the tangent function gives

$$\tan(x_1 \pm x_2) = \frac{\sin(x_1 \pm x_2)}{\cos(x_1 \pm x_2)},$$

so

$$\tan(x_1 \pm x_2) = \frac{\sin x_1 \cos x_2 \pm \cos x_1 \sin x_2}{\cos x_1 \cos x_2 \mp \sin x_1 \sin x_2}.$$

If we divide the numerator and denominator by $\cos x_1 \cos x_2$ we obtain

$$\tan(x_1 \pm x_2) = \frac{\dfrac{\sin x_1 \cos x_2}{\cos x_1 \cos x_2} \pm \dfrac{\cos x_1 \sin x_2}{\cos x_1 \cos x_2}}{\dfrac{\cos x_1 \cos x_2}{\cos x_1 \cos x_2} \mp \dfrac{\sin x_1 \sin x_2}{\cos x_1 \cos x_2}},$$

which simplifies to

$$\tan(x_1 \pm x_2) = \frac{\tan x_1 \pm \tan x_2}{1 \mp \tan x_1 \tan x_2}.$$

This is an interesting identity but not one that needs to be committed to memory. It is easily derived from the sine and cosine formulas and has the weakness that it cannot be applied when either $\cos x_1 = 0$ or $\cos x_2 = 0$. There is a similar sum and difference identity for the cotangent, but it is better to derive it. (If you are curious, you will find it in the exercises at the end of the chapter.)

EXAMPLE 2   Determine $\tan(\pi/12)$.

Solution   We saw in Example 1 that $\pi/12 = \pi/3 - \pi/4$, so we could use the tangent identity we just derived to obtain

$$\tan\frac{\pi}{12} = \frac{\tan(\pi/3) - \tan(\pi/4)}{1 + \tan(\pi/3)\tan(\pi/4)} = \frac{\sqrt{3}-1}{1+\sqrt{3}\cdot 1} = \frac{\sqrt{3}-1}{\sqrt{3}+1}.$$

Alternatively, we could use the value of $\sin(\pi/12)$ from Example 1 and the cosine difference formula to compute

$$\cos\frac{\pi}{12} = \cos\frac{\pi}{3}\cos\frac{\pi}{4} + \sin\frac{\pi}{3}\sin\frac{\pi}{4} = \frac{1}{2}\cdot\frac{\sqrt{2}}{2} + \frac{\sqrt{3}}{2}\frac{\sqrt{2}}{2} = \frac{\sqrt{2}}{4}(1+\sqrt{3}).$$

Then

$$\tan\frac{\pi}{12} = \frac{\sin\frac{\pi}{12}}{\cos\frac{\pi}{12}} = \frac{\frac{\sqrt{2}}{4}(\sqrt{3}-1)}{\frac{\sqrt{2}}{4}(1+\sqrt{3})} = \frac{\sqrt{3}-1}{\sqrt{3}+1}.$$

This second method might seem longer, but it has the advantage of simplicity. We need only know the definition of the tangent function and the identities for the sum and difference of the sine and cosine functions.   ∎

The double- and half-angle formulas are also consequences of the sum formulas for the sine and cosine functions. For the sine function we have the identity

$$\sin 2x = \sin(x+x) = \sin x \cos x + \sin x \cos x,$$

so

$$\sin 2x = 2\sin x \cos x.$$

For the cosine we have

$$\cos 2x = \cos x \cos x - \sin x \sin x,$$

which can be rewritten using the Pythagorean Identity in a variety of ways:

$$\cos 2x = (\cos x)^2 - (\sin x)^2 = (\cos x)^2 - (1 - (\cos x)^2) = 2(\cos x)^2 - 1$$

or

$$\cos 2x = (\cos x)^2 - (\sin x)^2 = 1 - (\sin x)^2 - (\sin x)^2 = 1 - 2(\sin x)^2.$$

These are called the *double-angle* formulas for the sine and cosine.

Double-Angle Formulas

$$\sin 2x = 2\sin x \cos x$$

$$\cos 2x = (\cos x)^2 - (\sin x)^2 = 2(\cos x)^2 - 1 = 1 - 2(\sin x)^2$$

EXAMPLE 3   Determine all the values of x in $[0, 2\pi]$ that satisfy each equation.

a. $\cos x = \sin 2x$ b. $\sin x = \cos 2x$ c. $1 = \sin x + \cos x$

Solution   a. We can use the double-angle formula for the sine to write

$$\cos x = \sin 2x = 2 \sin x \cos x,$$

which implies that

$$0 = \cos x (2 \sin x - 1).$$

The solutions to this equation occur when

$$\cos x = 0, \qquad \text{that is,} \qquad x = \frac{\pi}{2} \quad \text{or} \quad x = \frac{3\pi}{2},$$

or when

$$\sin x = \frac{1}{2}, \qquad \text{that is,} \qquad x = \frac{\pi}{6} \quad \text{or} \quad x = \frac{5\pi}{6}.$$

b. We can use the double-angle formula for the cosine in the form that involves only the sine function to write

$$\sin x = \cos 2x = 1 - 2(\sin x)^2,$$

which implies that

$$0 = 2(\sin x)^2 + \sin x - 1 = (\sin x + 1)(2 \sin x - 1).$$

The solutions to this equation occur when

$$\sin x = -1, \qquad \text{that is,} \qquad x = \frac{3\pi}{2},$$

or when

$$\sin x = \frac{1}{2}, \qquad \text{that is,} \qquad x = \frac{\pi}{6} \quad \text{or} \quad x = \frac{5\pi}{6}.$$

Often a graphing device can be used to visualize a solution that can then lead us to the desired algebraic solution.

c. The equation $1 = \sin x + \cos x$ doesn't initially seem to belong with the others, but the graph of $f(x) = \sin x + \cos x$ shown in Figure 3.47 indicates that we have the graph of an expanded and translated sine function. The graph also implies that $f(x) = 1$ on $[0, 2\pi]$ precisely when $x = 0$, $x = \pi/2$, or $x = 2\pi$.

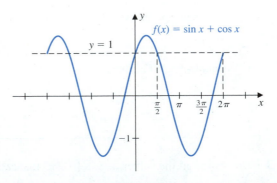

FIGURE 3.47

To verify these results algebraically, we first square both sides of the identity to give

$$1 = (\sin x + \cos x)^2 = (\sin x)^2 + (\cos x)^2 + 2 \sin x \cos x.$$

The Pythagorean and double-angle identities can be used to simplify this to

$$1 = 1 + 2 \sin x \cos x = 1 + \sin 2x,$$

or to

$$0 = \sin 2x.$$

To have $\sin 2x = 0$, we must have $2x$ a multiple of π; that is, x must be a multiple of $\pi/2$. At first glance, then we might conclude that the solutions are

$$x = 0, \qquad x = \frac{\pi}{2}, \qquad x = \pi, \qquad x = \frac{3\pi}{2}, \qquad \text{and} \qquad x = 2\pi.$$

Remember, though, that we squared the equation in the first step and that this might introduce *extraneous* solutions, that is, solutions to the final equation that are not solutions to the original equation. In fact, it did since the only values of x that satisfy the original equation, $1 = \sin x + \cos x$, are

$$x = 0, \qquad x = \frac{\pi}{2}, \qquad \text{and} \qquad x = 2\pi,$$

which agrees with our graphical solution. ■

The *half-angle* formulas come from the double-angle formulas involving $\cos 2x$ when x is replaced by $x/2$. Since $\cos 2x = 1 - 2(\sin x)^2$,

$$\cos x = 1 - 2 \left(\sin \frac{x}{2} \right)^2,$$

which implies that

$$\left(\sin \frac{x}{2} \right)^2 = \frac{1 - \cos x}{2},$$

For the cosine formula we have $\cos 2x = 2(\cos x)^2 - 1$, so

$$\cos x = 2 \left(\cos \frac{x}{2} \right)^2 - 1.$$

Solving for $\left(\cos \frac{x}{2} \right)^2$ gives the following result.

Half-Angle Formulas

For any real number x we have

$$\left(\sin \frac{x}{2} \right)^2 = \frac{1 - \cos x}{2} \qquad \text{and} \qquad \left(\cos \frac{x}{2} \right)^2 = \frac{1 + \cos x}{2}$$

or, equivalently, by replacing x with $2x$,

$$(\sin x)^2 = \frac{1 - \cos 2x}{2} \qquad \text{and} \qquad (\cos x)^2 = \frac{1 + \cos 2x}{2}.$$

EXAMPLE 4   Determine $\sin(3\pi/8)$, $\cos(3\pi/8)$, and $\tan(3\pi/8)$.

Solution   The half-angle formulas give

$$\left(\sin\frac{3\pi}{8}\right)^2 = \frac{1}{2}\left(1 - \cos\frac{3\pi}{4}\right) = \frac{1}{2}\left(1 + \frac{\sqrt{2}}{2}\right)$$

and

$$\left(\cos\frac{3\pi}{8}\right)^2 = \frac{1}{2}\left(1 + \cos\frac{3\pi}{4}\right) = \frac{1}{2}\left(1 - \frac{\sqrt{2}}{2}\right).$$

Since $0 < 3\pi/8 < \pi/2$, both the sine and cosine are positive, so

$$\sin\frac{3\pi}{8} = \sqrt{\frac{1}{2}\left(1 + \frac{\sqrt{2}}{2}\right)} = \sqrt{\frac{2 + \sqrt{2}}{4}} = \frac{\sqrt{2 + \sqrt{2}}}{2}$$

and

$$\cos\frac{3\pi}{8} = \sqrt{\frac{1}{2}\left(1 - \frac{\sqrt{2}}{2}\right)} = \sqrt{\frac{2 - \sqrt{2}}{4}} = \frac{\sqrt{2 - \sqrt{2}}}{2}.$$

The tangent is now easily determined to be

$$\tan\frac{3\pi}{8} = \frac{\sin\frac{3\pi}{8}}{\cos\frac{3\pi}{8}} = \frac{\sqrt{2 + \sqrt{2}}}{\sqrt{2 - \sqrt{2}}} = \sqrt{\frac{2 + \sqrt{2}}{2 - \sqrt{2}}}. \qquad\blacksquare$$

The following example illustrates one of the many identities that can be derived using the half-angle formula.

EXAMPLE 5   Show that the identity

$$\tan\frac{x}{2} = \frac{1 - \cos x}{\sin x}$$

holds when $0 < x < \pi$.

Solution   The double-angle formulas imply that

$$\tan\frac{x}{2} = \frac{\sin\frac{x}{2}}{\cos\frac{x}{2}} = \frac{\sqrt{\frac{1 - \cos x}{2}}}{\sqrt{\frac{1 + \cos x}{2}}} = \sqrt{\frac{1 - \cos x}{1 + \cos x}}.$$

Since we want the numerator to be $1 - \cos x$, we multiply the numerator and denominator within the square root by $1 - \cos x$, which in effect "rationalizes the numerator." This gives

$$\tan\frac{x}{2} = \sqrt{\frac{(1 - \cos x)^2}{(1 + \cos x)(1 - \cos x)}} = \sqrt{\frac{(1 - \cos x)^2}{1 - (\cos x)^2}} = \sqrt{\frac{(1 - \cos x)^2}{(\sin x)^2}}$$

$$= \pm\frac{1 - \cos x}{\sin x}.$$

But for $0 < x < \pi$ we have $0 < x/2 < \pi/2$, so

$$\cos\frac{x}{2} > 0, \qquad \sin\frac{x}{2} > 0, \qquad \text{and} \qquad \sin x > 0.$$

Since we also have $1 - \cos x \geq 0$, both sides of the identity must be positive and

$$\tan\frac{x}{2} = \frac{1 - \cos x}{\sin x}$$ ∎

The half-angle formulas are frequently needed in calculus to change powers of the sine or cosine functions into sums and differences in a manner illustrated in the following example.

EXAMPLE 6   Express $(\sin x)^4$ as a sum that involves only constants and sine and cosine functions to the first power.

Solution   The object is to reduce the power on the sine function, so we first use the half-angle formula for the sine to write

$$(\sin x)^4 = \left((\sin x)^2\right)^2 = \left(\frac{1 - \cos 2x}{2}\right)^2 = \frac{1}{4}\left(1 - 2\cos 2x + (\cos 2x)^2\right).$$

Now we use the half-angle formula for the cosine to rewrite $(\cos 2x)^2$ in terms of $\cos 4x$. This gives

The algebraic technique outlined in this example is useful in calculus, for example, in the context of integration.

$$(\sin x)^4 = \frac{1}{4}\left(1 - 2\cos 2x + \left(\frac{1 + \cos 4x}{2}\right)\right),$$

which simplifies to

$$(\sin x)^4 = \frac{3}{8} - \frac{1}{2}\cos 2x + \frac{1}{8}\cos 4x.$$ ∎

You might argue that the "simplified" expression for $(\sin x)^4$ in Example 6 looks more complicated than the original. In a sense you would be correct, and it might be better to have described the process as changing $(\sin x)^4$ into a "standard" form. As we will now see, simple products of sines and cosines can also be expressed in this form.

If we add the two formulas for the sine of a sum and difference,

$$\sin(x_1 + x_2) = \sin x_1 \cos x_2 + \cos x_1 \sin x_2,$$

$$\sin(x_1 - x_2) = \sin x_1 \cos x_2 - \cos x_1 \sin x_2,$$

and divide the result by 2 we obtain the identity

$$\sin x_1 \cos x_2 = \frac{1}{2}(\sin(x_1 - x_2) + \sin(x_1 + x_2)).$$

In a similar manner, the basic cosine identities

$$\cos(x_1 + x_2) = \cos x_1 \cos x_2 - \sin x_1 \sin x_2$$

$$\cos(x_1 - x_2) = \cos x_1 \cos x_2 + \sin x_1 \sin x_2$$

can be added to produce

$$\cos x_1 \cos x_2 = \frac{1}{2}(\cos(x_1 - x_2) + \cos(x_1 + x_2)).$$

or subtracted to produce

$$\sin x_1 \sin x_2 = \frac{1}{2}(\cos(x_1 - x_2) - \cos(x_1 + x_2)).$$

Taken together, we have the following formulas.

These identities are primarily needed for some of the topics in differential equations, a mathematics course that follows calculus.

Formulas for Products of Sines and Cosines

$$\sin x_1 \cos x_2 = \frac{1}{2}(\sin(x_1 - x_2) + \sin(x_1 + x_2))$$

$$\cos x_1 \cos x_2 = \frac{1}{2}(\cos(x_1 - x_2) + \cos(x_1 + x_2))$$

$$\sin x_1 \sin x_2 = \frac{1}{2}(\cos(x_1 - x_2) - \cos(x_1 + x_2))$$

EXAMPLE 7   Write $(\sin 2x)^2(\cos 3x)^2$ as a sum of sine and cosine functions.

Solution   We first use the half-angle formulas to rewrite the product as

$$(\sin 2x)^2(\cos 3x)^2 = \left(\frac{1 - \cos 4x}{2}\right)\left(\frac{1 + \cos 6x}{2}\right)$$

$$= \frac{1}{4}(1 - \cos 4x + \cos 6x - \cos 4x \cos 6x).$$

Then we use the cosine product formula to rewrite this as

$$(\sin 2x)^2(\cos 3x)^2 = \frac{1}{4}\left(1 - \cos 4x + \cos 6x - \frac{1}{2}(\cos(4x - 6x) + \cos(4x + 6x))\right)$$

$$= \frac{1}{4} - \frac{1}{4}\cos 4x + \frac{1}{4}\cos 6x - \frac{1}{8}\cos(-2x) - \frac{1}{8}\cos 10x.$$

Since the cosine is an even function, this can be simplified to

$$(\sin 2x)^2(\cos 3x)^2 = \frac{1}{4} - \frac{1}{8}\cos 2x - \frac{1}{4}\cos 4x + \frac{1}{4}\cos 6x - \frac{1}{8}\cos 10x. \quad \blacksquare$$

EXERCISE SET 3.5

In Exercises 1–6, use a sum or difference formula to determine the value of the trigonometric function.

1. $\sin\left(\frac{\pi}{3} - \frac{5\pi}{4}\right)$

2. $\cos\left(\frac{5\pi}{6} + \frac{\pi}{4}\right)$

3. $\sin\left(\frac{7\pi}{12}\right)$

4. $\sin\left(-\frac{\pi}{12}\right)$

5. $\tan\left(\frac{\pi}{12}\right)$

6. $\cot\left(-\frac{\pi}{12}\right)$

In Exercises 7–10, use a double-or half-angle formula to determine the value of the trigonometric function.

7. $\sin\left(\frac{\pi}{12}\right)$

8. $\cos\left(\frac{\pi}{8}\right)$

9. $\cos\left(\dfrac{3\pi}{8}\right)$ **10.** $\sin\left(\dfrac{5\pi}{12}\right)$

In Exercises 11–12, use the given information to find each value.

a. $\cos 2t$ b. $\sin 2t$ c. $\cos t/2$ d. $\sin t/2$

11. $\cos t = 3/5$, where $0 < t < \pi/2$

12. $\tan t = 5/12$, where $\sin t < 0$

In Exercises 13–18, use the addition and subtraction formulas to verify the given formula.

13. $\sin(t + \pi) = -\sin t$ **14.** $\cos(t + \pi) = -\cos t$

15. $\sin\left(t + \dfrac{\pi}{2}\right) = \cos t$ **16.** $\cos\left(t + \dfrac{\pi}{2}\right) = -\sin t$

17. $\sin(\pi - t) = \sin t$ **18.** $\cos(\pi - t) = -\cos t$

In Exercises 19–22, use a half-angle formula to rewrite the given expression so that it involves the sum or difference of only constants and sine and cosine functions to the first power.

19. $(\cos 2x)^2$ **20.** $(\sin 3x)^2$

21. $(\cos x)^4$ **22.** $(\sin 3x)^2(\cos 3x)^2$

In Exercises 23–26, rewrite each product as a sum or difference.

23. $\sin 6t \cos 5t$ **24.** $\cos 5t \sin 8t$

25. $\cos 2t \cos 3t$ **26.** $\sin 5t \sin 7t$

In Exercises 27–30, rewrite each sum or difference as a product.

27. $\sin 2t + \sin 3t$ **28.** $\sin 4t - \sin 6t$

29. $\cos 5t + \cos 2t$ **30.** $\cos t - \cos 3t$

In Exercises 31–34, find all values of x in the interval $[0, 2\pi]$ that satisfy the given equation.

31. $\sin 2x = \sin x$ **32.** $\sin 2x = \cos x$

33. $2(\cot x)^2 + (\csc x)^2 - 2 = 0$

34. $\tan x + \cot x = \dfrac{2}{\sin 2x}$

In Exercises 35–42, verify the identities.

35. $(1 - (\cos x)^2)(\sec x)^2 = (\tan x)^2$

36. $\cot x + \tan x = \sec x \csc x$

37. $\cot x - \tan x = 2 \cot 2x$

38. $(\sin x + \cos x)^2 = 1 + \sin 2x$

39. $\cos x = \sin x \sin 2x + \cos x \cos 2x$

40. $(\tan x)^2 - (\sin x)^2 = (\tan x)^2(\sin x)^2$

41. $\sec x - \cos x = \sin x \tan x$

42. $\cos x(\cot x + \tan x) = \csc x$

In Exercises 43–48, use a graphing device to sketch the graphs of f and g, and use the graphs to determine if $f(x) = g(x)$ is an identity.

43. $f(x) = (\sin x - \cos x)^2$; $g(x) = 1 - \sin 2x$

44. $f(x) = 2\left(\cos \dfrac{x}{2}\right)^2 - 1$; $g(x) = \cos x$

45. $f(x) = \dfrac{\sin 2x}{1 + \cos 2x}$; $g(x) = \tan x$

46. $f(x) = \dfrac{2 \cot x}{1 + (\cot x)^2}$; $g(x) = \sin 2x$

47. $f(x) = (\sin x - \cos x)^2$; $g(x) = 1$

48. $f(x) = \tan\left(\dfrac{x}{2}\right)$; $g(x) = \dfrac{1 + \cos x}{\sin x}$

3.6 RIGHT-TRIANGLE TRIGONOMETRY

Geometric and computational problems generally require that the trigonometric functions be defined in terms of angle measures. In Section 3.1 we saw how the measure of an angle is related to the length of the arc of the unit circle that is generated by the angle. Since we will be needing this relationship frequently in this section, let us briefly review the details.

An angle is said to be in *standard position* when its initial side lies along the positive x-axis and its vertex is at the origin. For example, the angle θ shown in Figure 3.48 is in standard position. This angle has radian measure t since the terminal side of θ intersects the unit circle at $P(t)$, and the length of the arc from $(1, 0)$ to $P(t)$ is t units.

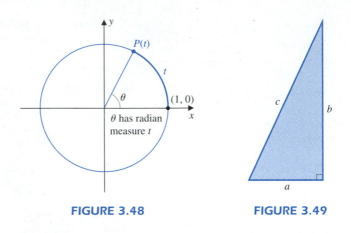

FIGURE 3.48   **FIGURE 3.49**

The trigonometric functions of an *angle* are defined using the radian measure. For an angle θ with radian measure t, we simply define $\sin\theta = \sin t$, $\cos\theta = \cos t$, and so on, for the other trigonometric functions. In a similar manner, we define the *reference angle* of θ to be the angle in $[0, \pi/2]$ whose radian measure is the reference number for t.

An angle with radian measure $\pi/2$ is called a *right angle*, and a triangle containing a right angle is a *right triangle*. (See Figure 3.49.) The sides of a right triangle satisfy the Pythagorean relationship $a^2 + b^2 = c^2$. Angles with radian measure θ, where $0 < \theta < \pi/2$, are called *acute* angles. So the nonright angles in a right triangle must be acute. An angle with radian measure $\theta > \pi/2$ is called an *obtuse* angle.

Values of the trigonometric functions of angles in a right triangle can be easily determined without referring to the unit circle. To see how this is done, consider the angle θ shown in the right triangle in Figure 3.50(a).

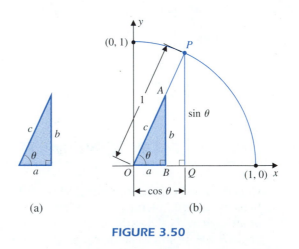

(a)   (b)

FIGURE 3.50

Superimpose an xy-coordinate system onto the right triangle with the vertex of θ at the origin and positive x-axis along the side of the triangle with length a. The side a of the triangle is called the *adjacent* side of the triangle relative to the angle θ, and the side b is called the *opposite* side. The side c is called the *hypotenuse* of the triangle.

Let P denote the intersection point of the unit circle with terminal side of θ. Figure 3.50(b) shows the situation when $0 < c < 1$. The situation when $c > 1$ is similar, and the algebra we do does not depend on the magnitude of c.

Relating the various sides of the similar triangles AOB and POQ, we have

$$\frac{\sin \theta}{1} = \frac{b}{c}, \qquad \frac{\cos \theta}{1} = \frac{a}{c}, \qquad \frac{\sin \theta}{\cos \theta} = \frac{b}{a},$$

$$\frac{1}{\sin \theta} = \frac{c}{b}, \qquad \frac{1}{\cos \theta} = \frac{c}{a}, \qquad \frac{\cos \theta}{\sin \theta} = \frac{a}{b}.$$

This gives the following values for the trigonometric functions of θ.

Trigonometric Functions of an Angle in a Right Triangle

For the angle θ in the triangle shown in Figure 3.51, we have

$$\sin \theta = \frac{b}{c}, \qquad \cos \theta = \frac{a}{c}, \qquad \tan \theta = \frac{b}{a},$$

$$\csc \theta = \frac{c}{b}, \qquad \sec \theta = \frac{c}{a}, \qquad \cot \theta = \frac{a}{b}.$$

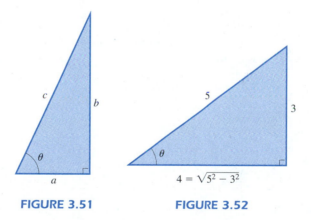

FIGURE 3.51   FIGURE 3.52

EXAMPLE 1   Suppose that an acute angle θ is known to have $\sin \theta = 3/5$. Determine the other trigonometric functions of this angle.

Solution   One strategy for solving this problem is to use the Pythagorean Identity $(\sin \theta)^2 + (\cos \theta)^2 = 1$ with the fact that $\sin \theta = \frac{3}{5}$ to find $\cos \theta$, and then determine the other values of the trigonometric functions as we did in Example 1 of Section 3.4. Instead of doing this, let us look at a right-triangle approach to solving this problem.

The right triangle in Figure 3.52 has been constructed with $\sin \theta = \frac{3}{5}$, and the angle lies in a right triangle, so it must be acute.

The Pythagorean Theorem implies that the horizontal side of the triangle is $\sqrt{5^2 - 3^2} = 4$. Once this is known, all the trigonometric functions of θ can be read directly from the triangle:

$$\sin \theta = \frac{3}{5}, \qquad \cos \theta = \frac{4}{5}, \qquad \tan \theta = \frac{3}{4},$$

$$\csc \theta = \frac{5}{3}, \qquad \sec \theta = \frac{5}{4}, \qquad \cot \theta = \frac{4}{3}. \qquad \blacksquare$$

The problem in Example 1 was simplified slightly because the angle was acute. If the angle is in the second, third, or fourth quadrants, we use the reference angle to compute the trigonometric functions, correcting the signs as necessary. For example, suppose that Example 1 had stated that θ is obtuse. In the second quadrant the cosine, tangent, cotangent, and secant are all negative, but the sine and cosecant remain positive. As a consequence we would have

$$\sin \theta = \frac{3}{5}, \qquad \cos \theta = -\frac{4}{5}, \qquad \tan \theta = -\frac{3}{4},$$

$$\csc \theta = \frac{5}{3}, \qquad \sec \theta = -\frac{5}{4}, \qquad \cot \theta = -\frac{4}{3}.$$

Measuring Angles Using Degrees

Although the natural method of expressing angles in calculus is radian measure, angles are also measured in degrees. An angle formed by rotating a ray counterclockwise from an initial position back to the initial position so that the terminal and initial sides coincide has, by definition, a measure of 360 degrees, written 360°. This angle also has radian measure 2π, as shown in Figure 3.53, so

$$360 \text{ degrees} = 2\pi \text{ radians}.$$

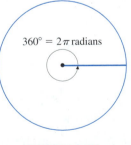

$360° = 2\pi$ radians

FIGURE 3.53

Conversion between Degrees and Radians

$$180° = \pi \text{ rad}, \qquad 1° = \frac{\pi}{180} \text{ rad}, \qquad 1 \text{ rad} = \frac{180°}{\pi}.$$

EXAMPLE 2   Express 135° and 405° in radians, and $5\pi/6$ in degrees.

Solution

$$135° = 135 \left(\frac{\pi}{180}\right) \text{ rad} = \frac{3\pi}{4} \text{ rad}$$

$$420° = 420 \left(\frac{\pi}{180}\right) \text{ rad} = \frac{7\pi}{3} \text{ rad}$$

$$\frac{5\pi}{6} \text{ rad} = \frac{5\pi}{6} \left(\frac{180}{\pi}\right)° = 150°.$$ ∎

The following examples show some ways that the angle-side relationships of the trigonometric functions can be used in applications. Since the first example uses angle measure in degrees, we present Table 3.8 to help with the trigonometric function conversion.

TABLE 3.8

Degrees	0	30	45	60	90
Radians	0	$\pi/6$	$\pi/4$	$\pi/3$	$\pi/2$
Sine	0	$1/2$	$\sqrt{2}/2$	$\sqrt{3}/2$	1
Cosine	1	$\sqrt{3}/2$	$\sqrt{2}/2$	$1/2$	0

EXAMPLE 3   A climber who wants to measure the height of a cliff is standing 35 ft from the base of the cliff. An angle of approximately 60° is formed by the lines joining the climber's feet with the top and bottom of the cliff, as shown in Figure 3.54. Use this information to approximate the height of the cliff.

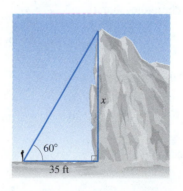

FIGURE 3.54

Solution   Let x denote the height of the cliff. Since 60° is equivalent to $\pi/3$ rad, we have $\tan 60° = \sqrt{3}$. As a consequence,

$$\sqrt{3} = \tan 60° = \frac{x}{35}, \quad \text{and} \quad x = 35\sqrt{3} \approx 60.6 \text{ ft.}$$ ∎

EXAMPLE 4   Two balls are against the rail at opposite ends of a 10-ft snooker table. The player must hit the ball on the left with the cue ball on the right without touching any of the other balls on the table. This is done by banking the cue ball off the bottom cushion, as shown in Figure 3.55. Where should the cue ball hit the bottom cushion, and what is the angle that its path makes with the bottom cushion?

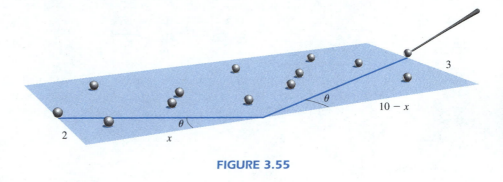

FIGURE 3.55

Solution   Let x denote the distance from the left bottom corner to the spot where the cue ball hits the cushion. Snooker players instinctively realize that when a ball hits the cushion, its angle of reflection is the same as its angle of incidence (unless there is spin on the ball, which we will assume is not the case). This means that for the shot to be successful, the triangles shown in Figure 3.55 must be similar. Hence we must have

$$\frac{2}{x} = \tan \theta = \frac{3}{10 - x},$$

which implies that

$$2(10 - x) = 3x.$$

Solving for x gives

$$20 - 2x = 3x, \qquad \text{and} \qquad x = \frac{20}{5} = 4.$$

Since $\tan \theta = 2/x$, we also have

$$\tan \theta = \frac{2}{4} = \frac{1}{2}.$$

This is not quite the satisfactory answer we would like since we have not come upon any angles whose tangent is $1/2$. The graph of the tangent function indicates that there is some number in $(0, \pi/2)$ whose tangent is $1/2$, and at this stage, this is the best we can do. In the next section we will find a more satisfactory method for representing this solution.   ∎

EXAMPLE 5

Example 5 is an optimization problem similar to the kind you will need to be able to set up, before finding a solution using techniques from calculus.

An engineer is designing a drainage canal that has a trapezoidal cross section, as shown in Figure 3.56. The bottom and sides of the canal are each L feet long, and the side makes an angle θ with the horizontal. Find an expression for the cross sectional area of the canal in terms of the angle θ. If the canal is S feet long, approximate the angle θ that will maximize the capacity of the canal.

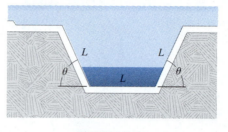

FIGURE 3.56

Solution   The cross section of the canal is shown in Figure 3.57. It has an area equal to the height times the average width of the top, w_1, and bottom, w_2; that is,

$$A = \frac{1}{2}(w_1 + w_2) \cdot h = \frac{1}{2}((L + 2a) + L) \cdot h.$$

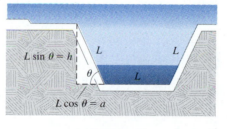

FIGURE 3.57

To express the area in terms of the angle θ, we have $a = L \cos \theta$ and $h = L \sin \theta$, so

$$A = \frac{1}{2}((L + 2L \cos \theta) + L) \cdot (L \sin \theta)$$

$$= \frac{1}{2}(2L + 2L \cos \theta)(L \sin \theta)$$

$$= L^2(1 + \cos \theta) \sin \theta.$$

The capacity of the canal is the volume V of water the canal can hold, so a canal S feet long has a capacity

$$V = A \cdot S = SL^2(1 + \cos \theta) \sin \theta \text{ ft}^3,$$

We cannot find the exact maximum value, but a graphing device allows us to estimate the maximum from the graph of the function.

and the capacity is maximized if θ maximizes

$$f(\theta) = \sin\theta + \cos\theta\sin\theta.$$

Figure 3.58 shows a computer-generated graph of $y = f(\theta)$.

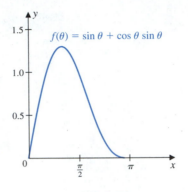

$f(\theta) = \sin\theta + \cos\theta\sin\theta$

FIGURE 3.58

Near the maximum we have

$$\theta \approx 1.047 \text{ rad}$$

$$= 1.047 \left(\frac{180}{\pi}\right)^{\circ}$$

$$\approx 59.99°.$$

The maximum capacity of the canal is approximately

$$V \approx SL^2(1 + \cos(1.047))\sin(1.047) = 1.299\ SL^2 \text{ cubic feet.} \qquad \blacksquare$$

EXERCISE SET 3.6

In Exercises 1–6, convert from degree measure to radian measure.

1. 60°

2. 40°

3. 225°

4. 150°

5. −72°

6. −270°

In Exercises 7–10, convert from radian measure to degree measure.

7. $\dfrac{3\pi}{4}$

8. $-\dfrac{\pi}{6}$

9. $\dfrac{2\pi}{3}$

10. $-\dfrac{11\pi}{6}$

In Exercises 11–14, find the value of the six trigonometric functions of the angle θ.

11.

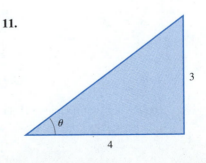

12.

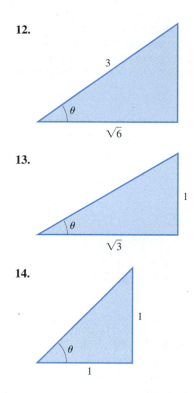

3

θ

$\sqrt{6}$

13.

1

θ

$\sqrt{3}$

14.

1

θ

1

In Exercises 15–18, find the value of x.

15.

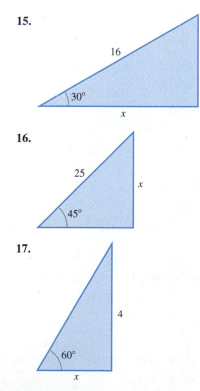

16

30°

x

16.

25

x

45°

17.

4

60°

x

18.

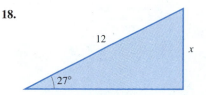

12

x

27°

In Exercises 19–24, refer to the right triangle in the figure. From the given information find any missing angles or sides.

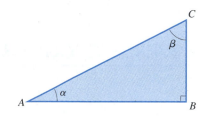

C

β

α

A B

19. $\alpha = 30°, \overline{BC} = 4$ **20.** $\alpha = 60°, \overline{AB} = 12$

21. $\beta = 45°, \overline{AC} = 5$ **22.** $\beta = 50°, \overline{BC} = 10$

23. $\alpha = 27.6°, \overline{BC} = 15.3$ **24.** $\beta = 52.6°, \overline{AC} = 12.3$

25. A climber who needs to estimate the height of a cliff stands at a point on the ground 40 ft from the base of the cliff and estimates that the angle from the ground to a line extending from the climber's feet to the top of the cliff is 70°. Assuming that the cliff is perpendicular to the ground, find the approximate height of the cliff.

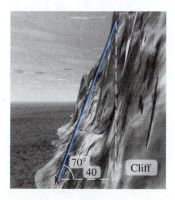

70°
40
Cliff

26. A pipe line is to be constructed between points A and B on opposite sides of a river, as shown in the following figure. To determine the amount of pipe needed using a theodolite at point A, a line perpendicular to AB is determined and point C is marked approximately 300 feet from A. At point A the theodolite is used to determine that the angle θ is 42°. Determine the amount of pipe needed to connect points A and B.

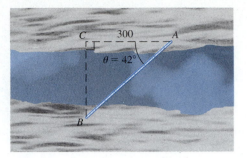

27. From the top of an observation tower a forest ranger spots an illegal campfire. The angle of depression made by the rangers line of sight and the campfire is 10.5°. If the tower is 80 ft high, how far is the campfire from the base of the tower?

3.7   INVERSE TRIGONOMETRIC FUNCTIONS

In Example 4 of Section 3.6 we needed to recover the value of θ from the fact that $\tan \theta = \frac{1}{2}$. This calls for an inverse function. But thinking back to Section 1.12 we know that:

A function has an inverse precisely when it is one-to-one.

Like other functions we have studied, the inverse trigonometric functions are used in several different contexts in calculus. In this section we review the basic definitions and properties in preparation for what you will see in your calculus course.

The trigonometric functions are all periodic, so they do not pass the Horizontal Line Test and cannot be one-to-one. The functions we will consider in this section are not the inverses of the ordinary trigonometric functions; instead they are inverses of the trigonometric functions after the domains have been restricted to intervals on which they

1. Are one-to-one, and
2. Have the same range as the ordinary trigonometric functions.

Before proceeding, however, let us review from Section 1.12 a few of the important facts about inverse functions.

Properties of Inverse Functions

Suppose that f is a one-to-one function. Then the inverse function f^{-1} is unique, and

(i) The domain of f^{-1} is the range of f.
(ii) The range of f^{-1} is the domain of f.
(iii) If x is in the domain of f^{-1} and y is in the domain of f, then

$$f^{-1}(x) = y \qquad \text{if and only if} \qquad f(y) = x.$$

(iv) $f(f^{-1}(x)) = x$ when x is in the domain of f^{-1}.
(v) $f^{-1}(f(x)) = x$ when x is in the domain of f.
(vi) The graph of $y = f^{-1}(x)$ is the reflection about the line $y = x$ of the graph of $y = f(x)$.

Although we have denoted the inverse function by f^{-1}, the inverse functions that are particularly important for applications have been given special notation. We will see that the inverse trigonometric functions fall into this category.

Now let us tackle the trigonometric functions and their inverses. We begin, as usual, with the sine function. It certainly is not one-to-one, since its graph, shown in Figure 3.59, crosses every horizontal line between -1 and 1 many times.

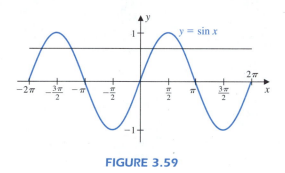

FIGURE 3.59

Suppose, however, that we consider the sine function restricted in domain to the interval $[-\pi/2, \pi/2]$. On this interval the sine function is one-to-one and assumes all the values in $[-1, 1]$. This function has an inverse, which we call the inverse sine, or *arcsine*, function.

The arcsine function is denoted either as arcsin or as $\sin^{-1}$. We will use only the arcsin notation to avoid possible confusion between the inverse sine function and the reciprocal of the sine function, $\csc x = (\sin x)^{-1}$.

The Arcsine Function

The **arcsine** function, denoted **arcsin**, has domain $[-1, 1]$ and range $[-\pi/2, \pi/2]$, and is defined by

$$\arcsin x = y \qquad \text{if and only if} \qquad \sin y = x.$$

The definition and properties of inverse functions imply the following.

Arcsine Properties

$$\sin(\arcsin x) = x \text{ when } x \text{ is in } [-1, 1]$$

and

$$\arcsin(\sin x) = x \text{ when } x \text{ is in } [-\pi/2, \pi/2].$$

The graph of $y = \arcsin x$ is the reflection about the line $y = x$ of the graph of the restricted sine function, as shown in Figure 3.60. Notice how steep the graph is at the ends of its domain. This corresponds to the flatness on the corresponding portions of the restricted sine function.

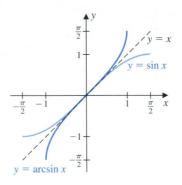

FIGURE 3.60

EXAMPLE 1   Find (a) $\arcsin \frac{1}{2}$, (b) $\arcsin \left(\sin \frac{\pi}{3} \right)$, and (c) $\arcsin \left(\sin \frac{3\pi}{4} \right)$.

Solution   a. We need to find a number in the interval $[-\frac{\pi}{2}, \frac{\pi}{2}]$ whose sine is $\frac{1}{2}$. Since $\sin \frac{\pi}{6} = \frac{1}{2}$ and $\frac{\pi}{6}$ is in $[-\frac{\pi}{2}, \frac{\pi}{2}]$, we have

$$\arcsin \frac{1}{2} = \frac{\pi}{6}.$$

b. Since $\frac{\pi}{3}$ is in $[-\frac{\pi}{2}, \frac{\pi}{2}]$, the result following the definition implies that

$$\arcsin \left(\sin \frac{\pi}{3} \right) = \frac{\pi}{3}.$$

c. This part is different from part (b), since $\frac{3\pi}{4}$ is not in the interval $[-\frac{\pi}{2}, \frac{\pi}{2}]$. What we must find is a number in $[-\frac{\pi}{2}, \frac{\pi}{2}]$ whose sine is the same as that of $\frac{3\pi}{4}$. But $\sin \frac{3\pi}{4} = \frac{\sqrt{2}}{2}$, and we also know that $\sin \frac{\pi}{4} = \frac{\sqrt{2}}{2}$. Since $\frac{\pi}{4}$ is in $[-\frac{\pi}{2}, \frac{\pi}{2}]$, we have

$$\arcsin \left(\sin \frac{3\pi}{4} \right) = \frac{\pi}{4}. \qquad \blacksquare$$

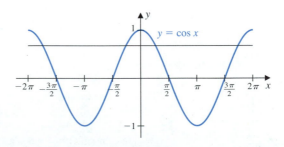

FIGURE 3.61

The inverses for the other trigonometric functions are defined by making domain restrictions similar to those made for the sine function. In the case of the cosine function, whose graph is shown in Figure 3.61, we restrict the domain to the interval $[0, \pi]$, since the function on this interval is one-to-one and assumes all the values in the range of the ordinary cosine function. For this restricted function we have an inverse, called the inverse cosine, or *arccosine*, function.

The Arccosine Function

The **arccosine** function, denoted **arccos**, has domain $[-1, 1]$ and range $[0, \pi]$ and is defined by

$$\arccos x = y \quad \text{if and only if} \quad \cos y = x.$$

The definition and properties of inverse functions imply the following.

Arccosine Properties

$\cos(\arccos x) = x$ when x is in $[-1, 1]$ and $\arccos(\cos x) = x$ when x is in $[0, \pi]$.

The graph of $y = \arccos x$ is the reflection about the line $y = x$ of the graph of the restricted cosine function, as shown in Figure 3.62.

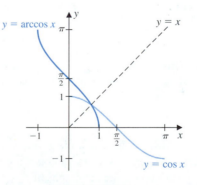

FIGURE 3.62

EXAMPLE 2 Find

a. $\cos\left[\arccos\left(-\tfrac{1}{2}\right)\right]$

b. $\sin\left[\arccos\left(-\tfrac{1}{2}\right)\right]$

c. $\arccos\left(\cos\tfrac{\pi}{3}\right)$

d. $\arccos\left[\cos\left(-\tfrac{\pi}{4}\right)\right]$.

Solution   a. The problem is clearer if you express this equation in words. Since the $\arccos\left(-\frac{1}{2}\right)$ is the number in $[0, \pi]$ whose cosine is $-\frac{1}{2}$, the cosine of this number must be $-\frac{1}{2}$. That is

$$\cos\left[\arccos\left(-\frac{1}{2}\right)\right] = -\frac{1}{2}.$$

b. We first apply the Pythagorean Identity to produce

$$\left(\sin\left[\arccos\left(-\frac{1}{2}\right)\right]\right)^2 + \left(\cos\left[\arccos\left(-\frac{1}{2}\right)\right]\right)^2 = 1.$$

Since

$$\cos\left[\arccos\left(-\frac{1}{2}\right)\right] = -\frac{1}{2},$$

we have

$$\left(\sin\left[\arccos\left(-\frac{1}{2}\right)\right]\right)^2 = 1 - \left(-\frac{1}{2}\right)^2 = \frac{3}{4}$$

and

$$\sin\left[\arccos\left(-\frac{1}{2}\right)\right] = \pm\frac{\sqrt{3}}{2}.$$

To determine which sign is appropriate, notice that all values of the arccosine lie in the interval $[0, \pi]$. On this interval the sine is never negative, so

$$\sin\left(\arccos\left(-\frac{1}{2}\right)\right) = \frac{\sqrt{3}}{2}.$$

c. This problem is also clearer if we express the equation in words. It states that we need to find the number in the interval $[0, \pi]$ whose cosine is the same as the cosine of $\frac{\pi}{3}$. Since $\frac{\pi}{3}$ is itself in the interval $[0, \pi]$, we have

$$\arccos\left(\cos\frac{\pi}{3}\right) = \frac{\pi}{3}.$$

d. This problem is slightly more complicated than part (c) since $-\frac{\pi}{4}$ is not in the interval $[0, \pi]$. However, the cosine function is even and $\pi/4$ is in $[0, \pi]$, so

$$\arccos\left(\cos\left(-\frac{\pi}{4}\right)\right) = \arccos\left(\cos\left(\frac{\pi}{4}\right)\right) = \frac{\pi}{4}. \qquad\blacksquare$$

A principal application of the inverse trigonometric functions in calculus is associated with the integrals of certain types. For this application we need the arcsine or the arccosine function but not both. It is the arcsine that is generally chosen.

Just as there is a close relationship between the graphs of the sine and cosine functions, there is a similar connection between the graphs of their inverse functions. In Figure 3.63 we see the graphs of $y = -\arccos x$ and $y = \arcsin x$. Notice that the shape of the graphs is the same and that the graphs differ only by a vertical shift of $\frac{\pi}{2}$ units. Hence we have

$$-\arccos x = \arcsin x - \frac{\pi}{2}$$

or, more simply,

$$\arcsin x + \arccos x = \frac{\pi}{2} \qquad \text{for all } x \text{ in } [-1, 1].$$

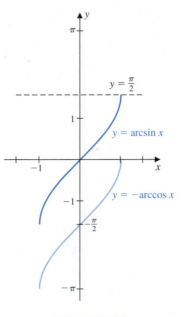

FIGURE 3.63

The arccosine function is not used extensively in calculus, but the arctangent function finds many applications.

The Arctangent Function

The **arctangent** function, denoted **arctan**, has domain $(-\infty, \infty)$ and range $(-\frac{\pi}{2}, \frac{\pi}{2})$ and is defined by

$$\arctan x = y \qquad \text{if and only if} \qquad \tan y = x.$$

The definition and properties of inverse functions imply that

---

Arctangent Properties

$\tan(\arctan x) = x$ when x is in $(-\infty, \infty)$ and $\arctan(\tan x) = x$ when x is in $(-\frac{\pi}{2}, \frac{\pi}{2})$.

The graph of $y = \arctan x$ is the reflection about the line $y = x$ of the graph of the restricted tangent function, as shown in Figure 3.64(a). Since the graph of $y = \tan x$ has vertical asymptotes at $x = -\frac{\pi}{2}$ and $x = \frac{\pi}{2}$, the graph of $y = \arctan x$ has horizontal asymptotes at $y = -\frac{\pi}{2}$ and $y = \frac{\pi}{2}$, as shown in Figure 3.64(b).

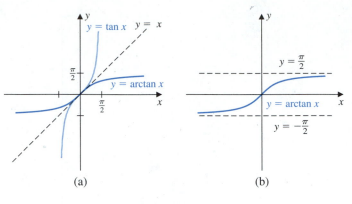

FIGURE 3.64

EXAMPLE 3   Find $\cos\left(\arctan\frac{12}{5} + \arcsin\frac{3}{5}\right)$.

Solution   This problem calls for the use of the sum formula for the cosine,

$$\cos(a + b) = \cos a \cos b - \sin a \sin b,$$

with $a = \arctan\frac{12}{5}$ and $b = \arcsin\frac{3}{5}$. Then we can rewrite the equation as

$$\cos\left(\arctan\frac{12}{5} + \arcsin\frac{3}{5}\right) = \cos\left(\arctan\frac{12}{5}\right)\cos\left(\arcsin\frac{3}{5}\right)$$

$$- \sin\left(\arctan\frac{12}{5}\right)\sin\left(\arcsin\frac{3}{5}\right)$$

We first note that $\sin\left(\arcsin\frac{3}{5}\right) = \frac{3}{5}$. The other values can be obtained from trigonometric identities, but we have chosen to find them by considering the triangles shown in Figure 3.65. In Figure 3.65(a) we have a right triangle with an angle θ satisfying

$$\theta = \arctan\frac{12}{5}, \qquad \text{that is,} \qquad \tan\theta = \frac{12}{5}.$$

The Pythagorean Theorem implies that the hypotenuse of this right triangle is $\sqrt{12^2 + 5^2} = 13$, so

$$\sin\theta = \sin\left(\arctan\frac{12}{5}\right) = \frac{12}{13} \qquad \text{and} \qquad \cos\theta = \cos\left(\arctan\frac{12}{5}\right) = \frac{5}{13}.$$

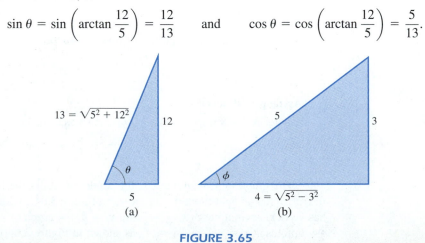

FIGURE 3.65

In a similar manner, the right triangle in Figure 3.65(b) shows an angle ϕ with

$$\phi = \arcsin \frac{3}{5}, \qquad \text{so} \qquad \sin \phi = \frac{3}{5}.$$

The Pythagorean Theorem implies that the base of this right triangle is

$$\sqrt{5^2 - 3^2} = 4, \qquad \text{so} \qquad \cos \phi = \cos \left(\arcsin \frac{3}{5} \right) = \frac{4}{5}.$$

Hence,

$$\cos \left(\arctan \frac{12}{5} + \arcsin \frac{3}{5} \right) = \frac{5}{13} \cdot \frac{4}{5} - \frac{12}{13} \cdot \frac{3}{5} = -\frac{16}{65}. \qquad \blacksquare$$

The inverse trigonometric functions occur frequently in calculus because the composition of an inverse trigonometric function followed by a trigonometric function produces some useful algebraic relationships. The following example shows two of these.

EXAMPLE 4   Show the following identities.

a. $\sin(\arccos x) = \sqrt{1 - x^2}$ and

b. $\cos(2 \arctan x) = \dfrac{1 - x^2}{1 + x^2}.$

Solution   a. The Pythagorean Identity implies that

$$(\sin(\arccos x))^2 + (\cos(\arccos x))^2 = 1,$$

and $\cos(\arccos x) = x$, so

$$1 = (\sin(\arccos x))^2 + x^2, \qquad \text{and} \qquad \sin(\arccos x) = \pm\sqrt{1 - x^2}.$$

Since the range of the arccosine function is $[0, \pi]$, where the sine function is non-negative, we have

$$\sin(\arccos x) = \sqrt{1 - x^2}.$$

Example 4 gives an example of how the inverse trigonometric functions in combination with the trigonometric functions can yield useful algebraic relationships which is exploited in calculus.

b. This identity is easiest to establish if we first introduce a variable, or parameter, to represent the argument of the cosine. Let

$$t = 2 \arctan x \qquad \text{which implies that} \qquad x = \tan \frac{t}{2}.$$

We can reexpress the latter equation using the half-angle formulas for the sine and cosine as

$$x = \tan \frac{t}{2} = \frac{\sin \frac{t}{2}}{\cos \frac{t}{2}} = \frac{\sqrt{(1 - \cos t)/2}}{\sqrt{(1 + \cos t)/2}} = \frac{\sqrt{1 - \cos t}}{\sqrt{1 + \cos t}}.$$

Squaring the right and left sides of this equation gives

$$x^2 = \frac{1 - \cos t}{1 + \cos t}, \qquad \text{or} \qquad x^2(1 + \cos t) = 1 - \cos t.$$

To solve for $\cos t$ we write

$$x^2 + x^2 \cos t = 1 - \cos t \qquad \text{and} \qquad x^2 \cos t + \cos t = 1 - x^2,$$

so

$$(x^2 + 1) \cos t = 1 - x^2.$$

Since $t = 2 \arctan x$ we have

$$\cos(2 \arctan x) = \cos t = \frac{1 - x^2}{1 + x^2}. \qquad \blacksquare$$

The arcsecant function is also used frequently in calculus and is defined by restricting the secant function to the same basic interval as its reciprocal, the cosine function. Then features of the arccosine function can be used to generate facts about the arcsecant.

Some calculus books use a different restriction on the secant function, the set $[0, \pi/2) \cup [\pi, 3\pi/2)$. Either definition is valid; we simply need to restrict the secant function to a set of real numbers where the function is one-to-one and where it assumes its entire range. Our interval restriction of a secant function has been chosen to agree as nearly as possible with the corresponding interval restriction with its reciprocal, the cosine function.

The Arcsecant Function

The **arcsecant** function, denoted **arcsec**, has domain $(-\infty, -1] \cup [1, \infty)$ and range $[0, \pi/2) \cup (\pi/2, \pi]$ and is defined by

$$\arcsec x = y \qquad \text{if and only if} \qquad \sec y = x.$$

The definition and properties of inverse functions imply the following.

Arcsecant Properties

$\sec(\arcsec x) = x$ when x is in $(-\infty, -1] \cup [1, \infty)$, and
$\arcsec(\sec x) = x$ when x is in $[0, \pi/2) \cup (\pi/2, \pi]$.

The graph of $y = \arcsec x$ is the reflection about the line $y = x$ of the graph of the restricted secant function, as shown in Figure 3.66. Since the graph of $y = \sec x$ has a vertical asymptote at $x = \pi/2$, the graph of $y = \arcsec x$ has a horizontal asymptote at $y = \pi/2$.

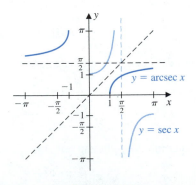

FIGURE 3.66

To determine the relationship between the arcsecant and the arccosine, consider $\cos(\operatorname{arcsec} x)$. If we introduce a variable t with $t = \operatorname{arcsec} x$, then $\sec t = x$. But $\sec t = 1/\cos t$, so this implies that $\cos t = 1/x$, and $t = \arccos(1/x)$. Hence we have

$$\operatorname{arcsec} x = \arccos \frac{1}{x}, \qquad \text{whenever } |x| \geq 1.$$

The inverse cosecant, denoted arccsc, and inverse cotangent, denoted arccot, functions are defined in a similar manner. Since they are not called on frequently in calculus, we do not consider them. Their basic properties are shown in Figure 3.67.

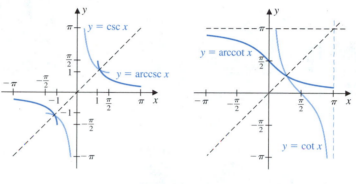

FIGURE 3.67

EXERCISE SET 3.7

In Exercises 1–12, find the exact value of the quantity, or explain why it is undefined.

1. $\arcsin(1/2)$

2. $\arccos(1/2)$

3. $\arcsin(1)$

4. $\arccos(-1)$

5. $\arctan(\sqrt{3})$

6. $\arctan(0)$

7. $\operatorname{arcsec}(-1)$

8. $\operatorname{arcsec}(0)$

9. $\operatorname{arccot}(-1)$

10. $\operatorname{arccot}(-\sqrt{3})$

11. $\arccos(2)$

12. $\operatorname{arcsec}(-2/\sqrt{3})$

In Exercises 13–28, find the exact value of each expression.

13. $\cos(\arccos(1/2))$

14. $\sin(\arccos(1/2))$

15. $\cos(\arcsin(\sqrt{2}/2))$

16. $\tan(\arcsin(\sqrt{3}/2))$

17. $\arcsin(\sin(\pi/2))$

18. $\arcsin(\sin(3\pi/4))$

19. $\arccos(\cos(2\pi/3))$

20. $\arccos(\cos(-\pi/4))$

21. $\arctan(\tan(-\pi/4))$

22. $\arctan(\tan(7\pi/6))$

23. $\cos(\arcsin(3/5))$

24. $\tan(\arccos(4/5))$

25. $\sin(\arccos(5/13))$

26. $\cos(\arctan(4))$

27. $\cos(\arcsin(3/5) + \arccos(4/5))$

28. $\tan(\arcsin(1/3) + \arccos(1/2))$

In Exercises 29–30, (a) solve each equation on the given interval, expressing the solution for x in terms of inverse trigonometric functions, and (b) use a calculator to approximate the solutions in part (a) to three decimal places.

29. $(\tan x)^2 - \tan x - 2 = 0$ on $(-\pi/2, \pi/2)$

30. $6(\cos x)^2 - \cos x - 5 = 0$ on $[\pi/2, \pi]$

In Exercises 31–36, verify the given identity.

31. $\arcsin(-x) = -\arcsin(x)$, where $|x| \leq 1$

32. $\arccos(-x) = \pi - \arccos(x)$, where $|x| \leq 1$

33. $\arccos x = \arcsin(\sqrt{1 - x^2})$, where $|x| \leq 1$

34. $\arctan x = \arcsin\left(\dfrac{x}{\sqrt{1 + x^2}}\right)$

35. $\tan(\arcsin x) = \dfrac{x}{\sqrt{1 - x^2}}$, where $|x| < 1$

36. $\cos(\arcsin x) = \sqrt{1 - x^2}$, where $|x| \leq 1$

37. A large picture measures a feet from top to bottom. The bottom of the picture is b feet above the eye level of an observer who is standing x feet from the wall on which the picture is hung. Show that the angle θ subtended by the picture at the eye of the observer is given by

$$\theta = \arctan\left(\frac{a + b}{x}\right) - \arctan\left(\frac{b}{x}\right)$$

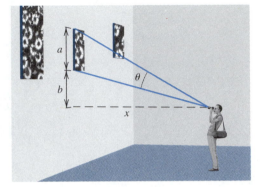

38. A lighthouse is 4 mi from a straight shoreline, as shown in the figure. If the light from the lighthouse is moving along the shoreline, express the angle θ formed by the beam of light and the shoreline in terms of the distance x.

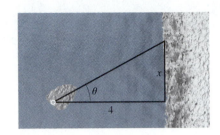

3.8   APPLICATIONS OF TRIGONOMETRIC FUNCTIONS

A Cessna Citation III business jet flying at 520 mi/h is directly over Logan, Utah, and heading due south toward Phoenix. Fifteen minutes later an F-15 Fighting Eagle passes over Logan traveling westward at 1535 mi/h. We would like to determine a function that describes the distance between the planes after the F-15 passes over Logan until it reaches the California border 20 min later.

Suppose we let $t = 0$ be the time in hours after the time the F-15 passes Logan, $x(t)$ be the distance in miles that the F-15 has traveled, and $y(t)$ be the distance in miles the Cessna has traveled at time t. Writing the 20 min as 1/3 h and taking into consideration the 1/4 hour lead time of the Cessna, we have

$$x(t) = 1535t \quad \text{and} \quad y(t) = \frac{1}{4}(520) + 520t, \quad \text{for } 0 \leq t \leq 1/3.$$

The situation is illustrated in Figure 3.68.

Since the triangle in the figure is a right triangle, the distance between the planes is

$$d(t) = \sqrt{[x(t)]^2 + [y(t)]^2} = \sqrt{(1535t)^2 + \left[520\left(t + \frac{1}{4}\right)\right]^2},$$

and when the F-15 reaches the California line the planes are $d(1/3) \approx 595$ mi apart.

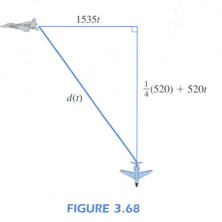

$1535t$

$\frac{1}{4}(520) + 520t$

$d(t)$

FIGURE 3.68

The reason for introducing the section with this example is to consider how dramatically the problem changes in the more likely situation when the paths of the planes are not perpendicular. Suppose, for example, that all the facts of the problem are the same except that the F-15 is heading over Logan on a course 24° west of south toward Nellis Air Force Base near Las Vegas. On this course it would cross the Nevada state line in approximately 14.7 min, or 0.245 h. Figure 3.69 illustrates this new situation, from which we can see that the triangle probably has no right angle.

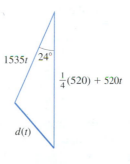

$1535t$ 24°

$\frac{1}{4}(520) + 520t$

$d(t)$

FIGURE 3.69

To handle this problem we need the Law of Cosines, of which the Pythagorean theorem is a special case. The Law of Cosines can be used to find the missing part of a non-right triangle provided that two sides and an angle are known, although it is easier to apply if the known angle is formed by the known sides.

Law of Cosines

Suppose that a triangle has sides of length a, b, and c and corresponding opposite angles α, β, and γ, as shown in Figure 3.70(a). Then

$$a^2 = b^2 + c^2 - 2bc \cos \alpha.$$

Since the sides and angles are arbitrarily labeled, the Law of Cosines also gives

$$b^2 = a^2 + c^2 - 2ac \cos \beta \qquad \text{and} \qquad c^2 = a^2 + b^2 - 2ab \cos \gamma.$$

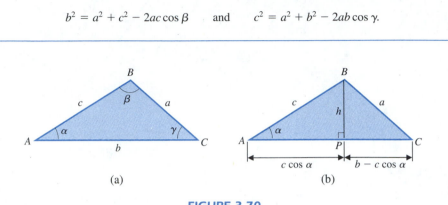

(a)   (b)

FIGURE 3.70

To see why the Law of Cosines holds, consider the triangles shown in Figure 3.70(b). We have labeled as P the point where the vertical from B intersects the base of the triangle and called this vertical distance h. Since APB and BPC are both right triangles,

$$c^2 = h^2 + (c \cos \alpha)^2 \qquad \text{and} \qquad a^2 = h^2 + (b - c \cos \alpha)^2.$$

Solving for h^2 in the first equation and substituting into the second produces the Law of Cosines:

$$a^2 = \left(c^2 - (c \cos \alpha)^2\right) + (b - c \cos \alpha)^2$$

$$= c^2 - (c \cos \alpha)^2 + b^2 - 2bc \cos \alpha + (c \cos \alpha)^2$$

$$= b^2 + c^2 - 2bc \cos \alpha.$$

Although we have illustrated the situation when α is an acute angle, the proof works equally well for obtuse angles.

We can now solve the problem of the aircraft whose paths are not at right angles to one another.

EXAMPLE 1   A Cessna Citation III business jet flying at 520 mi/h is directly over Logan, Utah, and heading due south to Phoenix. Fifteen minutes later an F-15 Eagle passes over Logan traveling toward Nellis Air Force Base near Las Vegas at 1535 mi/h on a course of 24° west of south. Determine a function that describes the distance between the planes after the F-15 passes over Logan until it reaches the Nevada border 0.245 h later.

Solution   Figure 3.71 illustrates the situation. The values of $x(t)$ and $y(t)$ are the same as in the opening example, namely,

$$x(t) = 1535t \qquad \text{and} \qquad y(t) = \frac{1}{4}(520) + 520t.$$

The Law of Cosines implies that in this situation

$$d(t) = \sqrt{[x(t)]^2 + [y(t)]^2 - 2x(t)y(t) \cos 24°}$$

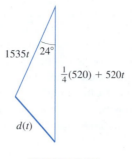

$1535t$ $24°$

$\frac{1}{4}(520) + 520t$

$d(t)$

FIGURE 3.71

In calculus you would probably be asked to determine the rate at which the distance $d(t)$ is changing.

for t between 0 and 0.245. When the F-15 reaches the Nevada line, the distance separating the planes is $d(0.245) \approx 176$ mi. ∎

EXAMPLE 2   A picture in an art museum is 5 ft high and hung so that its base is 8 ft above the ground. Find the viewing angle $\theta(x)$ of a 6-ft viewer standing x feet from the wall.

Solution   The situation is illustrated in Figure 3.72. Since APB and APC are both right triangles, we have $AC = \sqrt{49 + x^2}$ and $AB = \sqrt{4 + x^2}$.

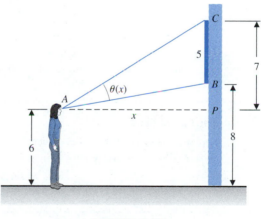

FIGURE 3.72

The Law of Cosines applied to $\triangle ABC$ implies that

$$25 = (4 + x^2) + (49 + x^2) - 2\sqrt{4 + x^2}\sqrt{49 + x^2} \cos \theta(x),$$

which gives

$$\cos \theta(x) = \frac{53 + 2x^2 - 25}{2\sqrt{4 + x^2}\sqrt{49 + x^2}},$$

so

$$\theta(x) = \arccos \frac{14 + x^2}{\sqrt{(4 + x^2)(49 + x^2)}}.$$

∎

You can expect to see problems like this Example as the first, and most difficult, step in a calculus problem.

Notice in Example 2 that as x becomes large, the constants in both the numerator and the denominator become less significant, and the fraction in the argument of the arccosine approaches 1. As a consequence, as x becomes large, $\theta(x)$ approaches zero. The argument of the arccosine also approaches 1 as x approaches zero, so $\theta(x)$ approaches zero in this case as well. The best view of the painting might be defined as occurring when $\theta(x)$ is a maximum. The graph of $y = \theta(x)$ shown in Figure 3.73 indicates that the maximum occurs at approximately $x = 3.7$ ft from the wall. Methods of calculus can be used to determine the exact value.

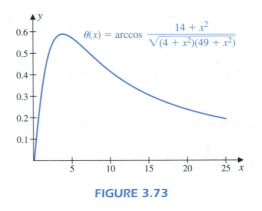

FIGURE 3.73

We have seen that the Law of Cosines can be used to find the missing parts of a triangle provided that two sides and an included angle are known. In other cases, for example, when we know two angles and a side, we need to use the Law of Sines.

The Law of Sines

Suppose that a triangle has sides of length a, b, and c and corresponding opposite angles α, β, and γ, as shown in Figure 3.74(a). Then

$$\frac{\sin \alpha}{a} = \frac{\sin \beta}{b} = \frac{\sin \gamma}{c}.$$

To see why the Law of Sines holds, we drop a perpendicular to the base, label the point of intersection P, and let h be the vertical distance from B to P, as shown in Figure 3.74(b).

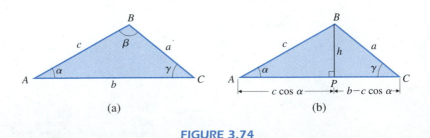

FIGURE 3.74

Since APB and BPC are both right triangles, we have

$$h = c \sin \alpha \qquad \text{and} \qquad h = a \sin \gamma.$$

Eliminating h in these equations gives

$$c \sin \alpha = a \sin \gamma, \qquad \text{or} \qquad \frac{\sin \alpha}{a} = \frac{\sin \gamma}{c}.$$

In a similar manner, we can show that $(\sin \beta)/b$ is also equal to these expressions.

EXAMPLE 3   The aircraft carrier Carl Vinson leaves the Pearl Harbor Naval shipyard and heads due west at 28 knots. A helicopter is 175 nautical miles from the carrier at 35° south of west. On what course should the helicopter travel at its cruising speed of 130 knots to intercept the aircraft carrier in the shortest time? How long will it take?

Solution   Figure 3.75(a) gives an illustration of the situation, assuming that the intersection point occurs at time t, which at present is unknown.

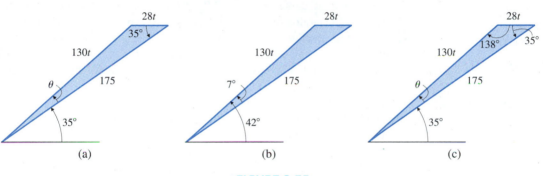

(a)   (b)   (c)

FIGURE 3.75

First we will find the angle θ that gives the course the helicopter should fly. The Law of Sines implies that

$$\frac{\sin \theta}{28t} = \frac{\sin 35°}{130t},$$

so

$$\sin \theta = \frac{28t}{130t} \sin 35° = \frac{28}{130} \sin 35° = 0.1235.$$

This gives $\theta = \arcsin 0.1235 \approx 7°$. The helicopter should, consequently, fly a course that is approximately $35 + 7 = 42°$ to the north of east, as shown in Figure 3.75(b).

To determine the time required to reach the carrier, we again use the Law of Sines, but now we involve the side that gives the original distance between the carrier and the helicopter, as shown in Figure 3.75(c). The remaining angle of the triangle has measure $180 - 35 - 7 = 138°$, so

$$\frac{\sin 138°}{175} = \frac{\sin 35°}{130t} \qquad \text{and} \qquad t = \frac{175 \sin 35°}{130 \sin 138°} \approx 1.154 \text{ hours,}$$

or approximately 1 h 9 min 14 s.   ∎

The Law of Sines is easier to apply than the Law of Cosines, but it can lead to some interesting situations when the given information consists of an angle and two sides, one of which is opposite the given angle. In Figure 3.76(a) we have shown a typical situation when we have been given an angle α, the length a of the side opposite of α, and the length c of an adjacent side. What happens depends on the given value a.

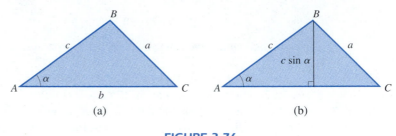

FIGURE 3.76

If we drop a perpendicular from B to the base, as shown in Figure 3.76(b) we find that the length of this vertical line segment is $c \sin \alpha$. If we are given $a < c \sin \alpha$, there is no triangle formed, as shown in Figure 3.77(a). If $a = c \sin \alpha$, there is precisely one triangle that meets the requirement, the right triangle shown in Figure 3.77(b). If $a \geq c$, there is also only one possibility, as shown in Figure 3.77(c). But if $c \sin \alpha < a < c$, we have an *ambiguous* case, since there are two possibilities, the triangles in Figure 3.77(d) that are labeled ABC and ABC'. Since the triangle $C'BC$ in this figure is isosceles, $\gamma' = \pi - \gamma$.

Generally the physical circumstances of the problem will dictate the required situation, but it is important to check the reasonableness of the solution to the problem involving this application of the Law of Sines.

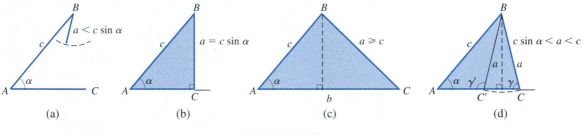

FIGURE 3.77

EXAMPLE 4   A campground lies at the west end of an east-west road in relatively flat but dense forest, and the starting point for a hike lies 30 km to the northeast of the campground. A hiker begins at the starting point and travels in the general direction of the campground, reaching the road after 25 km. Approximately how far is the campground?

Solution   The situation facing the hiker is illustrated in Figure 3.78, where the 45° angle and the 30-km distance are assumed to be correct, but the 25-km distance is probably somewhat inaccurate. We assume that $\triangle ABC'$ gives the correct solution to the problem since traveling along line BC would mean that the hiker is badly offcourse.

Our first step is to use the Law of Sines to determine the angle γ'. Since

$$\frac{\sin 45°}{25} = \frac{\sin \gamma'}{30}, \qquad \text{we have} \qquad \sin \gamma' = \frac{30 \sin 45°}{25} = \frac{3\sqrt{2}}{5}.$$

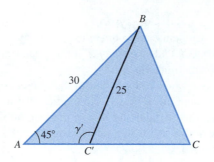

FIGURE 3.78

From this we find that either

$$\gamma' = \arcsin \frac{3\sqrt{2}}{5} \approx 58° \quad \text{or} \quad \gamma' \approx \pi - 58° = 122°.$$

The larger angle is correct unless the hiker is badly lost, so the angle at B is $180° - 122° - 45° = 13°$. Using the Law of Sines again gives us the length of the base AC':

$$\frac{\sin 45°}{25} = \frac{\sin 13°}{AC'}, \quad \text{so} \quad AC' = \frac{25 \sin 13°}{\sin 45°} \approx 7.95 \text{ km.}$$

If the hiker was lost but did measure the distance correctly, the distance to the camp could be as much as

$$AC = \frac{25 \sin(180° - 58° - 45°)}{\sin 45°} = \frac{25 \sin 77°}{\sin 45°} \approx 34.5 \text{ km.} \quad ■$$

Simple Harmonic Motion

Oscillating motion is common in physical problems. For example, if we ignore friction, a weight attached to a spring that is displaced from its equilibrium position will vibrate in a oscillatory manner according to a formula of the form

$$y(t) = A \sin \omega_0 t + B \cos \omega_0 t,$$

where ω_0 is a *spring constant* and $y(t)$ describes the displacement at time t. The spring constant determines the *frequency* of the motion, which is $\omega_0/(2\pi)$. (See Figure 3.79.)

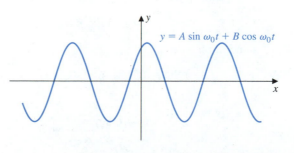

FIGURE 3.79

The use of the term
harmonic in music is
no coincidence.
Musical tones are
described
mathematically using
equations for
harmonic motion.

The same type of oscillatory motion is common in electrical circuits, musical instruments, and numerous other physical applications. All these phenomena exhibit what is called *harmonic* motion, and are described using combinations of sine and cosine functions.

One of the features of the sine or cosine function that finds regular use in physical applications follows from the fact that linear combinations of these functions that have the same argument can be combined into a single sine or cosine function with the same period. Specifically, for every real number x and nonzero constants A and B we have the following relationship.

Sine Combination Formula

$$A \sin x + B \cos x = C \sin(x + \delta), \quad \text{where} \quad C^2 = A^2 + B^2 \quad \text{and} \quad \delta = \arctan \frac{B}{A}.$$

This relationship can be seen from examining the right triangle shown in Figure 3.80.

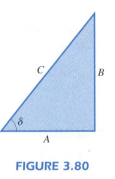

FIGURE 3.80

In this triangle we have

$$A = C \cos \delta \quad \text{and} \quad B = C \sin \delta,$$

so

$$A \sin x + B \cos x = C \cos \delta \sin x + C \sin \delta \cos x = C(\sin x \cos \delta + \sin \delta \cos x)$$
$$= C \sin(x + \delta).$$

In addition,

$$A^2 + B^2 = (C \cos \delta)^2 + (C \sin \delta)^2 = C^2 \left((\cos \delta)^2 + (\sin \delta)^2\right) = C^2,$$

and

$$\frac{B}{A} = \frac{C \sin \delta}{C \cos \delta} = \tan \delta \quad \text{so} \quad \delta = \arctan \frac{B}{A}.$$

There is a similar relationship involving the cosine function:

Cosine Combination Formula

$$A \sin x + B \cos x = C \cos(x - \delta), \qquad \text{where} \qquad C^2 = A^2 + B^2 \quad \text{and} \quad \delta = \arctan \frac{A}{B}.$$

EXAMPLE 5   Suppose that we have a spring with the spring constant k, the mass of the system is m, and the system has an initial position y_0 and velocity v_0, as shown in Figure 3.81. Then some basic rules of physics tell us that the motion of the mass at the end of the spring is given by

$$y(t) = v_0 \sqrt{\frac{m}{k}} \sin \sqrt{\frac{k}{m}} t + y_0 \cos \sqrt{\frac{k}{m}} t.$$

Sketch the graph of the motion of a spring-mass system that has a mass $m = 16$ kg, a spring constant $k = 1$ N/m, an initial position $y_0 = 3$ m, and an initial velocity 1 m/s.

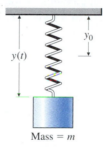

y_0

$y(t)$

Mass $= m$

FIGURE 3.81

Solution   Since $\sqrt{m/k} = \sqrt{16} = 4$, the motion of the system is given by the equation

$$y(t) = 4 \sin \frac{t}{4} + 3 \cos \frac{t}{4},$$

which can be reexpressed, using the sine combination formula, as

$$y(t) = \sqrt{4^2 + 3^2} \sin \left(\frac{t}{4} + \arctan \frac{3}{4} \right) = 5 \sin \left(\frac{t}{4} + \arctan \frac{3}{4} \right)$$

$$= 5 \sin \frac{1}{4} \left(t + 4 \arctan \frac{3}{4} \right).$$

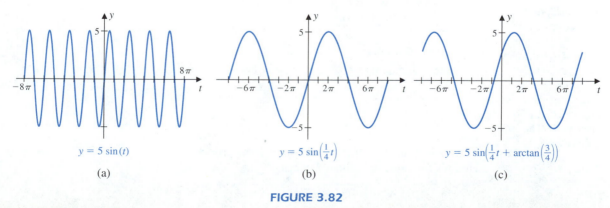

$y = 5 \sin(t)$

(a)

$y = 5 \sin\left(\frac{1}{4}t\right)$

(b)

$y = 5 \sin\left(\frac{1}{4}t + \arctan\left(\frac{3}{4}\right)\right)$

(c)

FIGURE 3.82

To sketch this graph we first sketch the graph of $y = 5 \sin t$, which is shown in Figure 3.82(a).

We expand this graph horizontally by a factor of 4 to produce the graph of $y = 5 \sin \frac{t}{4}$ shown in Figure 3.82(b). The final graph is the shift to the left by $4 \arctan(3/4) \approx 2.57$, as shown in Figure 3.82(c). This graph has amplitude 5, frequency $(1/4)/(2\pi) = 1/(8\pi)$, and phase shift $4 \arctan(3/4)$.   ■

Heron's Formula

The final application of the trigonometric functions is a bit of a misnomer. It isn't really an application of trigonometry, but it uses trigonometric identities in its derivation.

If you ask most people for a formula for the area of a triangle, you will be given "take one-half the base times the altitude." But this is a rather awkward formula to use when given the lengths of the sides of a triangle. Consider, for example, the triangle shown in Figure 3.83 whose sides have length 7, 9, and 12.

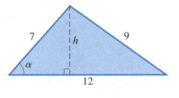

FIGURE 3.83

To determine the altitude from the horizontal base we need to know the appropriate angle α. This can be determined from the Law of Cosines, which implies that

$$9^2 = 7^2 + 12^2 - 2(7 \cdot 12) \cos \alpha \qquad \text{so} \qquad \alpha = \arccos \frac{49 + 144 - 81}{2(84)} = \arccos \frac{2}{3}.$$

Once α is known, we can find the altitude $h = 7 \sin \alpha$, and the area is $\mathcal{A} = \frac{1}{2}(12)h = 42 \sin \alpha$.

There is a easier formula to apply in this situation, one that has been known for over 2000 years. The perimeter of a triangle is the sum of the lengths of its sides and the *semiperimeter*, denoted s, is half this value. Heron's formula for the area of a triangle uses this semiperimeter.

A printed proof of this formula was first given by Heron in his book *Metrica* around 100 A.D., but the result was probably due to the great Archimedes, who lived more than 300 years earlier.

Heron's Formula

A triangle with sides of length a, b, and c has area given by

$$\mathcal{A} = \sqrt{s(s-a)(s-b)(s-c)}, \qquad \text{where} \qquad s = \frac{1}{2}(a+b+c).$$

The triangle in Figure 3.83 has a semiperimeter $s = (7 + 9 + 12)/2 = 14$, so Heron's formula implies that the area is

$$\mathcal{A} = \sqrt{14(14-7)(14-9)(14-12)} = \sqrt{980} = 14\sqrt{5}.$$

The application of Heron's formula might be easy, but the derivation is anything but. It is good practice for algebra, however, so let us attack it with vigor. We begin with the triangle shown in Figure 3.84, and the familiar formula of area being half the base times the altitude.

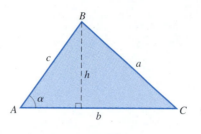

FIGURE 3.84

We can write

$$\mathcal{A}^2 = \left(\frac{1}{2}bh\right)^2 = \left(\frac{1}{2}bc\sin\alpha\right)^2 = \frac{1}{4}b^2c^2(\sin\alpha)^2.$$

But $(\sin\alpha)^2 = 1 - (\cos\alpha)^2$, so

$$\mathcal{A}^2 = \frac{1}{4}b^2c^2(1 - (\cos\alpha)^2) = \frac{1}{4}b^2c^2(1 + \cos\alpha)(1 - \cos\alpha)$$

$$= \frac{1}{4}(bc + bc\cos\alpha)(bc - bc\cos\alpha).$$

Now the Law of Cosines comes into play, since

$$a^2 = b^2 + c^2 - 2bc\cos\alpha \qquad \text{implies that} \qquad bc\cos\alpha = \frac{1}{2}(b^2 + c^2 - a^2).$$

Hence

$$\mathcal{A}^2 = \frac{1}{4}\left(bc + \frac{1}{2}(b^2 + c^2 - a^2)\right)\left(bc - \frac{1}{2}(b^2 + c^2 - a^2)\right)$$

$$= \frac{1}{16}(2bc + b^2 + c^2 - a^2)(2bc - b^2 - c^2 + a^2).$$

By regrouping and factoring we can rewrite this as

$$\mathcal{A}^2 = -\frac{1}{16}(a^2 - (b^2 + 2bc + c^2))(a^2 - (b^2 - 2bc + c^2))$$

$$= -\frac{1}{16}(a^2 - (b + c)^2)(a^2 - (b - c)^2).$$

Factoring this last formula gives

$$\mathcal{A}^2 = -\frac{1}{16}\left((a + (b + c))(a - (b + c))\right)\left((a - (b - c))(a + (b - c))\right)$$

$$= \frac{1}{16}(a + b + c)(b + c - a)(a + c - b)(a + b - c).$$

If we now add and subtract a, b, and c, respectively, in each of the last three factors we get

$$A^2 = \frac{1}{16}(a + b + c)(a + b + c - 2a)(a + b + c - 2b)(a + b + c - 2c)$$

$$= \frac{1}{16}(2s)(2s - 2a)(2s - 2b)(2s - 2c)$$

$$= s(s - a)(s - b)(s - c)$$

and, finally,

$$\mathcal{A} = \sqrt{s(s - a)(s - b)(s - c)}.$$

EXERCISE SET 3.8

In Exercises 1–6, let the angles of a triangle be α, β, and γ, with opposite sides of lengths a, b, and c, respectively. Use the Law of Cosines to find the remaining parts of the triangle.

1. $\alpha = 35°$; $b = 15$; $c = 25$

2. $\gamma = 105°$; $a = 16$; $b = 22$

3. $\beta = 30°$; $a = 25$; $c = 32$

4. $\alpha = 60°$; $b = 50$; $c = 35$

5. $a = 12$; $b = 22$; $c = 15$

6. $a = 3$; $b = 5$; $c = 7$

In Exercises 7–14, let the angles of a triangle be α, β, and γ, with opposite sides of lengths a, b, and c, respectively. Use the Law of Sines to find the remaining parts of the triangle.

7. $\alpha = 50°$; $\beta = 76°$; $c = 100$

8. $\alpha = 60°$; $\beta = 56°$; $a = 25$

9. $\alpha = 65°$; $\gamma = 50°$; $b = 10$

10. $\alpha = 30°$; $\gamma = 135°$; $a = 30$

11. $\beta = 100°$; $\gamma = 30°$; $c = 20$

12. $\beta = 72°$; $\gamma = 55°$; $b = 7.5$

13. $a = 6$; $c = 2\sqrt{3}$; $\alpha = 60°$

14. $b = \sqrt{3}$; $c = \sqrt{2}$; $\gamma = 45°$

15. Show that there is no triangle satisfying the conditions $a = 3$, $b = 10$, and $\alpha = 25.4°$.

16. Solve for the missing parts of the triangle satisfying $a = 4$, $b = 5$, and $\alpha = 53°$.

17. Is there a triangle that satisfies $a = 3$, $b = 10$, and $\alpha = 54°$?

18. Find the area of the triangle with sides of lengths 10 cm, 14 cm, and 18 cm.

19. A gas pipeline is to be constructed between towns A and B. The engineers have two alternatives. They can connect A and B directly, but then they must build the pipeline through a swamp, or they can build the pipeline from town A to town C, which is 3 mi directly west of A, and then to town B, which is 2 mi directly northwest of C. The cost of construction through the swamp from A to B is $125,000 per mile and the cost to go through C is $100,000 per mile.

   a. Which alternative should the engineers select?

   b. At what cost per mile for construction through C would there be no price difference in the two alternatives?

20. A surveyor needs to determine the distance across a pond and makes the measurements shown in the figure. What is the distance across the pond?

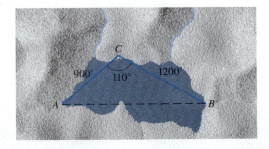

21. Two ships leave port at 10:00 A.M. One travels at a bearing of N62°E (62° north of east) at 20 mi/h and the second at a bearing of S75°E at 25 mi/h. How far apart are the ships at 12:00 A.M.?

22. In tracking the relative location of two aircraft, a controller determines that the distance from the station to the first aircraft is 150 mi and the distance to the second is 100 mi. If the angle between the two aircraft is 50°, how far apart are the two planes?

23. The National Forest Service maintains observation towers to check for the outbreak of forest fires. Suppose two towers are at the same elevation, one at point A and another 10 mi due west at a point B (see the figure). The ranger at A spots a fire in the northwest whose line of sight makes an angle of 63° with the line between the towers, and contacts the ranger at B. This ranger locates the fire along a line of sight that makes a 50° angle with the line between the towers. How far is the fire from the tower at B?

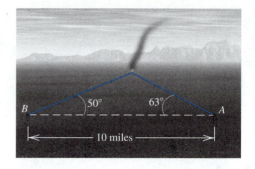

24. The lengths of the sides of a triangular parcel of land are approximately 200 ft, 300 ft, and 450 ft. If land is valued at $2000 per acre, what is the value of the parcel of land?

REVIEW EXERCISES FOR CHAPTER 3

In Exercises 1–6, find (a) $P(t)$, the terminal point on the unit circle determined by t, (b) the reference number for t, and (c) the values of the six trigonometric functions of t.

1. $t = \dfrac{\pi}{3}$

2. $t = \dfrac{11\pi}{6}$

3. $t = \dfrac{5\pi}{4}$

4. $t = \dfrac{8\pi}{3}$

5. $t = -\dfrac{21\pi}{4}$

6. $t = -\dfrac{23\pi}{3}$

In Exercises 7–10, find the values of the trigonometric functions from the given information.

7. $\cos t = \dfrac{3}{5}$, where $\dfrac{3\pi}{2} < t < 2\pi$.

8. $\sin t = -\dfrac{1}{2}$, where $\cos t < 0$.

9. $\tan t = \dfrac{1}{4}$, where $0 < t < \dfrac{\pi}{2}$.

10. $\sec t = -4$, where $\dfrac{\pi}{2} < t < \pi$.

In Exercises 11–16, find all values of x in $[0, \pi]$ that satisfy the given equation.

11. $\cos \dfrac{x}{3} = \dfrac{1}{2}$

12. $\sin 4x = -\dfrac{\sqrt{3}}{2}$

13. $(\tan x)^3 - 4 \tan x = 0$

14. $2 \tan x - 3 \cot x = 0$

15. $\cot x - \csc x = 1$

16. $\cos 2x + \sin x = 0$

In Exercises 17–20, determine whether the function is even, odd, or neither.

17. $f(x) = (\sin x)^3$

18. $f(x) = \sin x \cos x$

19. $f(x) = x(\sin x)^3$

20. $f(x) = x^5 + \cos x$

In Exercises 21–30, sketch one period of the graph of the function.

21. $y = 5 \sin \dfrac{1}{2}x$

22. $y = 2 \cos 2\pi x$

23. $y = -3 \cos 2x$

24. $y = 2 + 4 \sin 4x$

25. $y = \cos(2x - \pi)$

26. $y = -2 \cos\left(3x - \dfrac{\pi}{2}\right)$

27. $y = \cot\left(x + \dfrac{\pi}{6}\right)$

28. $y = -\tan\left(x - \dfrac{\pi}{2}\right)$

29. $y = \sec 4\pi x$

30. $y = -2 \csc\left(x - \dfrac{\pi}{4}\right)$

In Exercises 31–34 find a sine function or a cosine function whose graph matches the given curve.

31.

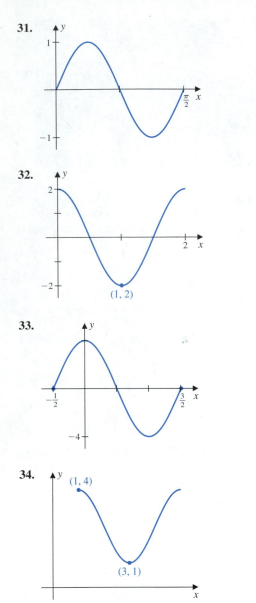

32.

33.

34.

In Exercises 35–36, use a graphing device to find solutions to the equation.

35. $\sin x = x^2$ **36.** $\cos x = x^3$

In Exercises 37–38, determine an appropriate viewing rectangle for the function and use it to sketch the graph.

37. $f(x) = 4\cos 125x$ **38.** $f(x) = x - 10\sin 25x$

In Exercises 39–42, use the addition and subtraction formulas to verify the given formula.

39. $\sin(t - \pi) = -\sin t$ **40.** $\cos(t - \pi) = -\cos t$

41. $\sin\left(\dfrac{3\pi}{2} - t\right) = -\cos t$

42. $\cos\left(\dfrac{3\pi}{2} - t\right) = -\sin t$

In Exercises 43–46, rewrite the expression so that it involves the sum or difference of only constants and sines and cosines to the first power.

43. $(\cos 3x)^2$ **44.** $(\sin 4x)^2$
45. $(\sin x)^4$ **46.** $(\sin 2x)^2(\cos 2x)^2$

In Exercises 47–50, rewrite each product as a sum or difference.

47. $\sin 4t \cos 5t$ **48.** $\cos 6t \sin 8t$
49. $\cos 2t \cos 4t$ **50.** $\sin 3t \sin 5t$

In Exercises 51–54, rewrite each sum or difference as a product.

51. $\sin 2t + \sin 6t$ **52.** $\sin 3t + \sin 7t$
53. $\cos 4t + \cos 2t$ **54.** $\cos 5t - \cos 3t$

In Exercises 55–58, verify the identities.

55. $(\cos x)^4 - (\sin x)^4 = \cos 2x$
56. $(\sin x - \cos x)^2 = 1 - \sin 2x$
57. $\dfrac{\sin x}{1 - \cos x} = \cot x + \csc x$
58. $\dfrac{\cos x}{1 - \tan x} + \dfrac{\sin x}{1 - \cot x} = \sin x + \cos x$

In Exercises 59–62, find the exact value of the quantity.

59. $\sin(\arctan(\sqrt{3}))$
60. $\cos\left(\arcsin\left(\dfrac{4}{5}\right)\right)$
61. $\sin\left(\arcsin\left(\dfrac{3}{5}\right) - \arcsin\left(\dfrac{5}{13}\right)\right)$
62. $\tan\left(\arccos\left(\dfrac{1}{4}\right) + \arcsin\left(\dfrac{1}{2}\right)\right)$

In Exercises 63–68, assume the angles of the triangle are α, β, and γ, with opposite sides of lengths a, b, and c, respectively. Use either the Law of Cosines or the Law of Sines to find the missing parts of the triangle.

63. $\alpha = 25°$; $b = 12$; $c = 20$
64. $\alpha = 30°$; $\beta = 100°$; $c = 25$

65. $a = 6; b = 8; c = 10$

66. $a = 65; \beta = 30°; c = 35$

67. $\beta = 76°; \gamma = 50°; b = 10.5$

68. $a = 8; c = \sqrt{3}; \gamma = 45°$

69. A gutter is to be made from a long strip of tin by bending up the sides, so that the base and the sides have the same length b. Express the area of a cross section of the gutter as a function of the angle θ that the sides make with the base.

70. Two ships A and B leave port at the same time, ship A traveling 45° NE and ship B traveling 45° SE. If ship A is moving at an average speed of 20 mi/h and ship B at an average speed of 35 mi/h, how far apart are the ships after 2 h? What is the bearing of ship B from ship A?

71. An airplane takes off from an airport and travels at a heading of 150°, measured clockwise from north. If the average speed is 380 mi/h, how far south and east is the plane from the airport after 2 h and 30 min?

CHAPTER 3: CALCULUS PREVIEW EXERCISES

In Exercises 1–4, show graphically that the following formulas hold.

1. $\sin(t + \pi) = -\sin t$

2. $\cos(t + \pi) = -\cos t$

3. $\sin\left(t + \dfrac{3\pi}{2}\right) = -\cos t$

4. $\cos\left(t + \dfrac{3\pi}{2}\right) = \sin t$

In Exercises 5–6, use a graphing device to sketch the graph of the function, and approximate the absolute maximum and minimum of the function on the specified interval.

5. $f(x) = x - \sin x; [-2, 3]$

6. $f(x) = \sin x + \cos x; [-\pi, \pi]$

In Exercises 7–10, use the formulas for the products of sines and cosines to verify the given sum-to-product formula.

7. $\sin x + \sin y = 2\sin\dfrac{x + y}{2}\cos\dfrac{x - y}{2}$

8. $\sin x - \sin y = 2\cos\dfrac{x + y}{2}\sin\dfrac{x - y}{2}$

9. $\cos x + \cos y = 2\cos\dfrac{x + y}{2}\cos\dfrac{x - y}{2}$

10. $\cos y - \cos x = 2\sin\dfrac{x + y}{2}\sin\dfrac{x - y}{2}$

11. Find all t in the interval $[0, 2\pi]$ such that $\sqrt{3}\cos t = 2 + \sin t$.

12. Use a graphing device to sketch the graph of $y = \sin x + \cos x$. Rewrite the equation of the curve $y = \sin x + \cos x$ in the form $y = a\sin(x + b)$.

13. Use a graphing device to sketch the graphs of the three functions on the same set of axes. Discuss the relationship of the three graphs.

    a. $f(x) = x, g(x) = -x, h(x) = x\sin x$

    b. $f(x) = \cos x, g(x) = -\cos x, h(x) = \cos x\cos 6\pi x$

    c. $f(x) = \sin\pi x, g(x) = -\sin\pi x,$
       $h(x) = \sin\pi x\sin 10\pi x$

14. Use a graphing device to sketch the graphs of $y = x\sin(1/x)$ and $y = x^2\sin(1/x)$. By zooming in repeatedly near the origin, discuss the differences in the behavior of the curves near the origin.

15. Use a graphing device to sketch the graphs of $y = x\sin(1/x)$, $y = x$ and $y = -x$ on the same set of axes for x in $[-10, 10]$. Find all points where $y = x\sin(1/x)$ intersects $y = x$ and $y = -x$.

16. Use a graphing device to plot the data points given in the table. Find a sine or cosine wave of the form $y = a\sin b(x + c) + d$ or $y = a\cos b(x + c) + d$ that fits the data points.

x	0	0.1	0.2	0.3	0.4	0.5	0.6	0.7	0.8	0.9
y	25	27	28	30	31	32	32	33	33	33
x	1.0	1.1	1.2	1.3	1.4	1.5	1.6	1.7	1.8	1.9
y	32	31	30	28	27	25	23	23	21	19
x	2.0	2.1	2.2	2.3	2.4	2.5	2.6	2.7	2.8	2.9
y	19	18	17	17	17	18	18	20	22	22

17. Discuss the effect of the positive constants a, b, and c on the curves $y = a\tan b(x + c)$. Where are the x-intercepts and the asymptotes?

18. Discuss the effect of the positive constants a, b, and c on the curves $y = a\sec b(x + c)$. Where do the local maxima and minima and the asymptotes occur?

19. Show the identity
$$\cot(x_1 \pm x_2) = \frac{\cot x_1 \cot x_2 \mp 1}{\cot x_2 \pm \cot x_1}$$

20. Suppose t_1 is an angle in quadrant I and t_2 is an angle in quadrant III with $\sin t_1 = 5/13$ and $\sin t_2 = -3/5$. Find $\sin(t_1 + t_2)$, $\sin(t_1 - t_2)$, $\cos(t_1 + t_2)$, and $\cos(t_1 - t_2)$.

21. Suppose t_1 is an angle in quadrant III with $\sin t_1 = -4/5$ and $\tan t_2 = 1/2$. Find $\tan(t_1 + t_2)$ and $\tan(t_1 - t_2)$.

22. Substitute $u = 2\sin t$, for $-\pi/2 \leq t \leq \pi/2$, in the expression $\sqrt{4 - u^2}$ and show the expression reduces to $2\cos t$.

23. Substitute $u = 3\tan t$, for $-\pi/2 \leq t \leq \pi/2$, in the expression $\sqrt{u^2 + 9}$ and show the expression reduces to $3\sec t$.

24. Make the indicated trigonometric substitution and simplify the expression, assuming that $a > 0$. (See Exercises 22 and 23.)

   a. $\sqrt{a^2 - u^2}$, $u = a\sin t$, $-\frac{\pi}{2} \leq t \leq \frac{\pi}{2}$

   b. $\sqrt{u^2 + a^2}$, $u = a\tan t$, $-\frac{\pi}{2} \leq t \leq \frac{\pi}{2}$

   c. $\sqrt{u^2 - a^2}$, $u = a\sec t$, $0 < t < \frac{\pi}{2}$

   d. $\dfrac{\sqrt{u^2 - a^2}}{u}$, $u = a\sec t$, $0 < t < \frac{\pi}{2}$

25. Express the area of the rectangle inscribed in a semicircle of radius r in terms of the angle θ shown in the figure.

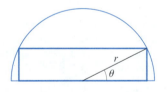

26. Suppose that points A and B are a distance of d meters apart and that the angles formed with the horizontal and the top of a hill are α and β, respectively, as shown in the figure. Show that

$$h = \frac{d}{\cot \alpha - \cot \beta}.$$

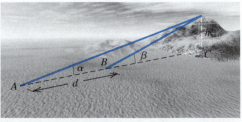

27. A reconnaissance plane is flying at an altitude of 33,000 ft above two ships A and B, as shown in the figure. The angle of depression from the plane to ship A is 32° and the angle of depression to ship B is 47°. Approximate the distance between the two ships, rounding your answer to the nearest 100 ft.

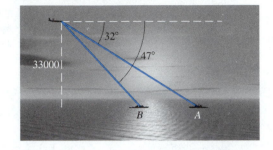

28. The accompanying figure shows an n-sided regular polygon (all sides are of equal length) inscribed in a circle of radius r. The polygon has been divided into n congruent triangles, each with central angle $2\pi/n$ radians. Show that the area of each triangle is

$$\frac{1}{2}r^2 \sin \frac{2\pi}{n}.$$

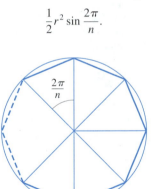

29. Rather than using inscribed polygons as in Exercise 28, circumscribed polygons can be used, as in the accompanying figure, where we have again divided the polygon

into n congruent triangles. Show that the area of each triangle is

$$r^2 \tan \frac{\pi}{n}.$$

30. A pilot needs to determine the bearing of city B from city A in order to deliver cargo. The pilot measures the distances between cities A and B and a third city C, which is due east of B, as shown in the figure. What bearing should the pilot fly?

31. To find the distance across a river a surveyor selects two reference points A and B, 300 ft apart on the same side of the river. The surveyor then selects a third reference point C on the opposite side of the river and determines the angles as shown in the figure. Approximate the distance from A to C.

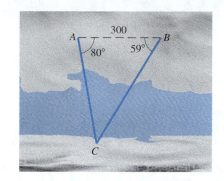

32. Approximate the area of the quadrilateral shown in the figure.

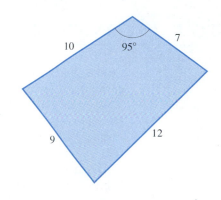

33. Use the Law of Cosines to show that in a triangle with angles α, β, and γ and opposite sides of lengths a, b, and c, respectively, we have

$$a = b \cos \gamma + c \cos \beta$$
$$b = c \cos \alpha + a \cos \gamma$$
$$c = a \cos \beta + b \cos \alpha$$

EXPONENTIAL AND LOGARITHM FUNCTIONS

CALCULUS CONNECTIONS

On the 19th of September in 1991, the frozen remains of a prehistoric man were found encased in ice near the border of Italy and Switzerland. The remains were clearly old, but how old were they? A common way to date objects in the likely age of this prehistoric man is to use a technique known as *radioactive carbon dating*. All living things contain carbon, which has a number of isotopes, predominant of which is the stable isotope $^{12}_{6}C$. Carbon in living beings also contains minute, but accurately measurable, amounts of the radioactive isotope $^{14}_{6}C$. The relative amounts of these two isotopes in living beings is well-known to scientists. It is assumed that the carbon is replenished from the atmosphere and that the relative amounts of the isotopes in living beings have remained constant throughout time.

When the organism dies, the replenishment of carbon ceases. Since the radioactive isotope $^{14}_{6}C$ decays with time, the proportion of that isotope decreases. It is known that the decay rate is such that half of a given amount will be lost after 5730 years. Based on this rate of decay, scientists can estimate the time that the organism died. This valuable technique has been used since the early 1950s, when it was proposed by the chemist Willard Libby, who won a Nobel Prize for his discovery.

This is only one of the many applications of calculus that requires *exponential* functions for their solution. In the example described here, the quantity $Q(t)$ of $^{14}_{6}C$ present at time t takes the form

$$Q(t) = Q_0 e^{-0.000121t},$$

where Q_0 is the estimated amount of the isotope in the sample at the time the man died. The function in this solution is the *natural exponential function* with base the irrational number e. Researchers estimated that 45% of the amount of the $^{14}_{6}C$ had decayed by the time the remains were found, and from this determined that the ancient man was about 5000 years old.

The radioactive-dating technique has limitations and is not without its detractors, but most scientists working in the area have found it to be a valuable tool for dating archaeological and anthropological materials.

4.1   INTRODUCTION

In Chapter 1 we saw that one-to-one functions have inverses, which reverse the function process. The function described by $f(x) = x^3$, for example, has the inverse $f^{-1}(x) = x^{1/3}$, and for every value of x we have both

$$f\big(f^{-1}(x)\big) = f\big(x^{1/3}\big) = \big(x^{1/3}\big)^3 = x$$

and

$$f^{-1}\big(f(x)\big) = f^{-1}\big(x^3\big) = \big(x^3\big)^{1/3} = x.$$

In Chapter 3 we saw that by restricting the domain of the trigonometric functions to a portion on which they are one-to-one, we could also define inverse trigonometric functions. For example, restricting the domain of the sine function to $[-\pi/2, \pi/2]$ allows us to define the arcsine function with the property that

$$\arcsin(\sin x) = x \qquad \text{for each } x \text{ in } [-\pi/2, \pi/2]$$

and

$$\sin(\arcsin x) = x \qquad \text{for each } x \text{ in } [-1, 1],$$

the range of the sine function. In this chapter we introduce another class of functions, the exponentials, and their inverses, the logarithms.

> The most important function-inverse function pair that you will study in calculus and see applied in the sciences is the natural exponential function and its inverse, the natural logarithm function.

One of the first topics discussed in algebra courses concerns the laws of exponents. These tell us that for every positive real number a and rational number $r = p/q$, where p and q have no common factors, we can define

$$a^r = a^{p/q} = \sqrt[q]{(a^p)} = \big(\sqrt[q]{a}\big)^p.$$

For example,

$$27^{2/3} = \big(\sqrt[3]{27}\big)^2 = 3^2 = 9 = \sqrt[3]{(3^2)^3} = \sqrt[3]{(3^3)^2} = \sqrt[3]{(27)^2}.$$

If a is positive and r_1 and r_2 are rational numbers, then we have the *arithmetic properties of exponents*

$$a^{r_1+r_2} = a^{r_1}a^{r_2}, \qquad a^{r_1-r_2} = \frac{a^{r_1}}{a^{r_2}} \qquad \text{and} \qquad \big(a_1^{r_1}\big)^{r_2} = a_1^{r_1 r_2}.$$

For a pair of positive real numbers a_1 and a_2 and a single rational number r, we also have

$$a_1^r a_2^r = (a_1 a_2)^r \qquad \text{and} \qquad \frac{a_1^r}{a_2^r} = \left(\frac{a_1}{a_2}\right)^r.$$

We want to define, for every positive number $a \neq 1$, an exponential function $f(x) = a^x$ that is valid for all real numbers x not just rational numbers. The definition should be such that the arithmetic properties of exponents still hold and such that when x is a rational number, the definition reduces to the root and power definition.

There are various equivalent ways of defining exponential functions, but the easiest is to take the limit of rational approximations. In Figure 4.1 we see a representation of $y = 3^x$ for rational numbers x. We want to extend the domain of this function to include all the real numbers, essentially filling the holes of the graph.

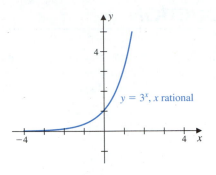

FIGURE 4.1

For example, suppose we want to define $3^{\sqrt{2}}$. Since $\sqrt{2}$ is irrational, we cannot define $3^{\sqrt{2}}$ in terms of roots and powers of 3. But we can use roots and powers of 3 to get arbitrarily accurate approximations to $3^{\sqrt{2}}$. Since the decimal expansion of $\sqrt{2}$ is

$$\sqrt{2} = 1.414213562\ldots,$$

we define $3^{\sqrt{2}}$ as the limit of the approximations

$$3^{1.4} = 3^{14/10} = 4.65553672\ldots,$$

$$3^{1.41} = 3^{141/100} = 4.70696500\ldots,$$

$$3^{1.414} = 3^{1414/1000} = 4.72769503\ldots,$$

$$3^{1.4142} = 3^{14142/10000} = 4.72873393\ldots,$$

$$3^{1.41421} = 3^{141421/100000} = 4.72878588\ldots,$$

We have again used an intuitive notion of "getting close," which is made precise in calculus with the concept of the limit.

and so on. To complete this process rigorously requires far more mathematical analysis than we have available to us, but it certainly appears that this limiting process is converging to a value that we can use to define $3^{\sqrt{2}}$ and that this value is close to 4.729.

In like manner we can define a^x whenever $a > 0$ and x is a real number. The specific value for various choices of a and x need not concern us at this time; what we want is the class of functions defined in this manner.

One exception to this process is the case $a = 1$. Since $1^r = 1$ for every rational number r, we also have $1^x = 1$ for every real number x, which gives the constant function $f(x) = 1^x \equiv 1$. This is a function that we have already considered, and it will be excluded from further discussion in this chapter. When we discuss exponential functions, we will mean that we have a function of the form $f(x) = a^x$, where a is a positive real number with $a \neq 1$.

4.2   THE NATURAL EXPONENTIAL FUNCTION

In the previous section we saw that for every positive number $a \neq 1$, called the *base*, we can define an *exponential function* of the form

$$f(x) = a^x,$$

whose domain is the set of all real numbers. The graphs of some exponential functions with base $a > 1$ are shown in Figure 4.2. Notice that each of these functions approaches the x-axis as $x \to -\infty$. They all pass through the point $(0, 1)$. In addition, they are always increasing, and the greater base a, the faster the rate of increase.

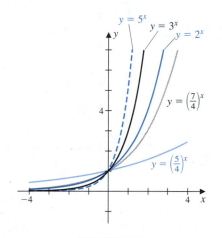

FIGURE 4.2

The graphs of some exponential functions with base $0 < a < 1$ are shown in Figure 4.3. Each function approaches the x-axis as $x \to \infty$. Each passes through $(0, 1)$, and each is always decreasing. The smaller the base a, the faster the rate of decrease.

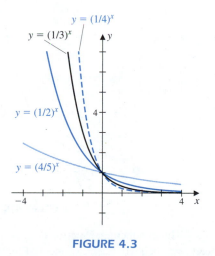

FIGURE 4.3

In the next example the graph of $y = 2^x$ and the graphing techniques from Chapter 1 are used to sketch the graphs of some modified exponential functions.

EXAMPLE 1   Use the graph of $y = 2^x$ to sketch the graphs of

a. $f(x) = 2^x - 3$ b. $g(x) = -3 \cdot 2^{(x-1)} + 1.$

Solution   a. The graph of $f(x) = 2^x - 3$ is obtained by shifting the graph of $y = 2^x$ downward 3 units, as shown in Figure 4.4. From the graph we can see that the domain of f is $(-\infty, \infty)$, the range is $(-3, \infty)$, and the graph has $y = -3$ as a horizontal asymptote.

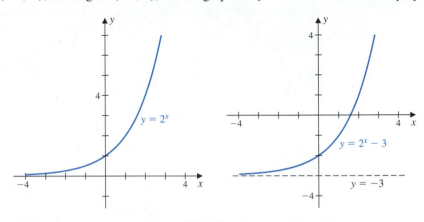

FIGURE 4.4

b. To sketch the graph of $g(x) = -3 \cdot 2^{(x-1)} + 1$, we first stretch the graph of $y = 2^x$ vertically by 3 units to obtain the graph of $y = 3 \cdot 2^x$, shown in Figure 4.5(a). Then we shift this graph one unit to the right to produce the graph of $y = 3 \cdot 2^{(x-1)}$, shown in Figure 4.5(b). The final graph of $g(x) = -3 \cdot 2^{(x-1)} + 1$ is obtained by reflecting this graph about the x-axis and then shifting it upward 1 unit, as shown in Figure 4.5(c).   ■

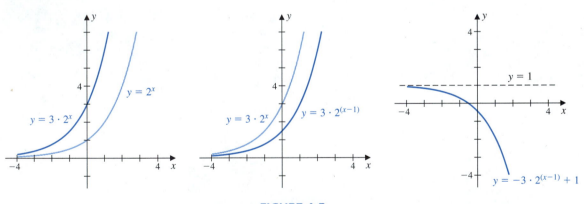

FIGURE 4.5

The exponential functions are even more closely related than their graphs would lead us to believe. Each exponential function can be converted into a scaled function of any single exponential function. Suppose, for example, that we want to write an arbitrary exponential function $f(x) = a^x$ in terms of an exponential function with base 2. The graph of $y = a^x$ satisfies the horizontal line test, so each of these exponential functions is one-to-one. Since the range of $f(x) = a^x$ is $(0, \infty)$, there is a unique real number k with $a = 2^k$. For this value of k we have

$$f(x) = a^x = \left(2^k\right)^x = 2^{kx}.$$

Every exponential function, then, is a scaled function of the exponential function with base 2. The choice of 2 as the fixed base was completely arbitrary; we could use as our base any positive number except 1.

The number we choose for the base of the **natural exponential function** is an irrational number denoted by the letter e. To see why this is the *natural* base requires some background work.

Throughout the book we have remarked that much of the study of calculus involves determining the limiting values of quotients of the form

$$\frac{f(x+h) - f(x)}{h} \qquad \text{as } h \to 0.$$

This number describes the slope of the curve at the point $(x, f(x))$.

For certain functions this limiting value is easy to determine. For example, if $f(x) = mx$ for some constant m, then

$$\frac{f(x+h) - f(x)}{h} = \frac{m(x+h) - mx}{h} = \frac{mx + mh - mx}{h} = \frac{mh}{h} = m,$$

regardless of the value of x and h. This is certainly reasonable since the graph of $f(x) = mx$ is a line with slope m.

If $f(x) = x^2$, we have

$$\frac{f(x+h) - f(x)}{h} = \frac{(x+h)^2 - x^2}{h} = \frac{x^2 + 2xh + h^2 - x^2}{h} = \frac{(2x+h)h}{h} = 2x + h,$$

which depends on the values of both x and h. As h approaches zero, the quotient approaches $2x$. Some illustrations of this result are shown in Figure 4.6.

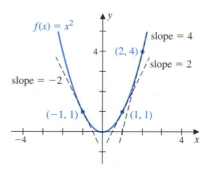

FIGURE 4.6

For an exponential function of the form $f(x) = a^x$, this quotient is

$$\frac{f(x+h) - f(x)}{h} = \frac{a^{x+h} - a^x}{h} = \frac{a^x a^h - a^x}{h} = a^x \left(\frac{a^h - 1}{h} \right).$$

The value of a that gives us the *natural* exponential function is the value for which

$$\frac{a^h - 1}{h} \to 1 \qquad \text{as } h \to 0.$$

So, the natural exponential function has the property that at each point on the graph, the slope of the graph is the same as the value of the function at the point, as shown in Figure 4.7.

The geometric interpretation of the *derivative* is the slope of the tangent line, or slope of the curve, at points on the curve. Understanding this geometric interpretation is key to understanding how the derivative is applied to many diverse problems in calculus.

A fundamental
property of the
exponential function
that is used
repeatedly in calculus
is the fact that the
slope of the graph at
a point of the
exponential function
is the value of the
function at that point.

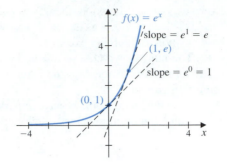

FIGURE 4.7

In Table 4.1 we see values of the quotient

$$\frac{a^h - 1}{h}$$

for various values of a and h.

TABLE 4.1

h	$a = 2$	$a = 3$	$a = 2.5$	$a = 2.75$	$a = 2.675$	$a = 2.7125$
0.1	0.717	1.161	0.960	1.064	1.034	1.049
0.01	0.695	1.105	0.921	1.017	0.989	1.003
0.001	0.693	1.099	0.917	1.012	0.984	0.998
0.0001	0.693	1.099	0.916	1.012	0.984	0.998

From this table it would appear that the number we seek is close to, but a little greater than the last entry, 2.7125. In fact, the unique value that satisfies our condition is the irrational number $e \approx 2.718281828459045\ldots$. That is, $f(x) = e^x$ has the property that

In calculus the
exponential function
that is of most
interest is the natural
exponential function.
In this section we
will concentrate on
the study of this
special function.

$$\frac{e^{x+h} - e^x}{h} = e^x \left(\frac{e^h - 1}{h} \right) \rightarrow e^x \cdot 1 = e^x, \qquad \text{as } h \rightarrow 0.$$

In addition, since

$$\frac{e^h - 1}{h} \approx 1 \qquad \text{for } h \text{ close to } 0$$

we have

$$e^h - 1 \approx h \qquad \text{and} \qquad e^h \approx 1 + h \qquad \text{for } h \text{ close to } 0.$$

These results imply that

$$(1 + h)^{1/h} \rightarrow e \qquad \text{as } h \rightarrow 0.$$

In Table 4.2 you can see that this expression does indeed appear to approach e as h approaches zero.

TABLE 4.2

h	0.1	0.01	0.001	0.0001	0.00001	0.000001
$(1 + h)^{1/h}$	2.593742	2.704814	2.716924	2.718146	2.718268	2.718281

We have once again used our intuitive notion of limit to argue that the number e can be realized as the value approached by $(1 + h)^{1/h}$ as h goes to 0. This is made precise in calculus.

Since $e \approx 2.71$ is between 2 and 3, the graph of $y = e^x$ lies between the graphs of $y = 2^x$ and $y = 3^x$, as shown in Figure 4.8. It passes thorough the points $(0, 1)$, $(1, e)$, and $(-1, 1/e)$. For any exponential function $f(x) = a^x$, a unique constant k exists with $f(x) = a^x = e^{kx}$. We will see in Section 4.3 how to precisely determine the specific constant k.

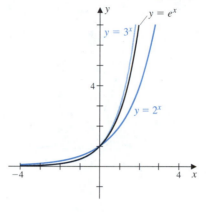

FIGURE 4.8

EXAMPLE 2   Sketch the graph of $f(x) = 2 - e^{x-1}$.

Solution   The graph of $y = e^{x-1}$ shown in Figure 4.9(a) is obtained by translating the graph of $y = e^x$ to the right 1 unit. The graph of $y = -e^{x-1}$ is the reflection about the x-axis of the graph of $y = e^{x-1}$, as shown in Figure 4.9(b). Translating this graph vertically upward two units gives the graph of $f(x) = 2 - e^{x-1}$, shown in Figure 4.9(c). ■

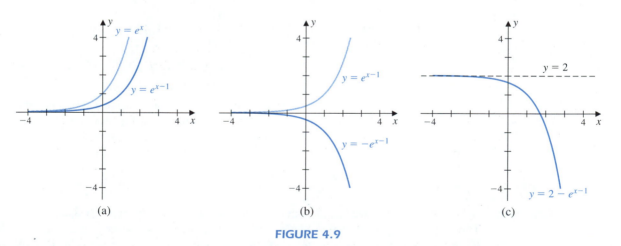

FIGURE 4.9

EXAMPLE 3   Use the graph of $y = e^x$ to determine the graph of $f(x) = e^x + e^{-x}$.

Solution   The arithmetic properties of exponentials imply that

$$e^{-x} = \frac{1}{e^x},$$

Combinations of exponential functions are quite easy to graph since they have little variational behavior.

so we first use the reciprocal graphing technique to determine the graph of $y = e^{-x}$. Both graphs pass through $(0, 1)$, and since e^x is always positive, so is e^{-x}. As $x \to \infty$, we have $e^x \to \infty$, and as $x \to -\infty$, we have $e^x \to 0$. So, as $x \to \infty$, we have $e^{-x} \to 0$, and as $x \to -\infty$, we have $e^{-x} \to \infty$. Finally, e^x is always increasing, so e^{-x} is always decreasing. As a consequence, the graph of $y = e^{-x}$ is a reflection about the y-axis of the graph of $y = e^x$, as shown in Figure 4.10(a).

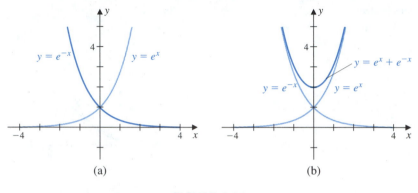

(a)   (b)

FIGURE 4.10

Next notice that we have $f(0) = e^0 + e^{-0} = 1 + 1 = 2$. Also,

$$\text{as } x \to \infty \qquad \text{we have} \qquad e^{-x} \to 0 \qquad \text{so} \qquad e^x + e^{-x} \to e^x,$$

and

$$\text{as } x \to -\infty \qquad \text{we have} \qquad e^x \to 0 \qquad \text{so} \qquad e^x + e^{-x} \to e^{-x}.$$

This result implies that the graph is similar to that shown in Figure 4.10(b). Notice that the graph is symmetric with respect to the y-axis since

$$f(-x) = e^{(-x)} + e^{-(-x)} = e^{-x} + e^x = f(x). \qquad \blacksquare$$

We cannot yet determine the exact value of the constant needed to change an arbitrary exponential function into a scaled natural exponential function, but the next example shows how we can use a graphing device to find a close approximation to the constant.

EXAMPLE 4   Approximate the value of k for which $2^x = e^{kx}$.

Solution   By letting $x = 1$, we see that the value of k that is required has the property that $2 = e^k$. The graph of $y = e^x$ shown in Figure 4.11(a) shows that it intersects $y = 2$ near 0.7. By zooming in on this area of the graph, we can refine our approximate value of k to 0.69, as shown in Figure 4.11(b). Figure 4.11(c) shows that the graphs of $y = 2^x$ and $y = e^{0.69x}$ are identical at this resolution. $\blacksquare$

We cannot yet solve $2^x = e^{kx}$ for k, but we can use a graphing device to estimate the value of k.

The natural exponential function has such wide application that modifications of its graph occur in many situations.

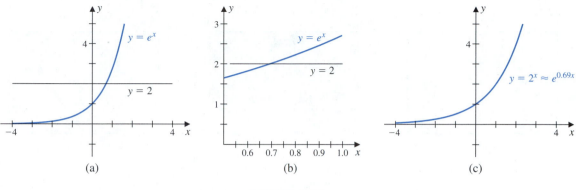

FIGURE 4.11

EXAMPLE 5   Sketch the graph of $f(x) = e^{-x^2}$.

Solution   Writing

$$f(x) = e^{-x^2} = \frac{1}{e^{x^2}},$$

we see that as x becomes large, $f(x)$ rapidly approaches 0. In addition, the graph is symmetric with respect to the y-axis, and $f(0) = 1$. The graph has the appearance of the *bell-shaped* curve shown in Figure 4.12. This a very familiar shape in statistical studies. ■

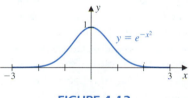

FIGURE 4.12

Graphing devices can be used to quickly sketch graphs in large viewing windows, allowing us to see the relative rates of growth of different functions.

We have seen that the natural exponential function grows rapidly for positive values of x. The next example gives a quantitative illustration of just how rapid this growth is.

EXAMPLE 6   Use a graphing device to compare the growth of $f(x) = e^x$ and $g(x) = x^4$ for large values of x.

Solution   If we use the viewing rectangle $[-5, 5] \times [-5, 5]$, we have the graphs shown in Figure 4.13(a). It appears from this figure that x^4 grows faster that e^x. In Figure 4.13(b) we have sketched the graph using the viewing rectangle $[0, 20] \times [0, 6800]$, which shows that there is another intersection point of the graphs and that e^x exceeds x^4 after this third intersection point. The final graph, in Figure 4.13(c), uses the viewing rectangle $[0, 20] \times [0, 100,000]$, in which it is clear that e^x far exceeds x^4 for values of x greater than 9.

An important feature of the natural exponential function is that it grows faster than x^n for any positive integer n.

By zooming in on the curves we can find that the three intersection points occur, when x is approximately -0.82, 1.43, and 8.61. ■

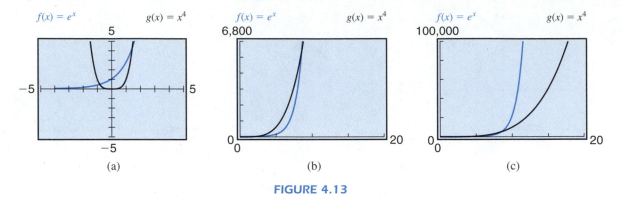

FIGURE 4.13

The natural exponential function is regularly required in physical situations, but calculus is needed for a full appreciation of its value. We give an introduction of this topic using compound interest as our model, but keep in mind that the applications of this behavior are much more far-reaching than this example suggests. In Section 4.4 we consider other applications.

Compound Interest

Suppose that we invest a sum A_0 in a savings account. How much we have in the account at the end of a time period depends on the interest rate i and on the number of times per year that the interest is compounded. In the past it was common to compound the interest only a few times a year, now most institutions compound interest daily. The more frequently the interest is compounded the faster the growth in the account, and the best of all possible situations occurs when the interest is compounded *continuously*. Let us examine the consequences of these various techniques.

If the interest is compounded yearly, with annual interest rate i, the amount in the account after 1 year is

$$A_y(1) = A_0(1 + i).$$

After 2 years we have the amount

$$A_y(2) = A_y(1)(1 + i) = A_0(1 + i)^2,$$

and so on. In general, we have

$$A_y(t) = A_0(1 + i)^t$$

in the account after t years.

If the interest is compounded semiannually, that is, twice a year, half the annual interest, $i/2$, is paid per period, and the number of periods doubles. Hence the amount after t years is

$$A_s(t) = A_0 \left(1 + \frac{i}{2}\right)^{2t}.$$

In a similar manner, monthly compounding produces the amount

$$A_m(t) = A_0 \left(1 + \frac{i}{12}\right)^{12t},$$

and daily compounding gives

$$A_d(t) = A_0 \left(1 + \frac{i}{365} \right)^{365t}.$$

In general, the interest i compounded n times a year produces

$$A_n(t) = A_0 \left(1 + \frac{i}{n} \right)^{nt}.$$

Suppose that we introduce the variable change $h = i/n$ into the formula for $A_n(t)$. Then $n = i/h$ and the arithmetic properties of exponents allow us to rewrite $A_n(t)$ as

$$A_n(t) = A_0(1 + h)^{(i/h)t} = A_0 \left((1 + h)^{1/h} \right)^{it}.$$

Recalling that

$$(1 + h)^{1/h} \longrightarrow e \qquad \text{as } h \to 0,$$

we have

$$A_n(t) \longrightarrow A_0 e^{it} \qquad \text{as } h \to 0.$$

That is, $A_n(t)$ approaches $A_0 e^{it}$ as the number $n = i/h$ of compounding periods increases without bound. This limiting term is said to be the amount when the interest is *compounded continuously* and is denoted by

$$A_c(t) = A_0 e^{it}.$$

It is larger than the amount produced by any compounding method but is quite close to that given by daily compounding, as the next example demonstrates.

EXAMPLE 7 Determine the value of a CD (Certificate of Deposit) in the amount of $1000 that matures in 6 years and pays 5% per year compounded annually, monthly, daily, and continuously.

Solution The decimal equivalent of 5% is 0.05, so the various compounding methods give

$$A_y(6) = 1000(1 + 0.05)^6 = \$1340.10,$$

$$A_m(6) = 1000 \left(1 + \frac{0.05}{12} \right)^{12(6)} = \$1349.02,$$

$$A_d(6) = 1000 \left(1 + \frac{0.05}{365} \right)^{365(6)} = \$1349.83,$$

and

$$A_c(6) = 1000 e^{0.05(6)} = \$1349.86. \qquad \blacksquare$$

Notice that there was very little difference in Example 7 between the amount produced by daily compounding and that given by continuously compounding. In addition, the exact compounding formulas are valid only for values of t that are integral multiples of the compounding period, whereas $A_c(t)$ is defined for all positive values of t. The next example shows how this can be useful for accurately approximating the future value in an account.

EXAMPLE 8   Determine the approximate length of time it takes an amount to double in value if it earns 9% compounded daily.

Solution   We first assume that instead of compounding daily, the account compounds continuously. Then the time, t, that it takes for the amount A_0 to double is

$$2A_0 = A_c(t) = A_0 e^{0.09t} \qquad \text{so} \qquad 2 = e^{0.09t}.$$

We cannot yet find the exact solution to this problem, but the graphs of $y = e^{0.09t}$ and $y = 2$ shown in Figure 4.14 indicate that we have $t \approx 7.7$.

We cannot yet find the solution to this example algebraically, but a graphing device can be used to see that a solution exists and to estimate its value.

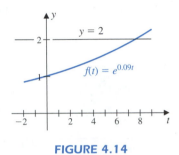

FIGURE 4.14

If we estimate conservatively, we might assume that it would take about 7.75 years, that is, 7 years 9 months, to double our investment at this rate. The exact value we would have after this time is

$$A_d(7.75) = A_0 \left(1 + \frac{0.09}{365}\right)^{365(7.75)} = 2.0086A_0,$$

so we are certainly in the ballpark.   ■

EXERCISE SET 4.2

In Exercises 1–12, sketch the graphs of the following.

1. $f(x) = -4^x$

2. $f(x) = 10^{-x}$

3. $f(x) = 2^x - 3$

4. $f(x) = 2^{x-3} - 4$

5. $f(x) = -e^x$

6. $f(x) = e^{2x}$

7. $f(x) = e^{x-2} + 1$

8. $f(x) = 3 - e^{-(x-1)}$

9. $f(x) = e^x - e^{-x}$

10. $f(x) = x + e^x$

11. $f(x) = e^{|x|}$

12. $f(x) = e^{-|x|}$

In Exercises 13–16, use a graphing device to approximate all solutions to the equation.

13. $e^{x-2} = x$

14. $e^x = x^2$

15. $e^{-x} = (x - 2)^2$

16. $xe^x = x^2 + 4x + 2$

In Exercises 17–20, use a graphing device to sketch the graphs of the functions and show intervals on which the function is increasing, intervals where it is decreasing, and local maximums and minimums.

17. $f(x) = xe^x$

18. $f(x) = \dfrac{e^x}{x}$

19. $f(x) = e^{-x^2 - x}$

20. $f(x) = e^{x^3 - x}$

21. Approximate the value of k for which $3^x = e^{kx}$.

22. a. Use a calculator to approximate the value of $(1 + 1/n)^n$ for $n = 1, 5, 10, 10^2, 10^3, 10^4, 10^5, 10^6$ and 10^7.

   b. Use a graphing device to plot $f(x) = (1 + 1/x)^x$ and $y = e$ in the viewing rectangle $[0, 30] \times [0, 3]$.

23. Use a graphing device to compare the rates of growth of the functions $f(x) = 2^x$ and $g(x) = x^5$ by graphing the two functions in the following viewing rectangles.

  a. $[-5, 5] \times [-5, 5]$

  b. $[0, 10] \times [0, 10^3]$

  c. $[0, 30] \times [0, 10^7]$

Approximate the solutions to $2^x = x^5$.

24. Use a graphing device to compare the rates of growth of $f(x) = e^x$ and $g(x) = x^{10}$ by graphing the functions together in several appropriate viewing rectangles. Approximate the solutions to $e^x = x^{10}$.

25. Use a graphing device to approximate all values of $x > 0$ for which $x^x < e^x$.

26. a. The hyperbolic cosine function is defined by

$$\cosh x = \frac{e^x + e^{-x}}{2}.$$

Use a graphing device to plot $f(x) = \cosh x$, $y = e^x/2$, and $y = e^{-x}/2$ on the same set of axes.

  b. The hyperbolic sine function is defined by

$$\sinh x = \frac{e^x - e^{-x}}{2}.$$

Use a graphing device to plot $f(x) = \sinh x$, $y = e^x/2$, and $y = -e^{-x}/2$ on the same set of axes.

27. Determine the value of a CD in the amount of $5000 that matures in 5 years and pays 6.5% per year compounded as indicated.

  a. Annually   **b.** Monthly

  c. Daily   **d.** Continuously

28. Suppose that $1000 is invested at 10% interest and the interest rate remains fixed for 8 years. Complete the following table.

Interest Compounded	Value after 8 years
Annually	
Semiannually	
Quarterly	
Monthly	
Weekly	
Daily	
Hourly	
Continuously	

4.3   LOGARITHM FUNCTIONS

We have seen that for each positive number $a \neq 1$, the exponential function $f(x) = a^x$ is one-to-one with domain $(-\infty, \infty)$ and range $(0, \infty)$. (See Figure 4.15.)

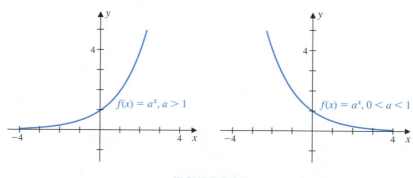

FIGURE 4.15

When $a > 1$ the function is always increasing, and the function is always decreasing when $0 < a < 1$. In either situation, the exponential function has an inverse, which we call the *logarithm function with base a*. It is defined by the following relationship.

The Logarithm Function with Base a

For each positive number $a \neq 1$ and x in $(0, \infty)$ we have

$$y = \log_a x \qquad \text{precisely when} \qquad x = a^y.$$

So

$$\log_a a^y = y \qquad \text{and} \qquad a^{\log_a x} = x$$

for each x in $(0, \infty)$ and each real number y.

EXAMPLE 1   Use the inverse relationship with the exponential functions to determine x if

a. $x = \log_3 81$ b. $\log_2(x^2 - 2x) = 3$.

Solution   a. The exponential-logarithm conversion implies that

$$x = \log_3 81 \Leftrightarrow 3^x = 81.$$

Since $81 = 3^4$, we have $x = 4$.

b. The conversion in this instance implies that

$$\log_2(x^2 - 2x) = 3 \Leftrightarrow x^2 - 2x = 2^3 = 8.$$

Solving this quadratic gives

$$0 = x^2 - 2x - 8 = (x - 4)(x + 2), \qquad \text{so } x = 4 \quad \text{or} \quad x = -2 \qquad \blacksquare$$

The graph of an inverse function is found by reflecting the graph of the function about the line $y = x$. So the graph of $y = \log_a x$ has the form shown in Figure 4.16(a) when $a > 1$ and has the form shown in Figure 4.16(b) when $0 < a < 1$. Notice, in particular, that for any value of a we have $\log_a 1 = 0$ and $\log_a a = 1$.

Notice how frequently this reflection property is used to find graphs of inverse functions.

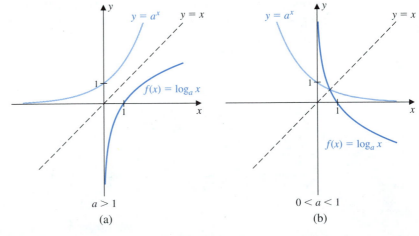

$a > 1$ $0 < a < 1$

(a)   (b)

FIGURE 4.16

For $a > 1$ we have

$$\log_a x \to -\infty \quad \text{as} \quad x \to 0^+ \qquad \text{and} \qquad \log_a x \to \infty \quad \text{as} \quad x \to \infty,$$

and for $0 < a < 1$ we have

$$\log_a x \to \infty \quad \text{as} \quad x \to 0^+ \qquad \text{and} \qquad \log_a x \to -\infty \quad \text{as} \quad x \to \infty.$$

EXAMPLE 2   Sketch the graphs of (a) $f(x) = \log_2 x$, and (b) $g(x) = \log_2(x - 1) + 3$.

Solution   a. We know that the general shape of the graph is as shown in Figure 4.16(a). Since

$$2 = 2^1, \qquad 4 = 2^2, \qquad 8 = 2^3, \qquad \text{and} \qquad \frac{1}{2} = 2^{-1}$$

the points $(1, 2)$, $(2, 4)$, $(3, 8)$ and $(-1, 1/2)$ lie on the graph of $y = 2^x$. As a consequence,

$$\log_2 2 = 1, \qquad \log_2 4 = 2, \qquad \log_2 8 = 3, \qquad \text{and} \qquad \log_2 \frac{1}{2} = -1,$$

and the points $(2, 1)$, $(4, 2)$, $(8, 3)$ and $(1/2, -1)$ lie on the graph of $f(x) = \log_2 x$, as shown in Figure 4.17.

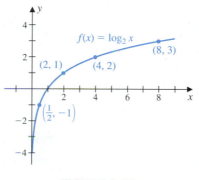

FIGURE 4.17

b. To sketch the graph of $g(x) = \log_2(x - 1) + 3$, we first translate the graph of $y = \log_2 x$ to the right 1 unit to obtain the graph of $y = \log_2(x - 1)$, as shown in Figure 4.18(a). Shifting the resulting graph 3 units upward gives the graph of $g(x) = \log_2(x - 1) + 3$, as shown in Figure 4.18(b). Notice that the graph has a vertical asymptote at $x = 1$, that the domain of g is the interval $(1, \infty)$, and the range is the set of all real numbers.

■

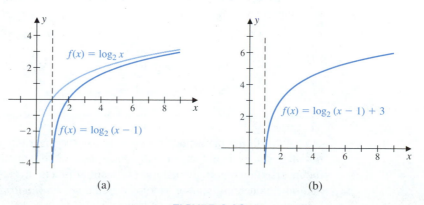

(a)   (b)

FIGURE 4.18

At the beginning of Section 4.1 we listed the arithmetic properties of exponents that hold for all positive numbers a and all real numbers r_1 and r_2:

$$a^{r_1+r_2} = a^{r_1}a^{r_2}, \qquad a^{r_1-r_2} = \frac{a^{r_1}}{a^{r_2}}, \qquad \text{and} \qquad (a^{r_1})^{r_2} = a^{r_1 r_2}.$$

Each of these rules has a logarithm equivalent that follows from the inverse relationship with the exponential functions. For example, if we let

$$\log_a x_1 = r_1 \qquad \text{and} \qquad \log_a x_2 = r_2,$$

then in exponential form these equations become

$$x_1 = a^{r_1} \qquad \text{and} \qquad x_2 = a^{r_2},$$

so

$$x_1 x_2 = a^{r_1} a^{r_2} = a^{r_1+r_2}.$$

But the inverse relationship between the logarithm and exponential function implies that we can rewrite this as

$$\log_a (x_1 x_2) = r_1 + r_2 = \log_a x_1 + \log_a x_2.$$

In a similar manner,

$$\frac{x_1}{x_2} = \frac{a^{r_1}}{a^{r_2}} = a^{r_1-r_2}$$

so

$$\log_a \left(\frac{x_1}{x_2} \right) = r_1 - r_2 = \log_a x_1 - \log_a x_2.$$

Finally, for each real number r we have

$$x_1^r = (a^{r_1})^r = a^{r_1 r} \Leftrightarrow \log_a x_1^r = r_1 r = \left(\log_a x_1 \right) r = r \log_a x_1.$$

In summary, we have the following arithmetic properties that are true for every logarithm function.

Arithmetic Properties of Logarithms

For each positive number $a \neq 1$, each pair of positive real numbers x_1, x_2, and each real number r we have

$$\log_a (x_1 x_2) = \log_a x_1 + \log_a x_2,$$

$$\log_a \left(\frac{x_1}{x_2} \right) = \log_a x_1 - \log_a x_2,$$

$$\log_a x_1^r = r \log_a x_1.$$

We will not have much occasion to use the logarithm functions that have a base a with $0 < a < 1$. In fact, we will concentrate on only one special logarithm function, the inverse function to the natural exponential function, $f(x) = e^x$. The inverse to the natural exponential function is known as the *natural logarithm function* and is written using the special notation $\ln x \equiv \log_e x$.

The Natural Logarithm Function

For each x in $(0, \infty)$ we have

$$y = \ln x \qquad \text{precisely when} \qquad x = e^y.$$

So

$$\ln e^y = y \qquad \text{and} \qquad e^{\ln x} = x$$

for each x in $(0, \infty)$ and each real number y.

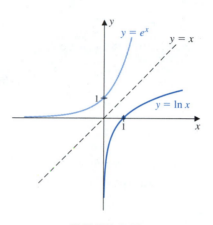

FIGURE 4.19

The arithmetic properties of logarithms are used in calculus to simplify complicated logarithm expressions.

The graph of the natural logarithm function is shown in Figure 4.19. The arithmetic logarithm rules imply that for each pair of positive real numbers x_1, x_2 and each real number r, we have the following properties.

Arithmetic Properties of Natural Logarithms

For each pair of positive real numbers x_1, x_2 and each real number r,

$$\ln(x_1 x_2) = \ln x_1 + \ln x_2, \qquad \ln\left(\frac{x_1}{x_2}\right) = \ln x_1 - \ln x_2, \qquad \text{and} \qquad \ln x_1^r = r \ln x_1.$$

EXAMPLE 3 Determine the values of x that satisfy each expression.

a. $\ln(x-1) + \ln(x-3) = 3 \ln 2$

b. $\ln \sqrt{\dfrac{1-x}{1+x}} = \dfrac{1}{2}$

Solution a. Since the equation can be written in the form

$$\ln(x-1) + \ln(x-3) - 3\ln 2 = 0,$$

we can use the arithmetic properties to rewrite the equation as

$$0 = \ln(x-1)(x-3) - \ln 2^3 = \ln \frac{(x-1)(x-3)}{8}.$$

Changing this equation to its equivalent exponential form and using the fact that $e^0 = 1$ produces

$$1 = \frac{(x-1)(x-3)}{8}, \qquad \text{or} \qquad 8 = (x-1)(x-3) = x^2 - 4x + 3.$$

As a consequence,

$$0 = x^2 - 4x - 5 = (x-5)(x+1), \qquad \text{so} \qquad x = 5 \quad \text{or} \quad x = -1.$$

But substituting $x = -1$ into the original equation gives a value at which the natural logarithm function is undefined, so $x = -1$ is an *extraneous* solution. The only true solution to the equation is $x = 5$. We can see that it is a solution since

$$\ln(5-1) + \ln(5-3) = \ln 4 + \ln 2 = \ln 2^2 + \ln 2 = 2\ln 2 + \ln 2 = 3\ln 2.$$

b. We can rewrite this equation as

$$\frac{1}{2} = \ln \sqrt{\frac{1-x}{1+x}} = \ln \left(\frac{1-x}{1+x} \right)^{1/2} = \frac{1}{2} \ln \frac{1-x}{1+x},$$

so

$$\ln \frac{1-x}{1+x} = 1 \qquad \text{and} \qquad \frac{1-x}{1+x} = e.$$

We can now solve this equation for x. Since

$$1 - x = (1+x)e = e + ex, \qquad \text{we have} \qquad 1 - e = x + ex = (1+e)x,$$

and

$$x = \frac{1-e}{1+e}.$$

To verify that this is indeed a solution, we have

$$\ln \sqrt{\frac{1 - \big((1-e)/(1+e)\big)}{1 + \big((1-e)/(1+e)\big)}} = \frac{1}{2} \ln \frac{2e/(1+e)}{2/(1+e)} = \frac{1}{2} \ln e = \frac{1}{2} \cdot 1 = \frac{1}{2}. \qquad \blacksquare$$

We can restrict our study of logarithms to the natural logarithms because of a relationship between exponential functions that we discovered in the Section 4.2. Recall that for any exponential function $f(x) = a^x$, there exists a number k with the property that

$$a^x = e^{kx}.$$

When we looked at this result in Section 4.2 we did not know how to find the exact value for k, although we did see how to use a graphing device to estimate k. With the arithmetic properties of the logarithm, we can determine the appropriate value of k.

Applying the natural logarithm to both sides of the equation $a^x = e^{kx}$ gives

$$\ln a^x = \ln e^{kx} \Leftrightarrow x \ln a = kx \ln e = kx \cdot 1 = kx.$$

Since this is true for all values of x,

$$k = \ln a.$$

This result permits us to express any exponential function $f(x) = a^x$ as

$$a^x = e^{(\ln a)x}.$$

This relationship between a general exponential function of the form $f(x) = a^x$ and the natural exponential function is critical for working with exponential functions in calculus. In fact, if you can remember only one fact about general exponential functions, this is the one you want.

The conversion formula gives the critical relationship between a^x and e^x. Make sure you understand this relationship in preparation for calculus.

General Exponential to Natural Exponential Conversion

For any positive real number a and every real number x, we have

$$a^x = e^{(\ln a)x}.$$

We can gain even more from this relationship. Suppose that we let $y = \log_a x$. Since

$$y = \log_a x \Leftrightarrow x = a^y = e^{(\ln a)y},$$

we have

$$\ln x = \ln \left(e^{(\ln a)y} \right) = (\ln a)y \ln e = (\ln a)y \cdot 1 = \ln a \log_a x.$$

As a consequence, every logarithm function is simply a multiple of the natural logarithm function.

General Logarithm to Natural Logarithm Conversion

For every positive real number $a \neq 1$ and every $x > 0$, we have

$$\log_a x = \frac{\ln x}{\ln a}.$$

This result implies that all logarithm functions are essentially equivalent, and we can choose any one of them for our basic function. If we have logarithm functions with any bases a and b, then the logarithmic conversion property implies that for every positive real number x we have

$$\log_b x = \frac{\ln x}{\ln b} = \frac{\ln x}{\ln b} \cdot \frac{\ln a}{\ln a} = \frac{\ln a}{\ln b} \cdot \frac{\ln x}{\ln a} = \frac{\ln a}{\ln b} \log_a x.$$

This shows that every logarithm function is a constant multiple of any other logarithm function.

For calculus applications the natural logarithm function is basic, but in some instances it is more useful to choose the logarithm with base 10, called the *common* logarithm. Because computers represent numbers using the binary system, which involves only the digits 0 and 1, it is also common in computer science to use the logarithm with base 2.

EXAMPLE 4 Suppose that $0 < a < 1$. Compare the graph of $y = \log_a x$ to the graph of $y = \log_{1/a} x$.

Solution Since $0 < a < 1$, we have $1 < 1/a$, and

$$\log_{1/a} x = \frac{\ln x}{\ln(1/a)} = \frac{\ln x}{\ln 1 - \ln a} = \frac{\ln x}{0 - \ln a} = -\frac{\ln x}{\ln a} = -\log_a x.$$

As a consequence, the graph of $y = \log_a x$ is just the reflection about the x-axis of the graph of $y = \log_{1/a} x$. Some specific examples are illustrated in Figure 4.20.   ∎

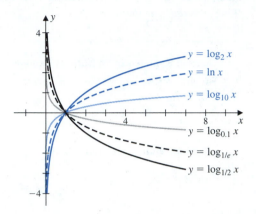

FIGURE 4.20

EXERCISE SET 4.3

In Exercises 1–12, evaluate the expression.

1. $\log_3 3^5$ **2.** $\log_2 32$

3. $\log_4 64$ **4.** $\log_8 8^{10}$

5. $\log_{10} 0.001$ **6.** $\log_{10} 10,000$

7. $\log_4 2$ **8.** $\log_9 3$

9. $e^{\ln 5}$ **10.** $5^{\log_5 6}$

11. $\ln e^{\pi}$ **12.** $e^{2 \ln \pi}$

In Exercises 13–20, use the properties of logarithms to simplify the expression so that the result does not contain logarithms of products, quotients, or powers.

13. $\ln x(x + 2)$ **14.** $\ln \dfrac{1}{x}$

15. $\log_2(2x - 1)^5$ **16.** $\log_3 \dfrac{x^4}{x + 1}$

17. $\ln \dfrac{3x^2}{(x + 3)^4}$ **18.** $\ln \dfrac{x\sqrt[3]{x^2}}{(x + 2)^3}$

19. $\log_3 \dfrac{(3x + 2)^{3/2}(x - 1)^3}{x\sqrt{x + 1}}$ **20.** $\ln \sqrt{x\sqrt{x + 1}}$

In Exercises 21–40, use the properties of logarithms to solve the equation for x.

21. $\log_3 x = 4$ **22.** $\log_2 x = 5$

23. $\log_2(3x - 4) = 3$ **24.** $\log_3(2 - x) = 2$

25. $\log_x 4 = 2$ **26.** $\log_x 3 = \dfrac{1}{3}$

27. $4^x = 3$ **28.** $5^{2x-1} = 2$

29. $\ln(2 - x) = 4$ **30.** $1 - \ln(3x + 2) = 0$

31. $\ln 2 + \ln(x + 1) = \ln(4x - 7)$

32. $2 \ln x = \ln 4 + \ln(x + 3)$

33. $\ln x + \ln(x - 1) = \ln 2$

34. $2 \ln x = \ln(7x - 6) - \ln 2$

35. $\ln(2x - 1) - \ln(x - 1) = \ln 5$

36. $2 \ln(x + 2) - \ln x = \ln 8$

37. $\log_3(2x^2 + 17x) = 2$ **38.** $\log_2(5x^2 - 8x) = 2$

39. $e^{2x} = 3^{x-4}$ **40.** $3e^{-x} = 4^{3x-1}$

In Exercises 41–50, sketch the graph of the function.

41. $y = \log_2(x - 3)$ **42.** $y = -\log_2(x + 4)$

43. $y = \log_3(x - 2) + 1$ **44.** $y = 2 - \log_2(x - 1)$

45. $y = \ln(x + 1)$ **46.** $y = -2\ln(x - 1)$

47. $y = \ln(-x)$ **48.** $y = \ln(3 - x)$

49. $y = |\ln x|$ **50.** $y = \ln |x|$

In Exercises 51–54, use a graphing device to sketch the graph of the function, showing asymptotes and the approximate location of local maximum and minimum points.

51. $f(x) = \ln(4 - x^2)$ **52.** $f(x) = x^2 - \ln x$

53. $f(x) = \dfrac{\ln x}{x}$ **54.** $f(x) = \ln|x^2 - 1|$

55. Let $f(x) = a + \ln x$ and $g(x) = \sqrt[n]{x}$ for a positive integer n and a real number a. Use a graphing device to sketch the graphs of f and g for different values of a and n, and determine which functions grow more rapidly as $x \to \infty$.

56. Determine the value of k for which $3^x = e^{kx}$, and compare this result to the approximation found in Exercise 21 of Section 4.2.

57. Determine the length of time it takes an amount to double in value if it earns 9% compounded continuously.

58. Determine the length of time it takes an initial investment to triple in value if it earns 10% compounded continuously.

4.4   EXPONENTIAL GROWTH AND DECAY

In this section you will see several real-world applications typical of the type of problem considered in calculus. Calculus is needed to get you started, but getting to the final solution requires only understanding the natural exponential and logarithm functions.

In many natural settings, quantities grow or decay at a rate that is approximately proportional to the amount of the quantity that is present. For example, the rate at which some cultures of bacteria and animal populations increase is proportional to the size of the population, at least over short time periods. The mass of a radioactive substance decays, or decreases with time, at a rate proportional to the mass. Using calculus it can be shown that when the rate of growth or decay of some quantity at a given instant is proportional to the amount present at that instant, then the quantity present at any time is related to the exponential function.

Let Q_0 denote the initial amount of a quantity and $Q(t)$ be the amount at time t. If

$$Q(t) = Q_0 e^{kt},$$

we say Q *grows exponentially* if $k > 0$, and *decays exponentially* if $k < 0$. The constant k is called the *constant of proportionality*.

In Section 4.2, we saw that if interest is compounded continuously at an interest rate i, then an initial investment of A_0 dollars is worth

$$A(t) = A_0 e^{it}$$

dollars after t years. This same type of growth occurs in many different settings.

EXAMPLE 1   A culture of cells is observed to triple in size in 2 days. Assuming exponential growth, how large will the culture be in 5 days?

Solution   Let the initial amount be denoted by Q_0. Since the size of the population triples after $t = 2$ days,

$$Q(2) = 3Q_0.$$

Using the formula for exponential growth, $Q(t) = Q_0 e^{kt}$, we have,

$$3Q_0 = Q(2) = Q_0 e^{2k}, \qquad \text{so} \qquad e^{2k} = 3.$$

Taking the natural logarithm of both sides of the last equation gives

$$2k = \ln 3, \qquad \text{and} \qquad k = \frac{\ln 3}{2} \approx 0.549.$$

As a consequence, the quantity at time t is

$$Q(t) = Q_0 e^{t(\ln 3)/2}.$$

After 5 days the size of the population is

$$Q(5) = Q_0 e^{5(\ln 3)/2} \approx Q_0 e^{2.747} \approx 15.6 Q_0,$$

or approximately 15.6 times its original size.   ∎

In Example 1, the size of the population at any time t was found to be

$$Q(t) = Q_0 e^{t(\ln 3)/2}.$$

It is at times useful to rewrite this expression using the inverse relation between the exponential and logarithmic functions. In this case we have

$$Q(t) = Q_0 e^{t(\ln 3)/2} = Q_0 \left(e^{\ln 3}\right)^{t/2} = Q_0 (3)^{t/2} = Q_0 \left((3)^{1/2}\right)^t = Q_0 \left(\sqrt{3}\right)^t.$$

EXAMPLE 2   Table 4.3 gives the population in the United States for the years 1930 through 1990. Assume the population changes by an amount proportional to the amount of population present. Use the population figures in (a) 1930 and 1940 and (b) 1970 and 1980 to predict the population in the years 1990 and 2000.

TABLE 4.3

Year	Population (in millions)
1930	123
1940	131
1950	150
1960	179
1970	203
1980	226
1990	250

Solution   a. Suppose that we let $Q(t)$ represent the population t years after 1930. Since the initial population is the 1930 population, $Q(0) = 123$. To find the proportionality constant k, we can use the population 10 years later, in 1940. That is,

$$131 = Q(10) = 123e^{10k}, \qquad \text{so} \qquad \frac{131}{123} = e^{10k}, \qquad \text{and} \qquad k = \frac{1}{10} \ln \frac{131}{123}.$$

These models of population growth are quite crude. You will probably see better models when you study calculus and differential equations.

Hence

$$Q(t) = 123e^{[(1/10)\ln(131/123)]t} \approx 123e^{0.00630t}.$$

The years 1990 and 2000 correspond to $t = 60$ and $t = 70$, respectively, so the populations are approximately

$$Q(60) \approx 180 \text{ million} \qquad \text{and} \qquad Q(70) \approx 191 \text{ million}.$$

To plot the original data points and this exponential approximation on the same set of axes, we shift the exponential function so that it is 0 when $t = 1930$. That is, we plot

$$y = 123e^{[(1/10)\ln(131/123)](t-1930)}$$

The graph and data points are shown in Figure 4.21. Notice that this function gives a poor approximation to the data after 1950.

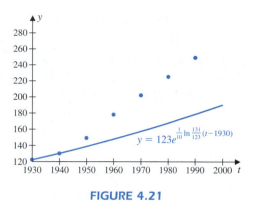

FIGURE 4.21

A graphing device can quickly plot the original data points and the function used to fit the data in this example. Figures 4.21 and 4.22 give an idea of how well the functions found in parts (a) and (b) predict the world population.

b. If we instead use the 1970 and 1980 figures to generate the exponential function and let $Q(t)$ now represent the population t years after 1970, we have

$$Q(t) = 203e^{[(1/10)\ln(226/203)]t} \approx 203e^{0.0107t}.$$

The new estimates for the populations for 1990 and 2000 are approximately

$$Q(20) \approx 252 \text{ million} \quad \text{and} \quad Q(30) \approx 280 \text{ million}.$$

Figure 4.22 shows the original data points along with the shifted graph

$$y = 203e^{[(1/10)\ln(226/203)](t-1970)}.$$

This function gives a much better approximation to the later data points, one that is more in line with the U. S. census prediction for the year 2000. It is, however, a poor model of the population before 1960. ∎

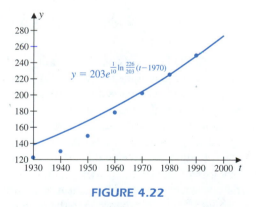

FIGURE 4.22

The growth of population is generally only approximated by an exponential function since the growth rate is dependent on additional conditions such as the availability of a food supply and pressures due to overcrowding. The next example concerns the decay of a radioactive substance. The exponential function gives a true quantitative picture in this situation.

EXAMPLE 3   The radioactive isotope strontium 90 has a half-life of 29.1 years, which is the time it takes for one-half of the original amount to decay to another substance.

a. How much strontium 90 will remain after 20 years from an initial amount of 300 kg?

b. How long will it take for 80% of the original amount to decay?

Solution   Let $Q(t)$ be the amount that remains after t years. Then $Q(0) = 300$, and, since the half-life is 29.1 years, $Q(29.1) = 150$. Using these two pieces of information, we can find the proportionality constant k. Since

$$Q(t) = Q_0 e^{kt} = 300e^{kt},$$

substituting $t = 29.1$ gives

$$150 = 300e^{29.1k}, \qquad \text{so} \qquad e^{29.1k} = \frac{1}{2}.$$

Thus,

$$29.1k = \ln \frac{1}{2} = \ln 1 - \ln 2 = -\ln 2, \qquad \text{so} \qquad k = -\frac{\ln 2}{29.1} \approx -0.0238,$$

and

$$Q(t) = 300e^{-[(\ln 2)/29.1]t} \approx 300e^{-0.0238t}.$$

a. Setting $t = 20$ gives

$$Q(20) \approx 300e^{-0.0238(20)} = 300e^{-0.476} \approx 300(0.621) = 186 \text{ kg.}$$

b. When 80% of the original amount has decayed, there will be 20%, or 60 kg, remaining. To find the time t when 60 kg remain we solve $Q(t) = 60$ for t. That is,

$$300e^{-0.0238t} = Q(t) = 60, \qquad \text{or} \qquad e^{-0.0238t} = 0.2.$$

Using the natural logarithm function to simplify this equation gives

$$-0.0238t = \ln 0.2, \qquad \text{or} \qquad t = -\frac{\ln 0.2}{0.0238} \approx -\frac{-1.61}{0.0238} = 67.6 \text{ years.} \quad ■$$

At the beginning of the chapter we discussed how scientists can determine the approximate age of ancient objects by a technique called radioactive dating. All living tissue contains carbon isotopes. Approximately 98.89% of this carbon consists of the stable isotope $^{12}_{6}\text{C}$, and most of the remainder is the stable isotope $^{13}_{6}\text{C}$, but a small and accurately measurable amount, about one part in one trillion, is the radioactive isotope $^{14}_{6}\text{C}$. The $^{14}_{6}\text{C}$ isotope decays over time to produce $^{14}_{7}\text{N}$.

Radioactive carbon dating makes the assumption that the percentage of various isotopes of carbon in all living things has remained constant throughout history. The date at which an ancient organism ceased to live can be estimated by comparing the current $^{14}_{6}\text{C}$ to $^{12}_{6}\text{C}$ proportion to the original proportion of these isotopes. The half-life of $^{14}_{6}\text{C}$ is about 5730 years.

EXAMPLE 4   Estimate the age of the Ice Man discussed in the opening of this chapter, assuming that 54.6% of the original amount of $^{14}_{6}\text{C}$ remained at the time of the discovery.

Solution   If $Q(t)$ represents the amount of $^{14}_{6}\text{C}$ present t years after the man died, then the amount originally present is $Q(0)$, and $Q(t)$ has the form

$$Q(t) = Q(0)e^{kt}.$$

Since the half-life of $^{14}_{6}C$ is 5730 years we have

$$\frac{1}{2}Q(0) = Q(5730) = Q(0)e^{5730k}.$$

Thus,

$$5730k = \ln\frac{1}{2} = \ln 1 - \ln 2 = -\ln 2, \qquad \text{so} \qquad k = -\frac{\ln 2}{5730} \approx -0.000121,$$

and

$$Q(t) = Q(0)e^{-0.000121t}.$$

Since 54.6% of the $^{14}_{6}C$ is currently present, we need to determine t so that

$$0.546Q(0) = Q(0)e^{-0.000121t}, \qquad \text{that is,} \qquad t = -\frac{\ln 0.546}{0.000121} \approx 5001.$$

Consequently, the Ice Man lived about 5000 years ago, around 3000 B.C.   ∎

EXERCISE SET 4.4

1. A bacteria culture starts with 1000 bacteria, and the population doubles every 4 h.

  a. Find an expression for the number of bacteria after t hours.

  b. Find the number of bacteria that will be present after 7 h.

  c. When will the population reach 20,000?

2. A bacteria culture starts with 500 bacteria and 5 h later has 4000 bacteria.

  a. Find an expression for the number of bacteria after t hours.

  b. Find the number of bacteria that will be present after 6 h.

  c. When will the population reach 15,000?

  d. How long does it take the population to double in size?

3. Under ideal conditions, a cell of the bacteria *Escherichia coli*, commonly found in human intestine, divides to create two cells in approximately 22 min. Assume the initial population is 200 cells.

  a. Find an expression for the number of cells after t minutes.

  b. Find the number of cells that will be present after 10 h.

  c. When will the population reach 10,000 cells?

4. The radioactive isotope thorium 234 has a half-life of approximately 578 h.

  a. If a sample has a mass of 50 mg, find an expression for the mass after t hours.

  b. How much will remain after 100 h?

  c. When will the initial mass decay to 10 mg?

  d. Use a graphing device to sketch the graph of the mass function.

5. If 200 g of a radioactive substance decays to 180 g in 2 years, find the half-life of the substance.

6. If a culture of bacteria doubles in size every 2 h, how long will it take to triple in size?

7. Find the half-life of a radioactive substance that decays by 3% in 5 years.

8. A certain radioactive substance has a half-life of 8 years.

  a. How much of a 200-g sample will remain after 15 years?

  b. How long will it take for 90% of the sample to decay?

9. The table gives estimates of the world population, in millions from 1950 to 1990, taken from the 1997 World Almanac.

Year	Population
1950	2513
1960	3027
1970	3678
1980	4478
1990	5321

a. Use the exponential model and population figures from 1950 and 1960 to predict the world population in the years 2000 and 2050.

b. Use the exponential model and the population figures from 1980 and 1990 to predict the world population in the years 2000 and 2050.

c. Use a graphing device to plot the original population data points and the exponential models given in (a) and (b).

10. Newton's Law of Cooling states that when an object at initial temperature T_0 is introduced into a medium of temperature T_m, where $T_m \neq T_0$, the rate at which the object changes temperature is proportional to the difference between the temperature of the object and the temperature of its surrounding medium. Using calculus it can be shown that the temperature of the object at time t is

$$T(t) = T_m + (T_0 - T_m)e^{kt}.$$

An outdoor thermometer reading $-3°C$ is brought into a room at $20°C$. One minute later the thermometer reads $5°C$. How long will it take to reach $19.5°C$?

11. A body was found floating face down in a lake. When the body was taken from the water at 11:50 A.M., its temperature was $66°F$. The temperature of the body when first found at 11:00 A.M. was $67°F$. The lake has a constant temperature of $62°F$ and the body of the victim was a normal $98.6°F$ before going into the water. Assuming that Newton's Law of Cooling applies, when did the victim die?

12. The rate of change of air pressure P with respect to altitude h is proportional to P, when the temperature is constant. Suppose at sea level the pressure is 1.01×10^5 pascals (Pa), and at altitude $h = 2$ km the pressure is 8.08×10^4 Pa. Find the atmospheric pressure at 4 km.

REVIEW EXERCISES FOR CHAPTER 4

In Exercises 1–12, sketch the graph of the function.

1. $f(x) = 2^{x-1} - 3$

2. $f(x) = 1 - 3^{2-x}$

3. $f(x) = e^{-x} + 1$

4. $f(x) = e^{x-2}$

5. $f(x) = 3e^{1-x}$

6. $f(x) = -2e^{x+1} + 1$

7. $f(x) = 2\ln x$

8. $f(x) = \ln(x - 2)$

9. $f(x) = 3 - \log_2(x + 1)$

10. $f(x) = \log_{10}(3 - x) + 2$

11. $f(x) = e^{-x^2 + 2x - 1}$

12. $f(x) = \ln x^{-2}$

In Exercises 13–20, evaluate the expression without using a calculator.

13. $\log_5 1$

14. $\log_{10} 0.000001$

15. $2^{\log_2 15}$

16. $\log_3 \dfrac{1}{81}$

17. $\log_9 3$

18. $\log_2 256$

19. $e^{3\ln 4}$

20. $\log_5 e^{-2\ln 5}$

In Exercises 21–24, rewrite the expression so that the result does not contain logarithms of products, quotients, or powers.

21. $\ln \dfrac{3x^2}{\sqrt{x - 1}}$

22. $\log_2 \left(\dfrac{x^2 - 1}{x^2 - 4} \right)$

23. $\log_{10} \dfrac{\sqrt{x + 1}\sqrt[3]{x - 1}}{x(x + 3)^{5/2}}$

24. $\ln \sqrt{\dfrac{x\sqrt{x + 1}}{x + 2}}$

In Exercises 25–28, rewrite the expression as a single logarithm.

25. $\ln x + \dfrac{1}{3}\ln(x + 1) - 2\ln(x - 1)$

26. $\dfrac{1}{2}\ln(2x + 1) + \ln(x - 1) - \ln(x^2 + 1)$

27. $3\ln(x^3 + 2) + \ln 5 - \dfrac{1}{2}\ln(x^5 - 1)$

28. $\dfrac{3}{2}\ln(x^2 - 2) - 2\ln(x + 1)$

In Exercises 29–36, determine the value of x without using a calculator.

29. $\ln(2x - 3) = 4$

30. $e^{3x-4} = 5$

31. $\ln(2x - 1) + \ln(3x - 2) = \ln 7$

32. $\ln(x - 1) - \ln(x - 3) = 1$

33. $3^x \cdot 5^{x-2} = 3^{4x}$

34. $3 \cdot 4^x = 2^{2x+1}$

35. $2e^x x^2 - e^x x = e^x$

36. $x \ln x - x = 0$

In Exercises 37–40, use a graphing device to approximate the solution to the inequality.

37. $e^x > x^4$

38. $\ln x < 2x - 3$

39. $e^{x-1} - 3 < x^5$

40. $\ln x^2 > x^3 - 2x^2 - x - 2$

41. Let $f(x) = x^2 e^{1-x^2}$. Use a graphing device to determine the intervals where the function is increasing, where it is decreasing, and any local maximums and minimums.

42. How long does it take for an amount of money to double if it is deposited at 6%, compounded continuously?

43. The radioactive isotope $^{225}_{92}\text{U}$ has a half-life of 8.8×10^8 years.

    a. How much of a 1-g sample will decay after 1000 years?

    b. How long will it take for 90% of the mass to decay?

44. A bacteria culture is known to grow at a rate proportional to the amount present. After 1 h, 1000 bacteria are present; after 4 h, 3000 bacteria are present.

    a. Find an expression for the number of bacteria at any time t.

    b. When will the population reach 20,000?

45. Determine the value of a CD in the amount of $10,000 that matures in 8 years and pays 10% per year compounded as indicated.

    a. Annually   **b.** Monthly

    c. Daily   **d.** Continuously

CHAPTER 4: CALCULUS PREVIEW EXERCISES

In Exercises 1–6, use a graphing device to sketch the graph of each pair of functions and determine when $f = g^{-1}$.

1. $f(x) = 2 \ln x; \quad g(x) = e^{x/2}$

2. $f(x) = \ln \dfrac{x}{2}; \quad g(x) = e^{2x}$

3. $f(x) = \ln|x|; \quad g(x) = e^{|x|}$

4. $f(x) = -\ln x; \quad g(x) = e^{-x}$

5. $f(x) = 1 + \ln x; \quad g(x) = e^{x-1}$

6. $f(x) = 2 \ln x; \quad g(x) = \dfrac{1}{2}e^x$

7. Use a graphing device to sketch the graph of $f(x) = a^x$ for different values of $a > 0$. Discuss the effect a has on the growth rate of a^x as $x \to \infty$.

8. Use a graphing device to sketch the graph of $f(x) = \log_a x$ for different values of $a > 0$. Discuss the effect a has on the growth rate of $\log_a x$ as $x \to \infty$.

9. Use a graphing device to compare the growth rates of $f(x) = x^n$ and $g(x) = a^x$ as $x \to \infty$. Use various positive values of n and a.

10. Use a graphing device to investigate how the number of intersections of the graphs of $y = a^x$ and $y = x^b$ is affected by the values of the positive integers a and b.

11. Use a graphing device to compare the long-term growth rates of the following functions as $x \to \infty$. Arrange them in order according to increasing growth rates:

$$\ln x, \qquad x^x, \qquad e^{3x}, \qquad x^{20},$$

$$x^{-4}, \qquad x^{10}e^{-x}, \qquad x^{1/20}, \qquad \frac{e^{6x}}{x^8}.$$

12. Explain how the graph of $y = 3e^{x-2}$ can be obtained from $y = e^x$ using only a horizontal translation.

13. Explain how the graph of $y = 3e^{x-2}$ can be obtained from $y = e^x$ using only vertical scaling.

14. Explain how the graph of $y = 3 + \ln 2x$ can be obtained from $y = \ln x$ using only horizontal scaling.

15. Explain how the graph of $y = 3 + \ln 2x$ can be obtained from the graph of $y = \ln x$ using only vertical translation.

16. Let $f(x) = (\ln x)^\alpha$ and $g(x) = x^\beta$, α and β any positive real numbers. Use a graphing device to sketch the graph of

$$y = \frac{f(x)}{g(x)}$$

for various values of α and β. Compare the growth rates of $f(x)$ and $g(x)$ as $x \to \infty$.

17. Use a graphing device to investigate the family of curves $y = a \ln(bx)$, where a is any real number and $b > 0$. How does the parameters a and b affect the curve?

18. Bankers often approximate the time it takes to double the amount of an investment made at a fixed interest rate by dividing the percent of the annual interest into 70. For example, $10,000 invested at 8.75% per year will become $20,000 in approximately $70/8.75 = 8$ years. Explain why this is a reasonable estimate.

19. Archaeologists find an animal bone and estimate that 78% of the original radioactive $^{14}_{6}\text{C}$ remains. Determine the age of the bone.

Chapter

CONIC SECTIONS, POLAR COORDINATES, AND PARAMETRIC EQUATIONS

5

CALCULUS CONNECTIONS

Much of calculus concerns the study of curves that occur in our physical world. By determining equations and functions that describe the curves, we can better examine their properties and those of the physical objects they represent. Take, as an example, some of the physical problems that are associated with the orbiting satellites that are a part of the global positioning system. These satellites transmit signals that permit objects to be located to a degree of accuracy only dreamed of a few years ago. The present applications of the system range from incredibly accurate ship and aircraft navigation to the scheduling of trucks on our highways to the allocation of personnel on some of the world's major airports. These applications have only scratched the surface, the effect of this precise locating system will probably not be fully realized for a number of years.

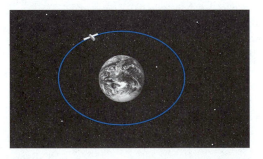

Satellites move in an elliptical orbit about Earth, just as Earth moves in an elliptical orbit about the sun. The signals that the satellites relay to Earth are collected in dishes that have a parabolic shape, and the satellite's location scheme is based on properties of hyperbolas. These three types of curves, the parabola, the ellipse, and the hyperbola, are the conic sections, curves we study in the first part of this chapter.

Curves that arise from rotating a circle about an object occur frequently in physical situations. The path of a tooth on a gear that is turning on another gear is one example, and a robot that rotates to paint or weld a car on an assembly line is another. Curves described in this manner are often difficult to represent using the familiar rectangular

coordinate system. For that matter, even the common circle cannot be represented by a function in the rectangular coordinate system. Rectangular equations to represent curves that spiral or overlap, such as the one shown in the figure, would be incredibly complicated.

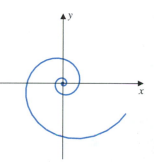

To better describe these curves we use the polar coordinate system, which involves a circular rather than rectangular method for describing points in the plane. We examine this system later in the chapter.

Some commonly occurring curves cannot be easily represented in a direct manner using either the rectangular or polar coordinate systems, but they can be described by introducing an additional variable called a *parameter*. A problem as simple as describing a specific point on a bicycle wheel as the bicycle moves down a road has this form. This type of representation is particularly important in physics, where the position of an object is often described using a time parameter. Paths of objects as they travel through space are

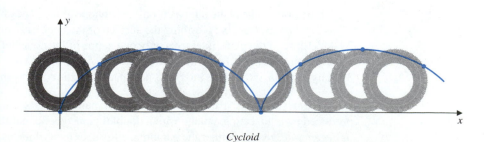

Cycloid

almost always of this form, but the discussion of curves in space will be postponed for a year or so, until you are in the final term of calculus. In this chapter we give just an introduction to the notion of parametric representation of curves.

5.1   INTRODUCTION

The general **quadratic equation** in x and y has the form

$$Ax^2 + Bxy + Cy^2 + Dx + Ey + F = 0,$$

where A, B, C, D, E, and F are constants. The graphs of these equations are called **conic sections**, or simply **conics**, since they were historically developed by considering the various curves generated when a double-napped cone (See Figure 5.1.) is cut by planes. These are the first natural extensions of the general equation of a line, which has the form $Ax + By + C = 0$, as we saw in Section 1.7.

FIGURE 5.1

The conic sections are basic graphs, some of whose special cases we have looked at often in this book. For example, the graphs of the parabola with equation $y = x^2$, and the circle having equation $x^2 + y^2 = 1$ are conic sections.

There are three basic distinct figures that result from the intersection of a plane with the double-napped cone:

1. A **parabola**, if the plane intersects only one nappe of the cone and is parallel to one of its *generators*, (the lines forming the surface of the cone), as shown in Figure 5.2(a).
2. An **ellipse**, if the plane intersects only one nappe of the cone but is not parallel to one of its generators, as shown in Figure 5.2(b).
3. A **hyperbola**, if the plane intersects both nappes of the cone, as shown in Figure 5.2(c).

Figures such as a point, circle, line, or pair of intersecting lines can also be produced by intersecting a plane with a double-napped cone, but these are only special, or *degenerate*, cases of either the parabola, ellipse, or hyperbola, as shown in Figure 5.3.

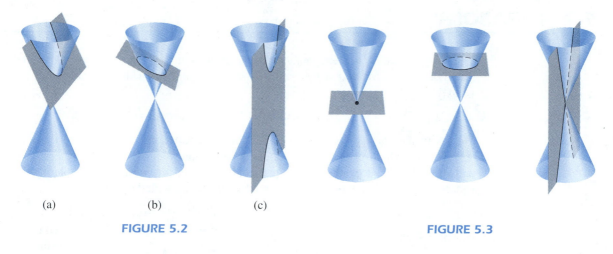

(a)   (b)   (c)

FIGURE 5.2   **FIGURE 5.3**

In calculus you will be expected to recognize equations describing parabolas, circles, ellipses, and hyperbolas and to be familiar with some of their applications. This chapter will give you a background in the basic properties and applications of these important curves.

This geometric approach to the study of conics leads to a number of important applications of these curves, which were first studied extensively by the ancient Greeks. The mathematician Apollonius, in the third century B.C., wrote eight volumes on the subject.

In the 16th century Galileo found that the path of a projectile fired upward at an angle is a parabola. In 1609, German mathematician and astronomer Johannes Kepler predicted that the planets and other objects in our solar system orbit the sun in elliptic orbits. Parabolas and hyperbolas are used in the construction of reflecting telescopes. The applications are extensive and encompass many areas, including engineering, physics, astronomy, architecture, and optics.

Our study will concentrate on the different types of curves that are generated by quadratic equations rather that on the intersection properties with the cone. In the first three sections we consider conics that occur as graphs of the quadratic equation

$$Ax^2 + Cy^2 + Dx + Ey + F = 0,$$

Most of the curves considered so far have been graphs of functions, but many applications in calculus involve curves that are not given this way. Polar coordinates are useful in plotting curves in the plane. Representing curves parametrically allows the description of many complicated curves.

obtained by setting $B = 0$ in the general quadratic equation. The more general situation where $B \neq 0$ is postponed to the final section of the chapter. In Section 5.2 we look at parabolas, curves produced when one of A or C is zero, but not both. Section 5.3 concerns ellipses, which occur when A and C are both nonzero and of the same sign. Section 5.4 considers the situation when A and C are nonzero but of different signs, which produces hyperbolas.

In this chapter we also study two additional methods for describing curves in the plane. The polar coordinate system is discussed in Section 5.5, and in Section 5.6 we describe the conics in the polar coordinate system. In Section 5.7 we study parametric equations. These new methods of representing curves allow us to visualize a greater variety of curves in the plane and lays the groundwork for describing the curves and surfaces in space you will need in multivariable calculus.

In the final section of the chapter we take up the case of the conic sections that are produced from the quadratic equation when $B \neq 0$ and show that the graph of the quadratic equation is obtained by a rotation of a conic in standard position.

5.2   PARABOLAS

The first time we encountered a parabola was in Section 1.4, where we considered the graph of $y = x^2$. In Section 1.8 we described more general quadratic functions of the form

$$y = f(x) = ax^2 + bx + c, \qquad \text{where } a \neq 0.$$

We found that the graphs of the general quadratic functions are similar to those of $y = x^2$ but might involve an elongation or compression (if $a \neq 1$) as well as horizontal and vertical shifts, depending on the particular values of the constants. Completing the square on the quadratic revealed its secrets, as illustrated with the graph of

$$f(x) = 2x^2 - 8x + 9 = 2(x^2 - 4x + 4) + 9 - 8 = 2(x - 2)^2 + 1$$

shown in Figure 5.4.

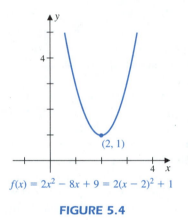

$$f(x) = 2x^2 - 8x + 9 = 2(x - 2)^2 + 1$$

FIGURE 5.4

The graphs of parabolas, then, have been studied quite thoroughly. What we will do in this section is show some important geometric properties of parabolas that feature in applications. To do this we introduce an alternative geometric definition for the parabola.

Parabola

A **parabola** is the set of points in a plane that are equidistant from a given point, called the **focal point**, and a given line, called the **directrix**, that does not contain the focal point.

The **axis** of the parabola is the line through the focal point and perpendicular to the directrix, and the point of intersection of the axis and the parabola is called the **vertex**. (See Figure 5.5.)

A *standard form* for the equation of a parabola is derived by superimposing an xy-coordinate system so that the axis of the parabola corresponds with the y-axis, and the vertex of the parabola, which is located midway between the focal point and the directrix, is at the origin. If the focal point is located at $(0, c)$, then the directrix has equation $y = -c$. The situation with $c > 0$ is shown in Figure 5.6(a), and that of $c < 0$ is shown in Figure 5.6(b).

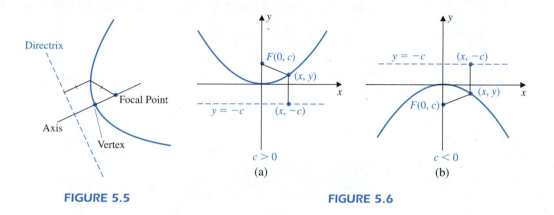

FIGURE 5.5   FIGURE 5.6

By the definition of the parabola, the distance from an arbitrary point $P(x, y)$ on the curve to the focal point $F(0, c)$ is the same as the distance from $P(x, y)$ to the directrix. From Figure 5.6 we see that this implies

$$\sqrt{(x - 0)^2 + (y - c)^2} = \sqrt{(x - x)^2 + (y + c)^2}.$$

Squaring both sides gives

$$x^2 + (y - c)^2 = (y + c)^2, \qquad \text{or} \qquad x^2 + y^2 - 2cy + c^2 = y^2 + 2cy + c^2.$$

This result simplifies to the equation

$$x^2 = 4cy, \qquad \text{or} \qquad y = \frac{1}{4c}x^2,$$

that is, a parabola in the usual form $y = ax^2$, where $a = 1/(4c)$.

From this derivation we can see that the basic parabola $y = x^2$, which we have seen so often, has its vertex at the origin and $c = 1/4$. Its focal point is at $(0, 1/4)$, and the equation of its directrix is $y = -1/4$, as shown in Figure 5.7.

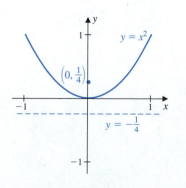

FIGURE 5.7

Standard-Position Parabolas

A parabola with focal point at $(0, c)$, vertex at $(0, 0)$, and directrix $y = -c$ is said to be in standard position with axis along the y-axis and has equation (see Figure 5.8(a))

$$y = \frac{1}{4c}x^2.$$

Similarly, a parabola with focal point at $(c, 0)$, vertex at $(0, 0)$, and directrix $x = -c$ is in standard position with axis along the x-axis and has equation (see Figure 5.8(b))

$$x = \frac{1}{4c}y^2.$$

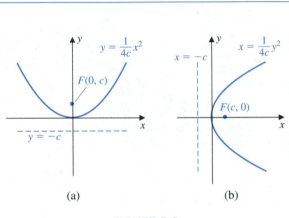

(a)

(b)

FIGURE 5.8

EXAMPLE 1   Find an equation of the parabola with focal point $(0, -2)$ and directrix $y = 2$.

Solution   The focal point lies along the y-axis. Since the vertex is the midpoint of the line segment from the focal point to the directrix, the vertex is at the origin. This implies that the parabola is in standard position with axis along the y-axis. Its graph is shown in Figure 5.9, and its equation is

$$y = \frac{1}{4(-2)}x^2 = -\frac{1}{8}x^2. \qquad \blacksquare$$

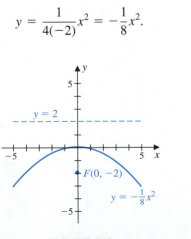

FIGURE 5.9

By using the graphing techniques we have so frequently employed, we can sketch the graph of any parabola whose directrix is parallel to one of the coordinate axes.

EXAMPLE 2 Find the focal point and directrix of the parabola with equation $2y^2 + 8y - x + 7 = 0$.

Solution We first complete the square on the y-variable so that we can compare this equation with the equation of a parabola that is in standard position. We have

$$2(y^2 + 4y + 4) - x + 7 - 2(4) = 0,$$

which gives

$$2(y + 2)^2 - x - 1 = 0, \quad \text{or} \quad x + 1 = 2(y + 2)^2.$$

We compare this with

$$x = 2y^2 = \frac{1}{4c}y^2,$$

which describes a parabola in standard position with axis along the x-axis, and since

$$\frac{1}{4c} = 2, \quad \text{we have} \quad c = \frac{1}{8}.$$

The standard-position parabola has focal point at $(1/8, 0)$ and directrix $x = -1/8$.

The graph of the parabola with equation $x + 1 = 2(y + 2)^2$ is the translation of the graph of $x = 2y^2$ to the left 1 unit and downward 2 units. Consequently, the focal point of the new parabola is at $(1/8 - 1, 0 - 2) = (-7/8, -2)$, and the directrix is $x = -1/8 - 1 = -9/8$, as shown in Figure 5.10. ■

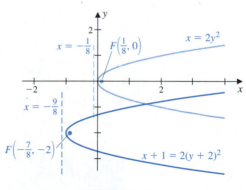

FIGURE 5.10

EXAMPLE 3 Determine the equation of the parabola with focal point at $(1, 5)$ and directrix $y = -1$.

Solution Since the vertex of the parabola lies midway between the focal point and the directrix, the distance from the vertex to the focal point is $c = 3$, and the vertex is at $(1, 2)$. The axis of the parabola is the line $x = 1$. Figure 5.11(a) shows the standard position parabola whose focal point is 3 vertical units above its vertex, and this parabola has equation

$$y = \frac{1}{4(3)}x^2 = \frac{1}{12}x^2.$$

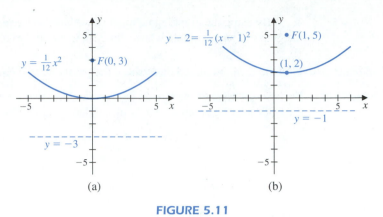

FIGURE 5.11

The parabola whose vertex is translated to $(1, 2)$ is shown in Figure 5.11(b) and has equation

$$y - 2 = \frac{1}{12}(x - 1)^2, \quad \text{or} \quad y = \frac{1}{12}(x - 1)^2 + 2.$$ ■

Parabolas have a distinctive reflective property. Sound, light, or other waves parallel to the axis of the parabola are reflected from the focus as shown in Figure 5.12.

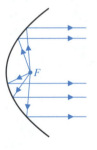

FIGURE 5.12

You will probably study surfaces of this type in your final term of calculus.

If a parabola is rotated about its axis to construct a surface called a *paraboloid*, then cross sections containing the axis are parabolic with a common focal point, as shown in Figure 5.13. In the case of a flashlight or search light, light emitted from the focus

FIGURE 5.13

is reflected in parallel rays, creating a concentrated beam of light. Similarly, reflecting telescopes have parabolic shapes since light or radio waves bouncing off the surface are reflected to the focus, where they are collected and amplified.

EXAMPLE 4

This is a type of problem you will likely be asked to set up and solve in calculus.

A satellite dish receiver has its amplifier in line with the edge of the dish, as shown in Figure 5.14(a). The diameter of the dish at the edge is 1 m. How deep is the dish?

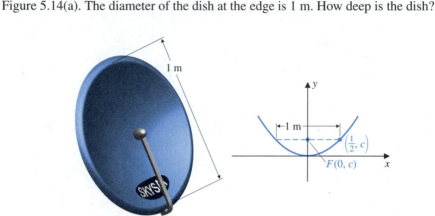

FIGURE 5.14

Solution   A cross section of the dish is shown in Figure 5.14(b), where the focal point is at $(0, c)$; the data in the problem imply that the point $(1/2, c)$ lies on the graph of the parabola. The equation of a parabola in standard position is

$$y = \frac{1}{4c}x^2,$$

so

$$c = \frac{1}{4c}\left(\frac{1}{2}\right)^2,$$

which implies that

$$c^2 = \frac{1}{16}, \quad \text{so} \quad c = \frac{1}{4}.$$

The distance from the focal point to the vertex and, hence, the depth of the dish is $\frac{1}{4}$ m = 25 cm. ■

EXERCISE SET 5.2

In Exercises 1–12, sketch the graph of the parabola, showing the vertex, focal point, and directrix.

1. $y = 2x^2$

2. $16y = 9x^2$

3. $y = -2x^2$

4. $9y = -16x^2$

5. $y^2 = 2x$

6. $9y^2 = -16x$

7. $x^2 + 4x + 4 = 2y$

8. $x^2 + 6x + 9 + 4y = 0$

9. $y^2 - 8y + 12 = 2x$

10. $y^2 + 6y + 6 - 3x = 0$

11. $2x^2 + 4x - 9y + 20 = 0$

12. $3x^2 - 12x + 4y + 8 = 0$

In Exercise 13–16, find an equation for the given parabola.

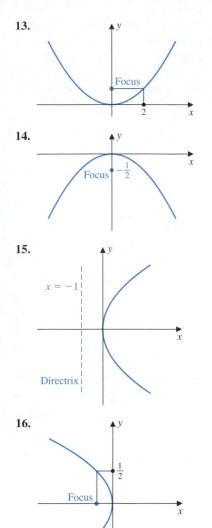

13.

14.

15.

16.

In Exercises 17–20, determine the equation of the parabola that satisfies the given conditions.

17. Focus at $(-2, 2)$, directrix $y = -2$.

18. Focus at $(-2, 2)$, directrix $x = 2$.

19. Vertex at $(-2, 2)$, directrix $x = 4$.

20. Vertex at $(-2, 2)$, focus at $(-2, 0)$.

In Exercises 21–24, find an equation of the parabola with axis parallel to the y-axis and vertex V that passes through the point P.

21. $V(0, 0)$, $(4, 6)$ **22.** $V(1, 0)$, $(5, 6)$

23. $V(0, 2)$, $(4, 8)$ **24.** $V(1, 2)$, $(5, 8)$

25–28. Find an equation of the parabola with axis parallel to the x-axis and vertex V that passes through the point P, where V and P are as given in Exercises 21–24.

29. Find a general form for the equation of a parabola with axis the y-axis and passing through $(1, 1)$.

30. A driving light has a parabolic cross section with a depth of 2 in. and a cross-section height of 4 in. Where should the light source be placed to produce a parallel beam of light?

31. A ball thrown horizontally from the top edge of a building follows a parabolic curve with vertex at the top edge of the building and axis along the side of the building. The ball passes through a point 100 ft from the building when it is a vertical distance of 16 ft from the top.

   a. How far from the building will the ball land if the building is 64 ft high?

   b. Recompute the answer if instead the ball is thrown from the top of the Sears Tower in Chicago, which has a height of 1450 ft.

5.3   ELLIPSES

An ellipse is an oval shaped curve that has the appearance of an elongated circle. In the 17th century it was discovered by Johannes Kepler, and proved by Isaac Newton, that the planets revolve about the sun in elliptical orbits.

There are many equivalent definitions for the ellipse. We give a definition that involves two fixed points and a given distance. In physics you are more likely to see a definition that uses a given point, a given line, and a number called the *eccentricity*, which describes how close the ellipse is to being a circle. This definition is considered in Section 5.6.

Ellipse

An **ellipse** is the set of points in a plane for which the sum of the distances from two fixed points is a given constant. The two fixed points are the **focal points** of the ellipse, the line passing through the focal points is called the **axis**, and the points of intersection of the axis and the ellipse are called the **vertices**. (See Figure 5.15(a).)

In vector calculus, we verify that Kepler's laws of planetary motion follow from Newton's law of motion. These laws revolutionized scientific thought in the 17th century.

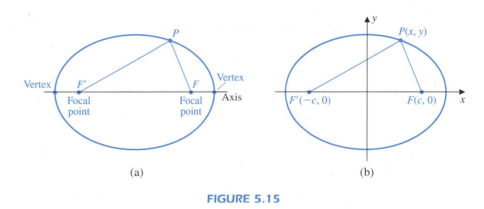

(a)   (b)

FIGURE 5.15

To determine standard equations for the ellipse, we position an xy-coordinate system with the x-axis along the axis of the ellipse and the origin midway between the focal points, as shown in Figure 5.15(b). The focal points have been given the coordinates $F(c, 0)$ and $F'(-c, 0)$. Since the distance from $F(c, 0)$ to $F'(-c, 0)$ is $2c$, the fixed constant specified in the definition must be greater than this value.

Let us call the fixed distance $2a$, where $a > c > 0$. Then the vertices have the coordinates $V(a, 0)$ and $V'(-a, 0)$ since

$$d(V, F) + d(V, F') = (a - c) + (a + c) = 2a$$

and

$$d(V', F) + d(V', F') = (a + c) + (a - c) = 2a.$$

In general, a point (x, y) is on the ellipse precisely when

$$\sqrt{(x - c)^2 + (y - 0)^2} + \sqrt{(x + c)^2 + (y - 0)^2} = d((x, y), (c, 0)) + d((x, y), (-c, 0))$$
$$= 2a,$$

that is, when

$$\sqrt{(x - c)^2 + y^2} = 2a - \sqrt{(x + c)^2 + y^2}.$$

Squaring both sides of this equation gives

$$(x - c)^2 + y^2 = 4a^2 - 4a\sqrt{(x + c)^2 + y^2} + (x + c)^2 + y^2,$$

which, after completing the multiplication and dividing by 4, simplifies to

$$a\sqrt{(x + c)^2 + y^2} = a^2 + cx.$$

Squaring again and canceling $2a^2cx$ from both sides of the equation gives

$$a^2x^2 + a^2c^2 + a^2y^2 = a^4 + c^2x^2,$$

which simplifies to

$$(a^2 - c^2)x^2 + a^2y^2 = a^4 - a^2c^2.$$

Since $a > c > 0$, we can divide both sides by $a^2 - c^2$ and a^2 to obtain

$$\frac{x^2}{a^2} + \frac{y^2}{a^2 - c^2} = 1.$$

For convenience we replace the term $a^2 - c^2$ with a new constant b^2, where $b > 0$, to produce the equation for the ellipse with axis lying along the x-axis:

$$\frac{x^2}{a^2} + \frac{y^2}{b^2} = 1, \qquad \text{where} \qquad b = \sqrt{a^2 - c^2}.$$

Notice that the squares on the x- and y-terms imply that the graph is symmetric about both the y- and the x-axes. The x-intercepts occur, as they must, at $(a, 0)$ and $(-a, 0)$, and the y-intercepts occur at $(0, b)$ and $(0, -b)$, where $b = \sqrt{a^2 - c^2} < a$. The *major axis* of an ellipse is the longest line segment joining two points on the ellipse. The major axis for this ellipse has length $2a$ and joins the points $(a, 0)$ and $(-a, 0)$, and the *minor* axis joins $(0, b)$ and $(0, -b)$.

An ellipse centered at the origin with axis along the x-axis is said to be in *standard position*. A similar type of equation occurs when the roles of the axes are interchanged. (See Figure 5.16.)

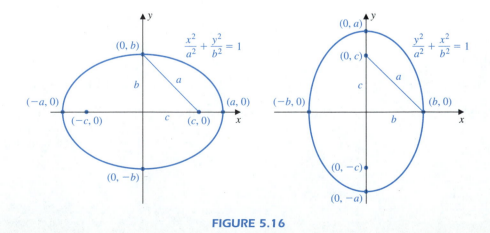

FIGURE 5.16

Standard-Position Ellipses

An ellipse with focal points $(c, 0)$ and $(-c, 0)$ and vertices $(a, 0)$ and $(-a, 0)$, where $a > c > 0$, is in *standard position* with axis along the x-axis and has equation

$$\frac{x^2}{a^2} + \frac{y^2}{b^2} = 1, \qquad \text{where} \qquad a > b = \sqrt{a^2 - c^2}.$$

Similarly, an ellipse with focal points $(0, c)$ and $(0, -c)$ and vertices $(0, a)$ and $(0, -a)$, where $a > c > 0$, is in *standard position* with axis along the y-axis and has equation

$$\frac{y^2}{a^2} + \frac{x^2}{b^2} = 1, \qquad \text{where} \qquad a > b = \sqrt{a^2 - c^2}.$$

EXAMPLE 1   Sketch the graph of the ellipse with equation $9x^2 + 16y^2 = 144$ and find its focal points.

Solution   Dividing both sides of the equation by 144 gives

$$\frac{x^2}{16} + \frac{y^2}{9} = 1.$$

The coefficient in the denominator of x^2 is larger than the coefficient in the denominator of y^2, so the ellipse is in standard position with axis along the x-axis. Since $a^2 = 16$ and $b^2 = 9$, we have $a = 4$ and $b = 3$. The axis intercepts are at $(4, 0)$, $(-4, 0)$, $(0, 3)$, and $(0, -3)$, as shown in Figure 5.17.

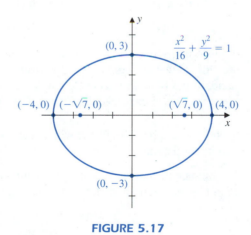

FIGURE 5.17

Since $b^2 = a^2 - c^2$, we have $c^2 = a^2 - b^2 = 16 - 9 = 7$, and the focal points are at $(\sqrt{7}, 0)$ and $(-\sqrt{7}, 0)$. ■

EXAMPLE 2   Find an equation of the ellipse in standard position that has a vertex at $(5, 0)$ and a focal point at $(3, 0)$.

Solution   The ellipse is in standard position, so both the vertices and the focal points are centered about the origin. As a consequence, the other vertex is at $(-5, 0)$, and the other focal point is at $(-3, 0)$. Since $a = 5$ and $c = 3$ we have

$$b = \sqrt{a^2 - c^2} = \sqrt{25 - 9} = \sqrt{16} = 4,$$

and the ellipse, shown in Figure 5.18, has equation

$$\frac{x^2}{25} + \frac{y^2}{16} = 1, \quad \text{or} \quad 16x^2 + 25y^2 = 400.$$

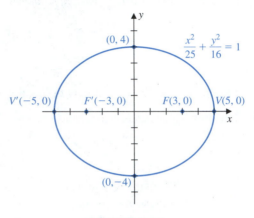

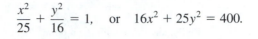

FIGURE 5.18

In the introduction we indicated that quadratic equations of the form

$$Ax^2 + Cy^2 + Dx + Ey + F = 0,$$

produce ellipses when A and C have the same sign—that is, when $AC > 0$. In fact, the ellipses that are produced are just vertical and horizontal translations of ellipses in standard position. The usual technique of completing the square is used to determine how the curve is translated, as illustrated in the following example.

EXAMPLE 3   Sketch the graph of the ellipse with equation $9x^2 - 72x + 4y^2 + 16y + 124 = 0$, and find its focal points and vertices.

Solution   To compare this equation with the equation of an ellipse in standard position, we first complete the square in both x and y. The equation can be written as

$$9(x^2 - 8x) + 4(y^2 + 4y) = -124,$$

so

$$9(x^2 - 8x + 16) + 4(y^2 + 4y + 4) = -124 + 9(16) + 4(4),$$

or

$$9(x - 4)^2 + 4(y + 2)^2 = 36.$$

Dividing by 36 gives

$$\frac{(x - 4)^2}{4} + \frac{(y + 2)^2}{9} = 1.$$

The equation

$$\frac{y^2}{9} + \frac{x^2}{4} = 1$$

describes the ellipse in standard position with axis along the y-axis, as shown in Figure 5.19(a). The vertices of this ellipse are at $(0, 3)$ and $(0, -3)$. Since

$$c^2 = a^2 - b^2 = 9 - 4 = 5,$$

the focal points of the ellipse in standard position occur at $(0, \sqrt{5})$ and $(0, -\sqrt{5})$.

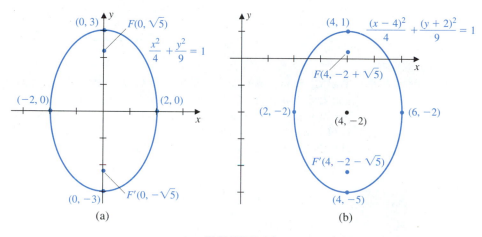

(a)

(b)

FIGURE 5.19

The graph of the original equation,

$$\frac{(y + 2)^2}{9} + \frac{(x - 4)^2}{4} = 1,$$

is simply a translation of the graph of the ellipse in standard position. It is translated downward 2 units and to the right 4 units. Hence, the focal points occur at $(4, -2 + \sqrt{5})$ and $(4, -2 - \sqrt{5})$, and the vertices are at $(4, 1)$ and $(4, -5)$, as shown in Figure 5.19(b).

■

Like the parabola, the ellipse also has an interesting reflection property. If a light or sound source is placed at one focus of a reflecting surface with elliptical cross sections, then the wave emitted from the source will be reflected off the surface to the other focus, as shown in Figure 5.20.

When a person stands at one of the focal points in a building with an elliptical ceiling, sound made at that point is reflected from the ceiling to the other focal point. Even whispers in a crowded room can be distinctly heard, and rooms with this property are called *whispering galleries*. These whispering galleries are no recent innovation. Statuary Hall in Washington, D.C., and the Mormon Tabernacle in Salt Lake City, Utah, were built over a century ago and use this principle, as does St. Paul's Cathedral in London, which was designed by the outstanding mathematician and architect Sir Christopher Wren in the decade after the great fire in 1666. In fact, the principles of this design were known to the Greek geometers of Euclid's time, more than 2300 years ago.

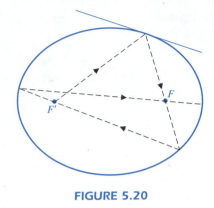

FIGURE 5.20

EXAMPLE 4   The Mormon Tabernacle in Salt Lake City is 250 ft long, 150 ft wide, and 80 ft high and has its longitudinal cross section in the shape of an ellipse. The conductor for performances stands at one of the focal points of the ellipse, and recording equipment is placed at the other. In this way the sound heard by the conductor corresponds very closely to the sound being recorded. Determine the location of these points.

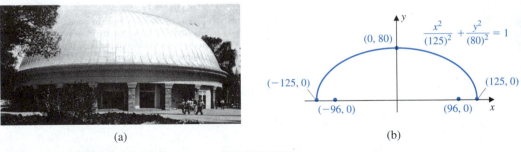

(a)   (b)

FIGURE 5.21

Solution   Figure 5.21(a) shows the Tabernacle, and Figure 5.21(b) shows the situation when an xy-coordinate system in superimposed on a cross section. The width of the building plays no part in the calculations, but the length and height are used to determine the lengths of the major and minor axes, 250 and 160 ft, respectively. The ellipse in standard position has equation

$$\frac{x^2}{(125)^2} + \frac{y^2}{(80)^2} = 1.$$

The focal points occur

$$c = \sqrt{a^2 - b^2} = \sqrt{(125)^2 - (80)^2} = \sqrt{9225} \approx 96 \text{ ft}$$

from the center of the building. So the conductor should be located approximately $125 - 96 = 29$ ft from one end of the building, and the recording equipment should be an equal distance from the other. ∎

The **eccentricity** of an ellipse tells how the ellipse differs from a circle. For an ellipse in standard position, the eccentricity is

$$e = \frac{c}{a} = \frac{\sqrt{a^2 - b^2}}{a},$$

with the usual definitions of a, b, and c. Since $a \geq b > 0$, we must have $0 \leq e < 1$, and $e = 0$ precisely when $a = b$, that is, when the ellipse is a circle. Figure 5.22 shows various ellipses with the same vertices. Notice that as the focal point approaches the origin the ellipse approaches a circle, and e approaches 0. The ellipse becomes increasingly elongated as the focal points approach the vertices, which occurs when e approaches 1 from the left.

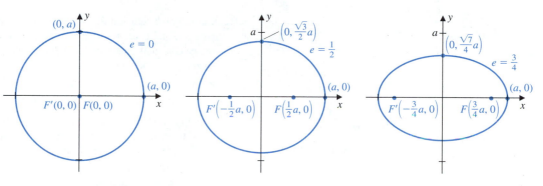

FIGURE 5.22

Ellipses play an important role in astronomy. In the early 17th century, Johannes Kepler used an extensive collection of data to predict that the orbits of the planets are ellipses with the sun at one focal point. The earth orbits the sun in a nearly circular orbit with eccentricity 0.0167, and the orbits of the other planets are also nearly circular, as you can see from the data in Table 5.1.

Even Pluto, the planet with the largest eccentricity, has a nearly circular orbit, since

$$0.2481 = \frac{\sqrt{a^2 - b^2}}{a} \qquad \text{implies that} \qquad b = 0.968a.$$

TABLE 5.1

Planet	Major Axis	Eccentricity	Planet	Major Axis	Eccentricity
Mercury	1.159×10^8 km	0.2056	Saturn	2.854×10^9 km	0.0543
Venus	2.162×10^8 km	0.0068	Uranus	5.741×10^9 km	0.0460
Earth	2.991×10^8 km	0.0167	Neptune	9.000×10^9 km	0.0082
Mars	4.557×10^8 km	0.0934	Pluto	1.182×10^{10} km	0.2481
Jupiter	1.556×10^9 km	0.0484			

On the other hand, the orbit of Halley's comet about the sun is an example of a very noncircular elliptic orbit. It has a major axis of 5.35×10^9 km and eccentricity of $e = 0.967$.

EXERCISE SET 5.3

In Exercises 1–10, sketch the graph of the ellipse and show the vertices and focal points.

1. $x^2 + \dfrac{y^2}{9} = 1$ **2.** $\dfrac{x^2}{16} + y^2 = 1$

3. $16x^2 + 25y^2 = 400$ **4.** $25x^2 + 16y^2 = 400$

5. $3x^2 + 2y^2 = 6$ **6.** $4x^2 + 3y^2 = 12$

7. $4x^2 + y^2 + 16x + 7 = 0$

8. $16x^2 + 9y^2 - 54y - 63 = 0$

9. $x^2 + 4y^2 - 2x - 16y + 13 = 0$

10. $4x^2 + 9y^2 - 16x + 90y + 97 = 0$

In Exercises 11–20, find an equation of the ellipse that satisfies the stated conditions.

11. x-intercepts at $(\pm 4, 0)$ and y-intercepts at $(\pm 3, 0)$

12. x-intercepts at $(\pm 2, 0)$ and y-intercepts at $(\pm 5, 0)$

13. Foci at $(\pm 2, 0)$ and vertices at $(\pm 3, 0)$

14. Foci at $(\pm 2, 0)$ and y-intercepts at $(0, \pm 2)$

15. Foci at $(0, \pm 1)$ and y-intercepts at $(0, \pm 2)$

16. Foci at $(3, 0)$ and $(1, 0)$ and a vertex at $(0, 0)$

17. Vertices at $(2, 2)$ and $(6, 2)$ and a focal point at $(5, 2)$

18. Vertices at $(3, 3)$ and $(3, -1)$ and passing through $(2, 1)$

19. Length of the major axis 5, length of the minor axis 3, and in standard position with foci on the x-axis

20. Foci at $(\pm 3, 0)$ and length of the major axis 8

21. The *latus rectum* of an ellipse is a line segment that passes through a focal point perpendicular to the major axis and joins the points on the ellipse (see the figure). Find the length of the latus rectum of the ellipse with equation $x^2/a^2 + y^2/b^2 = 1$ when $a > b > 0$.

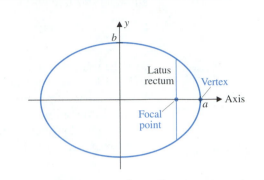

22. Consider the equation $Ax^2 + Cy^2 + Dx + Ey + F = 0$, where A and C are positive constants. Find conditions on the constants that ensure that this equation describes each of the following.

   a. An ellipse   **b.** A circle

   c. A single point   **d.** No points

23. Olympic Stadium in Montreal, shown in the figure, is constructed in the shape of an ellipse with major and minor axes 480 and 280 m, respectively. Find the equation of this ellipse.

24. Halley's comet is named to honor Edmund Halley (1656–1742). In 1682 Halley determined the orbit of this comet and predicted that it would return 76 years later. The orbit is elliptical with a focal point at the sun, its major axis is approximately 5.39×10^9 km, and its minor axis is approximately 1.36×10^9 km. How close does this comet pass to the sun?

5.4   HYPERBOLAS

Although the shapes of ellipses and hyperbolas are very different, their definitions are quite similar. Both of these families of curves are defined using two fixed points and a fixed distance. In the case of an ellipse, the fixed distance is the sum of the distances between the two fixed points, whereas the hyperbola uses the difference of these distances.

Hyperbola

A **hyperbola** is the set of points in a plane for which the magnitude of the difference between the distances from two fixed points is a given constant. The two fixed points are the **focal points** of the hyperbola, the line passing through the focal points is called the **axis**, and the points of intersection of the axis and the hyperbola are called the **vertices**. (See Figure 5.23(a).)

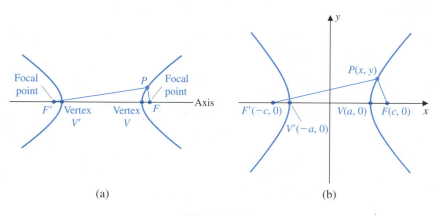

(a) (b)

FIGURE 5.23

To determine standard equations for the hyperbola, we proceed in the same manner as we did for the ellipse. An xy-coordinate system is placed with the x-axis along the axis of the hyperbola and the origin midway between the focal points, as shown in Figure 5.23(b). As in the case of the ellipse, the focal points have been given the coordinates $F(c, 0)$ and $F'(-c, 0)$, the vertices are located at $V(a, 0)$ and $V'(-a, 0)$, where $a \neq c$, and the fixed difference is $2a$. The vertices for the ellipse were located outside the focal points, but for the hyperbola the vertices are located between the focal points. This can be seen by observing that since both a and c are positive,

$$2a = |d(V, F) - d(V, F')| = d(V, F') - d(V, F) = c + a - |c - a|,$$

and

$$a - c = -|c - a|.$$

Since $a \neq c$, the right side of this last equation is negative, and so is the left side, which implies that $a < c$.

To derive an equation for the hyperbola, let us consider a point (x, y) on the hyperbola. To make the calculations more closely parallel those of the ellipse, we will assume that $x > 0$. Symmetry with respect to the y-axis ensures that the equation derived also holds when $x < 0$.

The point (x, y) is on the graph precisely when

$$\left| \sqrt{(x + c)^2 + (y - 0)^2} - \sqrt{(x - c)^2 + (y - 0)^2} \right|$$

$$= |d((x, y), (-c, 0)) - d((x, y), (c, 0))| = 2a$$

Since $x > 0$ and $c > 0$ we have $(x + c)^2 > (x - c)^2$, and the absolute values are not needed. So

$$\sqrt{(x + c)^2 + y^2} = 2a + \sqrt{(x - c)^2 + y^2}.$$

Squaring both sides of this last equation gives

$$(x + c)^2 + y^2 = 4a^2 + 4a\sqrt{(x - c)^2 + y^2} + (x - c)^2 + y^2,$$

which, after completing the multiplication and dividing by 4, simplifies to

$$-a\sqrt{(x - c)^2 + y^2} = a^2 - cx.$$

Squaring again and canceling $-2a^2cx$ from both sides of the equation gives

$$a^2x^2 + a^2c^2 + a^2y^2 = a^4 + c^2x^2,$$

the same equation as we found for the ellipse. This simplifies in the same way as it did for the ellipse to

$$\frac{x^2}{a^2} + \frac{y^2}{a^2 - c^2} = 1,$$

or, more appropriately in this case, since $a < c$,

$$\frac{x^2}{a^2} - \frac{y^2}{c^2 - a^2} = 1.$$

If we replace $c^2 - a^2$ with a new constant b^2, where $b > 0$, we produce the equation for a hyperbola with axis lying along the x-axis:

$$\frac{x^2}{a^2} - \frac{y^2}{b^2} = 1, \qquad \text{where} \qquad b = \sqrt{c^2 - a^2}.$$

Notice that the squares on the x- and y-terms imply that the graph is symmetric to both the x- and y-axes. The x-intercepts occur at the vertices $(a, 0)$ and $(-a, 0)$, but there are no y-intercepts since

$$y^2 = \frac{b^2}{a^2}(x^2 - a^2),$$

which implies that the equation is valid only for $x^2 \geq a^2$; that is, when $x \geq a$ or $x \leq -a$.

A hyperbola with equation in this form is said to be in *standard position* with axis along the x-axis. A similar type of equation occurs when the roles of the axes are reversed. (See Figure 5.24.)

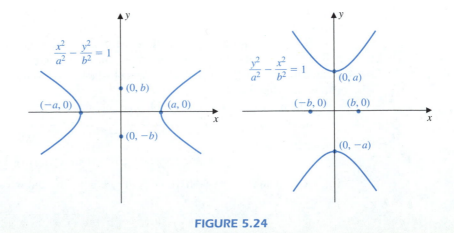

FIGURE 5.24

Standard-Position Hyperbolas

A hyperbola with focal points $(c, 0)$ and $(-c, 0)$ and vertices $(a, 0)$ and $(-a, 0)$, where $c > a > 0$, is in *standard position* with axis along the x-axis and has equation

$$\frac{x^2}{a^2} - \frac{y^2}{b^2} = 1, \qquad \text{where} \qquad b = \sqrt{c^2 - a^2}.$$

Similarly, a hyperbola with focal points $(0, c)$ and $(0, -c)$ and vertices $(0, a)$ and $(0, -a)$, where $c > a > 0$, is in *standard position* with axis along the y-axis and has equation

$$\frac{y^2}{a^2} - \frac{x^2}{b^2} = 1, \qquad \text{where} \qquad b = \sqrt{c^2 - a^2}.$$

The standard-position equations can be used to see the end behavior of the hyperbola. Consider the situation in the first quadrant for a hyperbola in standard position with axis along the x-axis. Solving for y in terms of x in

$$\frac{x^2}{a^2} - \frac{y^2}{b^2} = 1, \qquad \text{where} \qquad x > 0 \quad \text{and} \quad y > 0,$$

gives

$$y = \sqrt{\frac{b^2}{a^2}(x^2 - a^2)} = \frac{b}{a}\sqrt{x^2 - a^2}.$$

As x becomes increasingly large, the constant term, $-a^2$, under the radical becomes of less importance, and

$$y = \frac{b}{a}\sqrt{x^2 - a^2} \approx \frac{b}{a}\sqrt{x^2} = \frac{b}{a}x.$$

Hence the graph of the hyperbola approaches the line $y = \frac{b}{a}x$ as x becomes large. This line is a *slant asymptote* to the graph of the hyperbola. Since the hyperbola is symmetric with respect to both coordinate axes, similar behavior occurs in each of the quadrants,

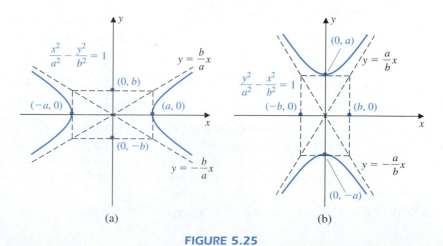

(a)   (b)

FIGURE 5.25

as shown in Figure 5.25. As a consequence, the hyperbola with equation

$$\frac{x^2}{a^2} - \frac{y^2}{b^2} = 1 \qquad \text{has asymptotes} \qquad y = \pm\frac{b}{a}x,$$

as shown in Figure 5.25(a).

In a similar manner, as shown in Figure 5.25(b),

$$\frac{y^2}{a^2} - \frac{x^2}{b^2} = 1 \qquad \text{has asymptotes} \qquad y = \pm\frac{a}{b}x.$$

EXAMPLE 1   Sketch the graph of the hyperbola with equation $16x^2 - 9y^2 = 144$.

Solution   The equation can be rewritten as

$$\frac{x^2}{9} - \frac{y^2}{16} = 1.$$

It is in standard position with axis along the x-axis, and we have $a = 3$ and $b = 4$. The asymptotes of the hyperbola are $y = \pm\frac{4}{3}x$. This information is sufficient to sketch the graph shown in Figure 5.26. Since $c = \sqrt{a^2 + b^2} = \sqrt{9 + 16} = \sqrt{25} = 5$, the focal points are at $(5, 0)$ and $(-5, 0)$. ∎

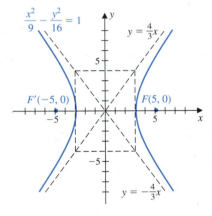

FIGURE 5.26

In Section 5.1 we indicated that quadratic equations of the form

$$Ax^2 + Cy^2 + Dx + Ey + F = 0$$

produce hyperbolas when A and C have opposite signs, that is, when $AC < 0$. The hyperbolas that are produced are just vertical and horizontal translations of hyperbolas in standard position. As in the case of the ellipse, we use completion of the square to determine how the curve is translated, which is illustrated in the following example.

EXAMPLE 2   Sketch the graph of the hyperbola with equation

$$y^2 - 2y - 9x^2 + 36x = 39.$$

Solution   The first step is to complete the square on the x- and y-terms so that we can relate this equation to one whose graph is in standard position. This process gives

$$(y^2 - 2y + 1) - 9(x^2 - 4x + 4) = 39 + 1 - 9(4) = 4,$$

so

$$(y - 1)^2 - 9(x - 2)^2 = 4, \quad \text{or} \quad \frac{(y - 1)^2}{4} - \frac{(x - 2)^2}{4/9} = 1.$$

This is a translation of the hyperbola in standard position with axis along the y-axis. Its equation is

$$\frac{y^2}{4} - \frac{x^2}{4/9} = 1.$$

It has vertices at $(0, 2)$ and $(0, -2)$, and asymptotes with equations

$$y = \pm \frac{2}{2/3} x = \pm 3x,$$

as shown in Figure 5.27(a).

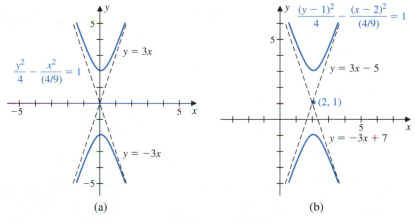

(a)   (b)

FIGURE 5.27

The graph of

$$\frac{(y - 1)^2}{4} - \frac{(x - 2)^2}{4/9} = 1$$

has the same shape, but it is shifted to the right 2 units and upward 1 unit, as shown in Figure 5.27(b). The asymptotes for the graph are also shifted, to become

$$y - 1 = \pm 3(x - 2).$$

That is,

$$y = 3x - 5 \quad \text{and} \quad y = -3x + 7. \quad \blacksquare$$

The global positioning system uses hyperbolic arcs to determine the position of objects. To describe how the positioning scheme works, suppose that we have three fixed points $P_1, P_2,$ and P_3 that emit signals simultaneously, as shown in Figure 5.28(a).

We wish to determine the location of a point P that has devices that can measure, very accurately, the time at which the signals reach P. Since the speed of the signals is constant, the times for the signals to reach P are directly proportional to the distances from P to the fixed points.

Suppose that the signal from P_1 is the first to reach P. Then there are constants c_1 and c_2 for which

$$d(P_2, P) - d(P_1, P) = c_1$$

and

$$d(P_3, P) - d(P_2, P) = c_2.$$

The first of these equations implies that P lies on the branch of a known hyperbola with foci at P_1 and P_2, as shown in Figure 5.28(b), and the second implies that P also lies on the branch of a known hyperbola with foci at P_3 and P_2. The intersection of the two hyperbolas is the location of P, as shown in Figure 5.28(c) The global positioning system has correction techniques that use signals from four or more satellites to average and correct the location error. The additional satellite is used to find the altitude of the point.

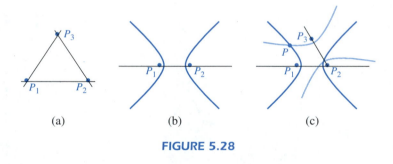

(a)   (b)   (c)

FIGURE 5.28

The hyperbola also has a reflection property. In the case of a hyperbola, light or sound emitted from one of the focal points is reflected off the surface along a line directly away from the other focal point. (See Figure 5.29.)

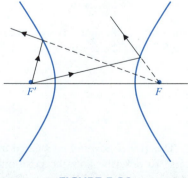

FIGURE 5.29

The hyperbola reflection property finds application in the construction of large telescopes, such as the Hale telescope at the Palomar Observatory in California. Within this telescope there is a Cassegrain configuration that consists of a hyperbolic mirror inserted between the parabolic reflector and the parabola's focal point. This hyperbola has one focal point coinciding with the focal point of the parabola and reflects the image back through a hole in the center of the parabolic mirror, as shown in Figure 5.30. From there it goes to the other focal point of the hyperbolic mirror, located beyond the vertex of the parabolic mirror.

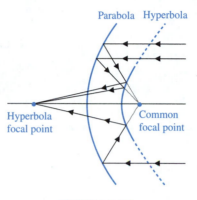

FIGURE 5.30

Eccentricity is also defined for hyperbolas. For a hyperbola in standard position, the eccentricity is

$$e = \frac{c}{a} = \frac{\sqrt{a^2 + b^2}}{a},$$

with our usual definitions of a, b, and c. Since $a < c$, we must have $e > 1$. Figure 5.31 shows various hyperbolas with the same vertex. Notice that as the focal point approaches the vertex, the hyperbola becomes increasingly narrow and e approaches 1 from the right.

As the focal point moves away from the vertex, the hyperbola becomes wider, approaching the pair of vertical lines $x = \pm a$.

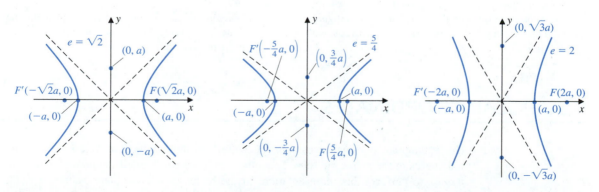

FIGURE 5.31

EXERCISE SET 5.4

In Exercises 1–12, sketch the graph of the hyperbola and show the vertices, focal points, and asymptotes.

1. $\dfrac{x^2}{4} - \dfrac{y^2}{9} = 1$

2. $\dfrac{x^2}{9} - \dfrac{y^2}{4} = 1$

3. $\dfrac{y^2}{4} - \dfrac{x^2}{9} = 1$

4. $\dfrac{y^2}{9} - \dfrac{x^2}{4} = 1$

5. $x^2 - y^2 = 1$

6. $y^2 - 4x^2 = 1$

7. $x^2 + 2x - 4y^2 = 3$

8. $9y^2 - 18y - 4x^2 = 27$

9. $3x^2 - y^2 = 6x$

10. $2y^2 + 8y = 9x^2$

11. $9x^2 - 4y^2 - 18x - 8y = 31$

12. $y^2 - 4x^2 - 2y - 16x = 19$

In Exercises 13–22, find an equation of the hyperbola that satisfies that stated conditions.

13. Foci at $(\pm 5, 0)$ and vertices at $(\pm 3, 0)$

14. Foci at $(0, \pm 13)$ and vertices at $(0, \pm 12)$

15. Foci at $(0, \pm 5)$ and vertices at $(0, \pm 4)$

16. Foci at $(\pm 13, 0)$ and vertices at $(\pm 5, 0)$

17. Foci at $(-1, 4)$ and $(5, 4)$ and a vertex at $(0, 4)$

18. Focus at $(2, 2)$ and vertices at $(2, 1)$ and $(2, -3)$

19. Vertex at $(0, \pm 2)$ and passing through $(3, 4)$

20. Vertices at $(2, \pm 2)$ and passing through $(8, 8)$

21. Vertices at $(\pm 3, 0)$ and equations of asymptotes $y = \pm 3x/4$

22. Foci at $(6, 1)$ and $(-2, 1)$ and equations of asymptotes $y = 3x/4 - 1/2$ and $y = -3x/4 + 5/2$

23. The *latus rectum* of a hyperbola is a line segment that passes through a focal point perpendicular to the axis and joins two points on the hyperbola (see the figure). Find the length of a latus rectum of the hyperbola with equation $x^2/a^2 - y^2/b^2 = 1$.

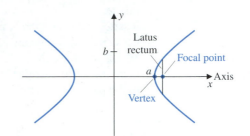

24. Consider the equation $Ax^2 - Cy^2 + Dx + Ey + F = 0$, where A and C are positive constants. Find conditions on the constants $A, C, D, E,$ and F that ensure that this equation describes each of the following.

   a. A hyperbola with axis parallel to the x-axis

   b. A hyperbola with axis parallel to the y-axis

   c. Intersecting lines

25. Three detection stations lie on an east–west line 1150 m apart. The eastmost station detects a sound from an object 2 s before the westmost station and 1 s before the station in the middle. Can the object emitting the noise be pinpointed? (Assume that the sound travels at 330 m/s.)

26. A company has two manufacturing plants that produce identical automobiles. Because of differing manufacturing and labor conditions in the plants, it costs $130 more to produce a car in plant A than in plant B. The shipping costs from both plants are the same, $1 per mile, as are the loading and unloading costs, $25 per car. State the criteria for determining from which plant a car should be shipped.

27. The hyperbolas

$$\frac{x^2}{a^2} - \frac{y^2}{b^2} = 1 \quad \text{and} \quad \frac{x^2}{a^2} - \frac{y^2}{b^2} = -1$$

are *conjugates* of each other. Find the conjugate to the hyperbola $9y^2 - 4x^2 + 36 = 0$, and describe how their graphs are related.

5.5   POLAR COORDINATES

The rectangular coordinate system has been used throughout the book to represent points and curves in the plane, but there is another common way, called the *polar coordinate system*, to perform this representation. In the rectangular coordinate system, a point in the plane is specified by an ordered pair (x, y) that describes the distances of the point from the *x*- and *y*-axes, as shown in Figure 5.32(a). The polar coordinate system represents a point in the plane using an ordered pair (r, θ), which describes a distance and direction from a fixed reference point, as shown in Figure 5.32(b).

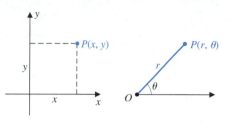

FIGURE 5.32

Many times in calculus the analysis of a problem can be simplified by changing from the rectangular coordinate system to a *polar* coordinate system. Being able to recognize when a change of coordinate systems is appropriate is an important tool.

To describe the polar coordinate system, we first choose a fixed point *O*, called the origin, or **pole**. We then draw a half-line, or *ray*, called the **polar axis**, originating at the pole and extending to the right. To determine polar coordinates of a point *P* in the plane, we use an angle θ formed by the polar axis and the line segment $\overline{OP}$ and the directed distance *r* from *O* to *P* along this line determined by the line segment $\overline{OP}$, as shown in Figure 5.33. The ordered pair (r, θ) is a set of **polar coordinates** of the point *P*.

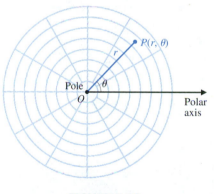

FIGURE 5.33

In the rectangular coordinate system each point in the plane has a unique representation, but this is not true in the polar coordinate system. The angles $\theta + 2n\pi$, for $n = 0, \pm 1, \pm 2, \ldots$, all have the same terminal side, so each point in the polar coordinate system has many representations. For example, $(1, \pi/6)$, $(1, 13\pi/6)$, and $(1, -11\pi/6)$ all represent the same point, as shown in Figure 5.34.

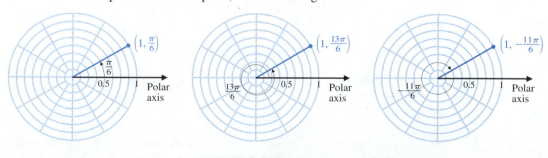

FIGURE 5.34

It is also convenient to allow r to be negative, with the understanding in this case that (r, θ) is the point $-r$ units from the origin in the direction *opposite* to that given by θ. For example, the points $(-2, \pi/3)$ and $(-3, 3\pi/4)$ are as shown in Figure 5.35.

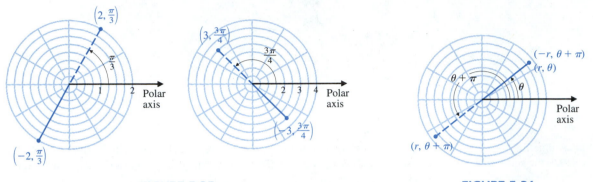

FIGURE 5.35   **FIGURE 5.36**

This also implies, for example, that the polar coordinates (r, θ) and $(-r, \theta + \pi)$ always represent the same point (see Figure 5.36).

We often need to convert between the rectangular and polar coordinate systems. To see how this is done, draw the two coordinate systems with a common origin and the positive x-axis coinciding with the polar axis. If a point P has polar coordinates (r, θ), then, as shown in Figure 5.37, the rectangular coordinates (x, y) are given by

$$x = r \cos \theta \qquad \text{and} \qquad y = r \sin \theta.$$

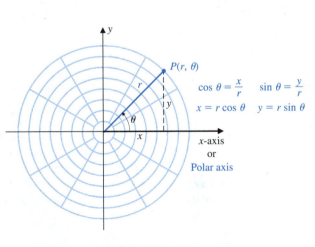

FIGURE 5.37

In addition, for a given set of rectangular coordinates $P(x, y)$, one set of polar coordinates for P is

$$\theta = \arctan \frac{y}{x} \qquad \text{and} \qquad r = \pm\sqrt{x^2 + y^2},$$

where the sign for r is chosen to ensure that the point is in the correct quadrant. This fails when $x = 0$, but in this case we have $\theta = \pi/2$ and $r = y$.

Relationship Between Rectangular and Polar Coordinates

a. A point with polar coordinates $P(r, \theta)$ has rectangular coordinates $P(x, y)$, where

$$x = r \cos \theta \quad \text{and} \quad y = r \sin \theta.$$

b. One set of polar coordinates for the point with rectangular coordinates $P(x, y)$ is given, when $x \neq 0$, by

$$\theta = \arctan \frac{y}{x} \quad \text{and} \quad r = \pm \sqrt{x^2 + y^2},$$

where the sign of r is chosen to ensure that the point is in the correct quadrant. When $x = 0$, one set of polar coordinates is $\theta = \pi/2$ and $r = y$.

EXAMPLE 1 Find rectangular coordinates of the point that has polar coordinates $(2, 2\pi/3)$.

Solution Since $r = 2$ and $\theta = 2\pi/3$, we have

$$x = r \cos \theta = 2 \cos \left(\frac{2\pi}{3} \right) = 2 \left(-\frac{1}{2} \right) = -1,$$

and

$$y = r \sin \theta = 2 \sin \left(\frac{2\pi}{3} \right) = 2 \left(\frac{\sqrt{3}}{2} \right) = \sqrt{3}.$$

Thus the point has rectangular coordinates $(-1, \sqrt{3})$, as shown in Figure 5.38. ■

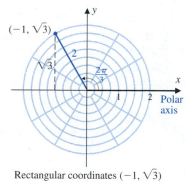

Rectangular coordinates $(-1, \sqrt{3})$
Polar coordinates $\left(2, \frac{2\pi}{3} \right)$

FIGURE 5.38

EXAMPLE 2 Find polar coordinates of the point that has rectangular coordinates $(1, -1)$.

Solution Since $x = 1$ and $y = -1$, we have

$$r = \pm \sqrt{x^2 + y^2} = \pm \sqrt{(1)^2 + (-1)^2} = \pm \sqrt{2},$$

and

$$\theta = \arctan \frac{y}{x} = \arctan \frac{-1}{1} = \arctan(-1) = -\frac{\pi}{4}.$$

Since the point is in the fourth quadrant, the polar coordinates for the point are $(\sqrt{2}, -\pi/4)$, as shown in Figure 5.39.

Of course, the polar coordinates we have found are not unique in describing this point. We could also use $(-\sqrt{2}, 3\pi/4)$ or $(\sqrt{2}, 7\pi/4)$, for example.   ■

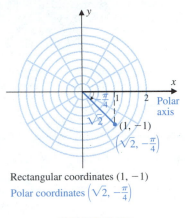

Rectangular coordinates $(1, -1)$
Polar coordinates $\left(\sqrt{2}, -\frac{\pi}{4}\right)$

FIGURE 5.39

Graphs of Polar Equations

A *polar equation* is an equation involving the polar variables r and θ. For example,

$$r = 2 + 2\cos\theta, \qquad r = \sin 2\theta, \qquad r = 3, \qquad \text{and} \qquad \theta = \frac{\pi}{2}$$

are all polar equations. The curve described by the equation is the collection of all points $P(r, \theta)$ that have *at least one* polar representation (r, θ) that satisfies the equation.

Quite often a polar equation is represented as $r = f(\theta)$, so r is a function of the independent variable θ. Computer algebra systems and most graphing calculators can sketch curves in polar coordinates and permit us to see curves in the plane that might otherwise be very difficult to represent. Some of these examples are shown in Figure 5.40. In this brief introduction to polar curves, we look just at a few of those that are commonly seen in calculus.

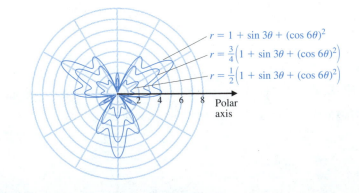

FIGURE 5.40

EXAMPLE 3   Sketch the graph of each polar equation.

a. $r = 3$ b. $\theta = \pi/3$

Solution   a. The graph consists of all points with r-coordinate 3, that is, all points that are 3 units from the origin. This graph is the circle with center at the origin and radius 3, as shown in Figure 5.41(a). The rectangular equation is $x^2 + y^2 = 9$.

b. The graph consists of all points with θ-coordinate $\pi/3$. Any point $(r, \pi/3)$ with $r > 0$ lies in the first quadrant. If $r < 0$, the point lies in the third quadrant. So the graph is the line passing through the origin making an angle of $\pi/3$ radians with the polar axis, as shown in Figure 5.41(b). If a point on this line has rectangular coordinates (x, y), then

$$\frac{y}{x} = \tan \theta = \tan \frac{\pi}{3} = \sqrt{3},$$

so the line has the rectangular equation $y = \sqrt{3}x$.   ■

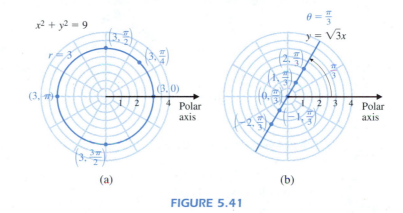

FIGURE 5.41

EXAMPLE 4   Sketch the graph of the polar equation $r = 2 \sin \theta$, and transform this polar equation to one in rectangular coordinates.

Solution   Table 5.2 lists sample values of θ and the corresponding values of r. Although θ can assume any real value, we need consider only values of θ between 0 and 2π since the sine function has period 2π.

TABLE 5.2

θ	0	$\frac{\pi}{6}$	$\frac{\pi}{4}$	$\frac{\pi}{3}$	$\frac{\pi}{2}$	$\frac{2\pi}{3}$	$\frac{3\pi}{4}$	$\frac{5\pi}{6}$	π
$r = 2 \sin \theta$	0	1	$\sqrt{2}$	$\sqrt{3}$	2	$\sqrt{3}$	$\sqrt{2}$	1	0
θ	$\frac{7\pi}{6}$	$\frac{5\pi}{4}$	$\frac{4\pi}{3}$	$\frac{3\pi}{2}$	$\frac{5\pi}{3}$	$\frac{7\pi}{4}$	$\frac{11\pi}{6}$	2π	
$r = 2 \sin \theta$	-1	$-\sqrt{2}$	$-\sqrt{3}$	-2	$-\sqrt{3}$	$-\sqrt{2}$	-1	0	

Connecting the points in Table 5.2 with a smooth curve gives the graph shown in Figure 5.42. The graph appears to be a circle with center on the ray $\theta = \pi/2$, 1 unit from the origin and radius 1.

The points obtained for θ between π and 2π are the same points as those determined for θ between 0 and π, so the total graph is obtained by plotting only values of θ between 0 and π.

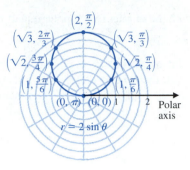

FIGURE 5.42

To verify that this graph is a circle, we can change the polar equation to a rectangular equation. It is easier to change r^2 to rectangular coordinates than it is r, so we first multiply both sides of the polar equation $r = 2 \sin \theta$ by r. This gives

$$r^2 = 2r \sin \theta.$$

Since $r^2 = x^2 + y^2$ and $y = r \sin \theta$ we have

$$x^2 + y^2 = 2y, \qquad \text{and} \qquad x^2 + y^2 - 2y = 0.$$

Completing the square on the y-terms gives

$$x^2 + (y - 1)^2 = 1.$$

This equation describes the circle with center $(0, 1)$ and radius 1. ■

The equations in Examples 3(a) and 4 are special cases of the family of circles shown in Figure 5.43 in the case $a > 0$.

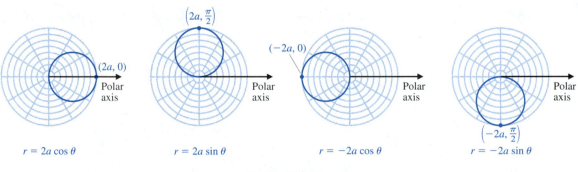

$r = 2a \cos \theta$ 　　　　 $r = 2a \sin \theta$ 　　　　 $r = -2a \cos \theta$ 　　　　 $r = -2a \sin \theta$

FIGURE 5.43

EXAMPLE 5   Sketch the graph of $r = 1 + \cos \theta$, and transform the equation to a rectangular equation.

Solution   Since $\cos(\theta) = \cos(-\theta)$ the graph is symmetric with respect to the x-, or polar, axis, so it is sufficient to consider only values of θ between 0 and π. The values in Table 5.3, which decrease from 2 to 0 as θ increases from 0 to π, allow us to plot the upper half of the graph shown in Figure 5.44(a). The lower half is simply the reflection of the upper half through the polar axis. The complete graph is shown in Figure 5.44(b).

TABLE 5.3

θ	0	$\frac{\pi}{6}$	$\frac{\pi}{4}$	$\frac{\pi}{3}$	$\frac{\pi}{2}$	$\frac{2\pi}{3}$	$\frac{3\pi}{4}$	$\frac{5\pi}{6}$	π
$r = 1 + \cos\theta$	2	$1 + \frac{\sqrt{3}}{2}$	$1 + \frac{\sqrt{2}}{2}$	$\frac{3}{2}$	1	$\frac{1}{2}$	$1 - \frac{\sqrt{2}}{2}$	$1 - \frac{\sqrt{3}}{2}$	0

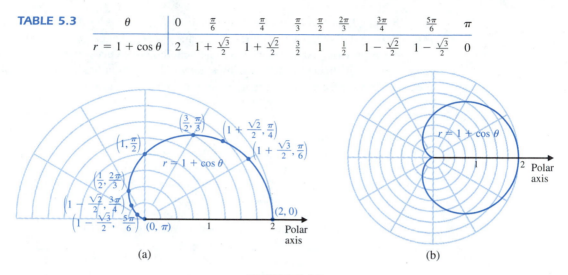

(a)   (b)

FIGURE 5.44

To transform the polar equation to a rectangular equation, we multiply both sides of $r = 1 + \cos\theta$ by r and use the relations $r^2 = x^2 + y^2$ and $x = r\cos\theta$. This process gives

$$r^2 = r + r\cos\theta,$$

so

$$x^2 + y^2 = r + x, \qquad \text{or} \qquad x^2 - x + y^2 = r.$$

Squaring both sides of the last equation gives

$$\left(x^2 - x + y^2\right)^2 = r^2,$$

or, since $r^2 = x^2 + y^2$,

$$\left(x^2 - x + y^2\right)^2 = x^2 + y^2.$$

This rectangular equation is much more complicated then the one found in Example 4 and would be difficult to graph without the polar equivalent. ■

The heart-shaped curve in Figure 5.44 is called a **cardioid**. It is a member of a family of curves called **limaçons** (pronounced *lim-a-sons*). The three types of limaçons are shown in Figure 5.45, where the graph is symmetric with respect to the *x*-axis—that

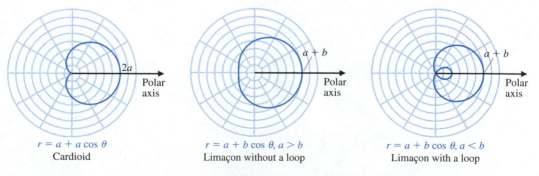

$r = a + a\cos\theta$ $r = a + b\cos\theta, a > b$ $r = a + b\cos\theta, a < b$
Cardioid   Limaçon without a loop   Limaçon with a loop

FIGURE 5.45

is, to the line described by $\theta = 0$. If the cosine is replaced with the sine, the graph of $r = a + b\sin\theta$ is symmetric with respect to the y-axis—that is, to the line $\theta = \pi/2$.

In the previous examples we have encountered polar curves that exhibited one or more types of symmetries. Recognizing symmetries can greatly simplify the graphing process. Three of the most important tests for symmetry are given below.

Tests for Symmetry

a. If replacing θ with $-\theta$ leaves the polar equation unchanged, then the graph is symmetric with respect to the polar, or x-axis. (See Figure 5.46(a).)

b. If replacing θ with $\pi - \theta$ leaves the polar equation unchanged, then the graph is symmetric with respect to the ray $\theta = \pi/2$, or y-axis. (See Figure 5.46(b).)

c. If replacing r with $-r$ leaves the polar equation unchanged, then the graph is symmetric with respect to the pole, or origin. (See Figure 5.46(c).)

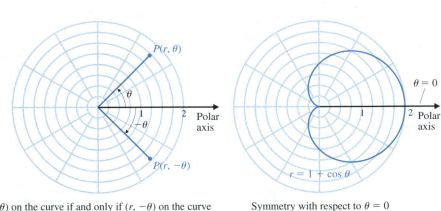

(r, θ) on the curve if and only if $(r, -\theta)$ on the curve
Symmetry with respect to $\theta = 0$

Symmetry with respect to $\theta = 0$

(a)

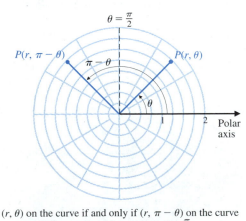

(r, θ) on the curve if and only if $(r, \pi - \theta)$ on the curve
Symmetry with respect to $\theta = \frac{\pi}{2}$

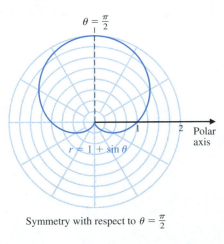

Symmetry with respect to $\theta = \frac{\pi}{2}$

(b)

FIGURE 5.46

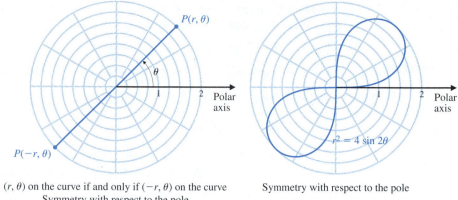

$P(r, \theta)$

$P(-r, \theta)$

(r, θ) on the curve if and only if $(-r, \theta)$ on the curve
Symmetry with respect to the pole

$r^2 = 4 \sin 2\theta$

Symmetry with respect to the pole

(c)

FIGURE 5.46

EXAMPLE 6   Sketch the graph of $r = \cos 2\theta$.

Solution   Since the cosine function is even, we have $\cos(\theta) = \cos(-\theta)$ for all values of θ, and the graph is symmetric with respect to the polar, or x-axis. The graph is also symmetric with respect to $\theta = \pi/2$, the y-axis, since

$$\cos 2(\pi - \theta) = \cos(2\pi - 2\theta) = \cos(-2\theta) = \cos(2\theta).$$

Because of this symmetry the graph is determined once we know the shape of the graph in the first quadrant. Table 5.4 lists values of θ in $[0, \pi/2]$ and the corresponding values of r.

TABLE 5.4

θ	0	$\frac{\pi}{8}$	$\frac{\pi}{6}$	$\frac{\pi}{4}$	$\frac{\pi}{3}$	$\frac{5\pi}{12}$	$\frac{\pi}{2}$
$r = \cos 2\theta$	1	$\frac{\sqrt{2}}{2}$	$\frac{1}{2}$	0	$-\frac{1}{2}$	$-\frac{\sqrt{3}}{2}$	-1

Since $r = 0$ when $\theta = \pi/4$, the graph approaches the pole along the line $\theta = \pi/4$, as shown in Figure 5.47(a). Symmetry with respect to the x-axis allows us to

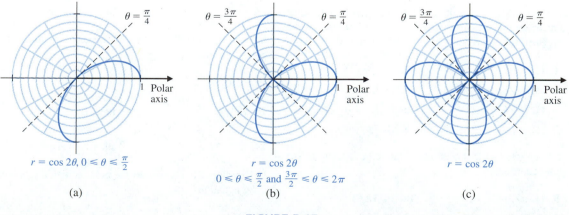

$\theta = \frac{\pi}{4}$

$r = \cos 2\theta, 0 \le \theta \le \frac{\pi}{2}$

(a)

$\theta = \frac{3\pi}{4}$ $\theta = \frac{\pi}{4}$

$r = \cos 2\theta$
$0 \le \theta \le \frac{\pi}{2}$ and $\frac{3\pi}{2} \le \theta \le 2\pi$

(b)

$\theta = \frac{3\pi}{4}$ $\theta = \frac{\pi}{4}$

$r = \cos 2\theta$

(c)

FIGURE 5.47

extend the graph as shown in Figure 5.47(b), and symmetry with respect to the y-axis gives the complete graph shown in Figure 5.47(c). The graph is called a *four-leafed rose*.   ■

EXAMPLE 7   Sketch the graphs of $r = -2 \sin \theta$ and $r = 2 + 2 \sin \theta$, and find the points of intersection.

Solution   Equating the two r values and solving for θ gives

$$-2 \sin \theta = 2 + 2 \sin \theta, \qquad 4 \sin \theta + 2 = 0, \qquad \text{and} \qquad \sin \theta = -\frac{1}{2},$$

so $\theta = 7\pi/6$ or $\theta = 11\pi/6$. Substituting these two values into either of the original equations gives $r = 1$, so points of intersection occur at $(1, 7\pi/6)$ and $(1, 11\pi/6)$.

The graphs are in Figure 5.48, which shows a third point of intersection at the origin. This point of intersection was not obtained from solving the two equations simultaneously. On $r = -2 \sin \theta$ the origin is represented by the coordinates $(0, 0)$, but on $r = 2 + 2 \sin \theta$ it is represented by the coordinates $(0, 3\pi/2)$.   ■

In calculus you will use integration to find the area bounded between two curves. The first step in the solution to this type of problem is to find the points of intersection of the curves.

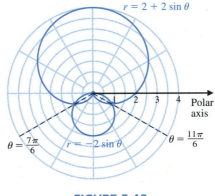

FIGURE 5.48

EXERCISE SET 5.5

In Exercises 1–8, plot the point with the given polar coordinates and give two other pairs of polar coordinates that represent the point, one with $r > 0$ and one with $r < 0$.

1. $(2, \pi/3)$

2. $(3, -\pi/4)$

3. $(5, 2\pi/3)$

4. $(-3, -\pi/4)$

5. $(1, 7\pi/6)$

6. $(-4, \pi/3)$

7. $(4, -\pi/6)$

8. $(8, 7\pi/4)$

In Exercises 9–14, convert the polar coordinates to rectangular coordinates.

9. $(1, \pi/3)$

10. $(4, 3\pi/2)$

11. $(0, 5\pi)$

12. $(5, 7\pi/3)$

13. $(-2, 5\pi/4)$

14. $(-3, -7\pi/6)$

In Exercises 15–20, convert the rectangular coordinates to polar coordinates.

15. $(2, 0)$

16. $(-2, 0)$

17. $(0, 4)$

18. $(0, -4)$

19. $(-2, 2)$

20. $(3\sqrt{3}, 3)$

In Exercises 21–24, convert the polar equation to a rectangular equation.

21. $r = 4$

22. $\theta = 3\pi/4$

23. $r = 2\cos\theta$

24. $r = \dfrac{1}{1 + 2\sin\theta}$

c. $r = 2 + \sin\theta$ **d.** $r = 3 + \sin\theta$

60. a. $r = \sin\theta$ **b.** $r = \cos\theta$

c. $r = \csc\theta$ **d.** $r = \sec\theta$

In Exercises 25–30, convert the rectangular equation to a polar equation.

25. $y = x$ **26.** $x - 2y = 4$

27. $x^2 + y^2 = 16$ **28.** $x^2 + y^2 = 2y$

29. $y = x^2$ **30.** $x^2 - y^2 = 1$

61. The lengthwise cross section of an apple can be reasonably approximated using a polar curve that we have seen in this section. What would be the form of the equation if the polar axis is along the stem of the apple with the pole at the base of the stem?

62. Find the coordinates of the bases of a baseball field if a polar coordinate system is placed on the field with the pole at home plate and the polar axis is aligned as indicated.

In Exercises 31–54, sketch the graph of the polar equation.

31. $r = 3$ **32.** $r = 5\pi/3$

33. $\theta = 5\pi/3$ **34.** $\theta = 3$

35. $r = 3\cos\theta$ **36.** $r = 6\sin\theta$

37. $r = -3\cos\theta$ **38.** $r = -6\sin\theta$

39. $r = 1 + \cos\theta$ **40.** $r = 1 + \sin\theta$

41. $r = 2 + \sin\theta$ **42.** $r = 2 + \cos\theta$

43. $r = 2 - 2\sin\theta$ **44.** $r = 1 - 2\sin\theta$

45. $r = 3\sin3\theta$ **46.** $r = 4\cos3\theta$

47. $r = 3\cos2\theta$ **48.** $r = 4\sin2\theta$

49. $r^2 = 16\cos2\theta$ **50.** $r^2 = 4\sin2\theta$

51. $r = \theta$ **52.** $r = 2^{-\theta}$

53. $r = e^{\theta}$ **54.** $r = \ln\theta$

a. Along the first base line

b. Parallel to the line between first and third base

Assume that the infield is in the shape of a square with sides of 90 ft.

In Exercises 63–66, find all points of intersection of the two curves.

63. $r = 2 + 2\cos\theta,\ r = -2\cos\theta$

64. $r = 1 + 2\cos\theta,\ r = 1$

65. $r = \sin2\theta,\ r = 1$

66. $r = 2 - 2\sin\theta,\ r = 2\sin\theta$

In Exercises 55–60, compare the graphs of the polar equations.

55. a. $r = 1 + \cos\theta$ **b.** $r = 1 - \cos\theta$

c. $r = \cos\theta - 1$ **d.** $r = -\cos\theta - 1$

56. a. $r = 1 + \sin\theta$ **b.** $r = 1 - \sin\theta$

c. $r = \sin\theta - 1$ **d.** $r = -\sin\theta - 1$

57. a. $r = \cos\theta$ **b.** $r = 1 + \cos\theta$

c. $r = 2 + \cos\theta$ **d.** $r = 3 + \cos\theta$

58. a. $r = \sin\theta$ **b.** $r = \sin2\theta$

c. $r = \sin3\theta$ **d.** $r = \sin4\theta$

59. a. $r = \sin\theta$ **b.** $r = 1 + \sin\theta$

67. Use a graphing device to sketch several curves from the family of curves.

$$r = 1 + \sin(n\theta) + \left(\cos(2n\theta)\right)^2, \qquad n \geq 1$$

68. Use a graphing device to sketch several curves from the family of curves.

$$r = (\sin m\theta)(\cos n\theta)$$

where m and n are positive integers. Discuss the effect of m and n on the curves.

69. Use a graphing device to compare the curves

$$r = \sin m\theta \qquad \text{and} \qquad r = |\sin m\theta|.$$

5.6   CONIC SECTIONS IN POLAR COORDINATES

Earlier in this chapter we found rectangular equations for parabolas in terms of a focus and a directrix and rectangular equations for ellipses and hyperbolas in terms of two foci. In many physical problems it is more useful to define all the conic sections in terms of a focus and a directrix and to use these to determine *polar* equations.

Conics Defined by Eccentricity

Let F be a fixed point, called the **focus**, l be a fixed line, called the **directrix**, and e, called the **eccentricity**, be a fixed positive number. The set of all points P in the plane such that

$$d(P, F) = ed(P, l),$$

is a conic section. Moreover, if

a. $e = 1$, the conic is a parabola;
b. $e < 1$, the conic is an ellipse;
c. $e > 1$, the conic is a hyperbola.

It is the polar form of the equation of an ellipse that is used in vector calculus when verifying that the orbit of a planet around the sun is an ellipse.

In Figure 5.49 we see that d is the distance from the directrix to the pole, so the distance from a point P on the graph to the directrix is $d + r \cos \theta$, and the distance from P to the focal point is r.

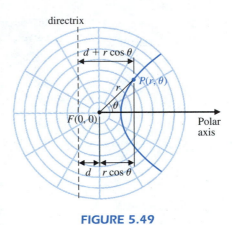

FIGURE 5.49

Hence we have

$$r = e(d + r \cos \theta).$$

Squaring both sides of this equation and changing to rectangular coordinates produces

$$x^2 + y^2 = e^2(d + x)^2 = e^2 d^2 + 2e^2 dx + e^2 x^2,$$

which is a quadratic equation in x and y, so it does describe a conic section. To discover which conic section it represents, we rewrite the equation as

$$(1 - e^2)x^2 - 2e^2 dx + y^2 = e^2 d^2.$$

If $e = 1$, the coefficient of x^2 is zero and we have a parabola. If $0 < e < 1$, the coefficient of x^2 is positive and we have an ellipse, whereas if $e > 1$, the coefficient of x^2 is negative and we have a hyperbola.

Solving the defining equation

$$r = e(d + r \cos \theta)$$

for r gives

$$r(1 - e \cos \theta) = ed, \quad \text{or} \quad r = \frac{ed}{1 - e \cos \theta}$$

when the focal point is at the pole and the directrix is vertical and to the left of the pole.

The following result describes the general situation for conics with a focus at a pole and the directrix parallel to a coordinate axis.

Polar Equations of Conic Sections

The graph of a polar equation of the form

$$r = \frac{ed}{1 \pm e \cos \theta} \quad \text{or} \quad r = \frac{ed}{1 \pm e \sin \theta}$$

is a conic section with eccentricity e. The graph is a parabola if $e = 1$, an ellipse if $e < 1$, and a hyperbola if $e > 1$. The focus is at the pole, and the directrix is d units from the pole.

It can be shown that eccentricity as defined here agrees with the earlier definition for ellipses and hyperbolas.

The form with the cosine in the denominator has a vertical directrix, and the form with the sine in the denominator has a horizontal directrix. When the denominator $1 \pm e \cos \theta$ has a plus sign, the directrix is to the right of the pole, and when it has a minus sign, the directrix is to the left of the pole. When the denominator $1 \pm e \sin \theta$ has a plus sign, the directrix is above the pole, and when it has a minus sign, the directrix is below the pole. These situations are illustrated in Figure 5.50.

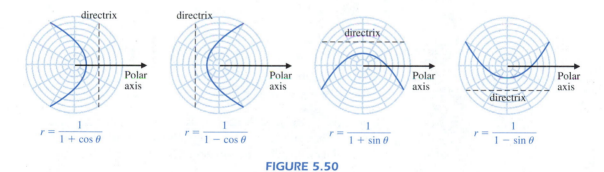

$$r = \frac{1}{1 + \cos \theta} \qquad r = \frac{1}{1 - \cos \theta} \qquad r = \frac{1}{1 + \sin \theta} \qquad r = \frac{1}{1 - \sin \theta}$$

FIGURE 5.50

Notice that a conic written in standard polar position has a focus at the pole, which is the origin in rectangular coordinates. In standard rectangular position the focus will not be at the origin, so *standard* position depends on the coordinate system being used.

EXAMPLE 1 Sketch the graph of the parabola whose polar equation is

$$r = \frac{2}{1 + \sin \theta}.$$

Solution In this equation $e = 1$ and $d = 2$. The focus of the parabola is at the origin, and the directrix is the horizontal line 2 units above the pole, that is, $y = 2$. The vertex of the parabola is midway between the focus and the directrix, so it has polar coordinates

$(1, \pi/2)$ and rectangular coordinates $(0, 1)$. The parabola opens downward and intersects with the x-axis at the points $(2, 0)$ and $(2, \pi)$, as shown in Figure 5.51. ■

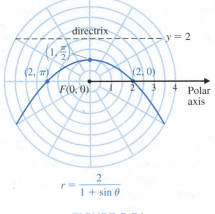

$$r = \frac{2}{1 + \sin \theta}$$

FIGURE 5.51

EXAMPLE 2   Sketch the graph of the conics defined by each equation.

a. $r = \dfrac{16}{5 + 3 \cos \theta}$

b. $r = \dfrac{10}{2 + 5 \sin \theta}$

Solution   a. The standard form for the conic section with focal point at the origin and directrix vertical and d units to the right of the focal point is

$$r = \frac{ed}{1 + e \cos \theta}.$$

Our equation is

$$r = \frac{16}{5 + 3 \cos \theta} = \frac{16}{5(1 + \frac{3}{5} \cos \theta)} = \frac{\frac{16}{5}}{1 + \frac{3}{5} \cos \theta},$$

so $e = 3/5$. Since $ed = 16/5$, we have $d = 16/3$. The eccentricity is less than 1, so the equation describes an ellipse whose vertices, which occur when $\theta = 0$ and $\theta = \pi$, are at the points with polar coordinates $(2, 0)$ and $(8, \pi)$. The graph is shown in Figure 5.52.

b. Writing the equation in standard form gives

$$r = \frac{10}{2 + 5 \sin \theta} = \frac{5}{1 + \frac{5}{2} \sin \theta}.$$

Since $e = \frac{5}{2} > 1$, the conic is a hyperbola with directrix above the x-axis at a distance

$$d = \frac{5}{e} = \frac{5}{\frac{5}{2}} = 2.$$

The vertices of the hyperbola occur when $\theta = \pi/2$ and $3\pi/2$, at the points with polar coordinates $(10/7, \pi/2)$ and $(-10/3, 3\pi/2)$. The portion of the hyperbola lying below the directrix can be obtained by plotting $(10/7, \pi/2)$ and the points of

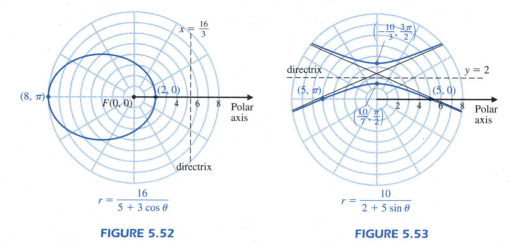

FIGURE 5.52

FIGURE 5.53

intersection with the x-axis, which occur at $(5, \pi)$ and $(5, 0)$. The portion of the hyperbola lying above the directrix is the reflection of the lower part about the horizontal line equidistant from $y = 10/7$ and $y = 10/3$, as shown in Figure 5.53.

EXAMPLE 3   The initial flight of the NASA space shuttle occurred when the Columbia was placed in elliptical earth orbit on April 12, 1981. The *apogee* (the maximum height above the earth) was 250 km and the *perigee* (the minimum height above the earth) was 238 km, with a focal point of the orbit located at the center of mass of the earth. Determine the eccentricity of the orbit.

Solution   We will assume for simplicity that the earth is spherical with its center of mass located at the center of the earth, as shown in Figure 5.54. We will also assume that the distance from the center of the earth to the surface is 6373 km (approximately 3960 mi).

This is another type of real-world application that is considered in calculus. There we can calculate the length of time it took a shuttle to orbit the earth.

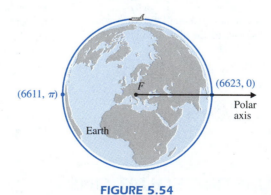

FIGURE 5.54

Since the orbit is elliptical, the form of the equation is

$$r = \frac{ed}{1 - e\cos\theta}, \qquad \text{where } e < 1.$$

The shuttle is at its apogee when $\theta = 0$, so

$$\frac{ed}{1 - e} = 6373 + 250 = 6623 \text{ km},$$

and the shuttle is at its perigee when $\theta = \pi$, so

$$\frac{ed}{1 + e} = 6373 + 238 = 6611 \text{ km}.$$

Solving these two equations for ed and equating them gives

$$6623(1 - e) = ed = 6611(1 + e),$$

so

$$e = \frac{12}{13234} \approx 0.9 \times 10^{-3}.$$

This very small value for the eccentricity reflects the fact that the orbit is nearly circular. At its perigee it is 6611 km from the center of the earth, only 12 km less than at its apogee. ■

EXERCISE SET 5.6

In Exercises 1–6, sketch the graph of the conic section, and find a corresponding rectangular equation.

1. $r = \dfrac{2}{1 + \cos \theta}$ **2.** $r = \dfrac{2}{4 + \cos \theta}$

3. $r = \dfrac{1}{1 + 2 \cos \theta}$ **4.** $r = \dfrac{3}{1 - \cos \theta}$

5. $r = \dfrac{3}{3 - \sin \theta}$ **6.** $r = \dfrac{3}{1 - 2 \sin \theta}$

In Exercises 7–10, find a polar equation of the conic with a focus at the origin and satisfying the given conditions.

7. $e = 2$, directrix $x = -4$

8. $e = 1/2$, directrix $y = -2$

9. $e = 1$, directrix $y = -1/4$

10. $e = 1/3$, directrix $r = \sec \theta$

11. Find a polar equation of the ellipse in standard polar position that has vertices at the points with polar coordinates $(1, 0)$ and $(3, \pi)$.

12. Find a polar equation of the parabola in standard polar position that has a vertex at the point with rectangular coordinates $(-6, 0)$.

13. The world's first orbiting satellite, *Sputnik I*, was launched in the Soviet Union on October 4, 1957. Its elliptical orbit reached a maximum height of 560 mi above the earth and a minimum height of 145 mi above the earth. Write a polar equation of the orbit, assuming that the pole is placed at the center of the earth, which is 3960 mi below the surface.

14. The earth moves in an elliptical orbit, with the sun at one focal point and an eccentricity $e = 0.0167$. The major axis of the elliptical orbit is approximately 2.99×10^8 km. Write a polar equation for this ellipse, assuming that the pole is at the sun.

5.7   PARAMETRIC EQUATIONS

We have represented curves using rectangular coordinates and polar coordinates. In each case the curve consists of all points in the plane that satisfy a single equation. An alternative method of describing a curve in the plane is to express the x- and y-coordinates of points on the curve separately as functions of a third variable t, called a *parameter*.

In *multivariable calculus* you will study surfaces in space. Parametric equations often allow for a better description of the surface. One of the uses of parametric surfaces is in generating three-dimensional computer graphics.

This gives a flexibility that is not available using strictly rectangular or polar equations, and is the basis for the method used to describe curves that lie in three-dimensional space.

Parametric Equations

The pair of equations

$$x = f(t) \quad \text{and} \quad y = g(t)$$

are called the **parametric** equations for the curve determined by the points $(x, y) = (f(t), g(t))$, as t assumes all the values in the common domain of f and g.

EXAMPLE 1 Describe and sketch the curve whose parametric equations are

$$x = f(t) = 2t + 1, \qquad y = g(t) = 4t^2 - 1, \qquad \text{for } -1 \le t \le 1.$$

Solution By choosing representative values of t in the interval $[-1, 1]$, we obtain the points on the graph shown in Figure 5.55(a). Instead of proceeding in this manner, however, let us try to eliminate the parameter and see if we can determine a direct relation between x and y.

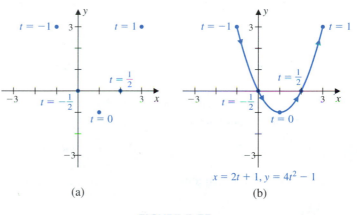

(a)   (b)

FIGURE 5.55

Solving for t in the first equation and substituting into the second gives

$$t = \frac{x - 1}{2}, \qquad \text{so} \qquad y = 4\left(\frac{x - 1}{2}\right)^2 - 1 = (x - 1)^2 - 1.$$

This is the equation of a parabola. As t varies from -1 to 1, points on the curve are traced from $(-1, 3)$ to $(3, 3)$. The x-intercepts occur at values of t making $y = 0$, so

$$y = 4t^2 - 1 = 0, \quad t^2 = \frac{1}{4}, \qquad \text{and} \qquad t = \pm\frac{1}{2}.$$

Substituting these values into the parametric equation for x gives the x-intercepts $(0, 0)$ and $(2, 0)$. The curve is shown in Figure 5.55(b). ∎

EXAMPLE 2   Find parametric equations for the line through $(1, 2)$ with slope $1/2$.

Solution   A rectangular equation for this line is

$$y - 2 = \frac{1}{2}(x - 1).$$

We can find one set of parametric equations by setting $t = x - 1$. Then

$$y - 2 = \frac{1}{2}t, \quad \text{so} \quad y = \frac{1}{2}t + 2.$$

The line shown in Figure 5.56 is then described by the parametric equations

$$x = t + 1, \qquad y = \frac{1}{2}t + 2, \qquad \text{for } -\infty < t < \infty.$$

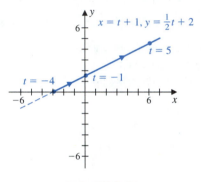

FIGURE 5.56

Parametric representations of curves are far from unique. Another set of parametric equations for the line is

$$x = t, \qquad y = \frac{1}{2}(t - 1) + 2, \qquad \text{for } -\infty < t < \infty.$$

The only conditions that need be fulfilled are that

$$y = \frac{1}{2}(x - 1) + 2$$

and that both x and y assume all real values.   ■

The parametric equations

$$x = t^2, \qquad y = \frac{1}{2}(t^2 - 1) + 2 = \frac{1}{2}(t^2 + 3), \qquad \text{for } -\infty < t < \infty$$

also satisfy the equation

$$y = \frac{1}{2}(x - 1) + 2$$

given in Example 2, but the curve described by these equations does not trace the entire line. Since $x \geq 0$ for all values of t, only the portion of the line in Figure 5.57 is traced.

Moreover, except for the point $(0, 3/2)$, the curve is traced twice, once for t in $(-\infty, 0)$ and again for t in $(0, \infty)$.

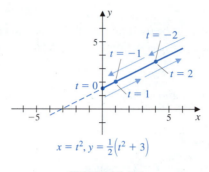

$$x = t^2, y = \tfrac{1}{2}\left(t^2 + 3\right)$$

FIGURE 5.57

EXAMPLE 3 Describe the curve given by the following parametric equations.

a. $x = \cos t, y = \sin t, 0 \le t \le 2\pi$ b. $x = \sin 2t, y = \cos 2t, 0 \le t \le 2\pi$

Solution a. Eliminating the parameter t gives

$$x^2 + y^2 = (\cos t)^2 + (\sin t)^2 = 1,$$

so all the points (x, y) lie on the unit circle. As t increases from 0 to 2π, the points $(x, y) = (\cos t, \sin t)$ make one complete revolution counterclockwise around the circle, starting at $(1, 0)$ and returning to the starting point when $t = 2\pi$. (See Figure 5.58(a).)

b. As in part (a), eliminating the parameter gives

$$x^2 + y^2 = (\sin 2t)^2 + (\cos 2t)^2 = 1,$$

so the parametric equations again trace the unit circle. But this time as t increases from 0 to 2π, the points $(x, y) = (\sin 2t, \cos 2t)$ start at $(0, 1)$ and make *two* complete revolutions *clockwise* around the circle, as shown in Figure 5.58(b), one for t in $[0, \pi]$ and the other when $t \in [\pi, 2\pi]$. ∎

When using a graphing device to sketch a curve given parametrically, be sure to observe how the curve is traced.

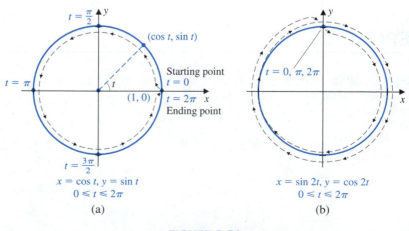

$$x = \cos t, y = \sin t$$
$$0 \le t \le 2\pi$$
(a)

$$x = \sin 2t, y = \cos 2t$$
$$0 \le t \le 2\pi$$
(b)

FIGURE 5.58

Polar coordinates were introduced in Section 5.5 and were used to describe conics in Section 5.6. The conversion formulas from polar to rectangular coordinates,

$$x = r \cos \theta, \qquad y = r \sin \theta,$$

can also be used to convert a polar equation of the form $r = f(\theta)$, for $\alpha \leq \theta \leq \beta$, into parametric equations. Simply replace r with $f(\theta)$ to get

$$x = f(\theta) \cos \theta, \qquad y = f(\theta) \sin \theta, \qquad \alpha \leq \theta \leq \beta.$$

For example, parametric equations of the cardioid

$$r = 1 + \sin \theta, \qquad 0 \leq \theta \leq 2\pi$$

are

$$x = (1 + \sin \theta) \cos \theta, \qquad y = (1 + \sin \theta) \sin \theta, \qquad 0 \leq \theta \leq 2\pi.$$

Computer algebra systems and most graphing calculators can graph curves given parametrically since they employ simple point-plotting methods. As a consequence, complicated curves, as in Figure 5.59, can easily be generated by computer, but would be very difficult to generate by hand.

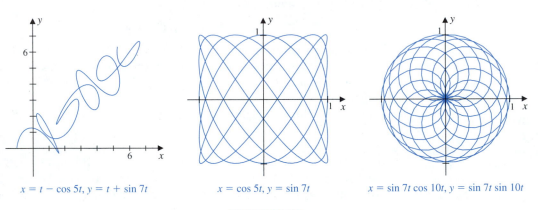

$x = t - \cos 5t, y = t + \sin 7t$ $\qquad$ $x = \cos 5t, y = \sin 7t$ $\qquad$ $x = \sin 7t \cos 10t, y = \sin 7t \sin 10t$

FIGURE 5.59

EXERCISE SET 5.7

In Exercises 1–14, sketch the graph of the curve described by the parametric equations, and find a corresponding rectangular equation.

1. $x = 3t, \quad y = t/2$

2. $x = 2t + 1, \quad y = 3t - 2$

3. $x = \sqrt{t}, \quad y = t + 1$

4. $x = t + 2, \quad y = -3\sqrt{t}$

5. $x = \sin t, \quad y = (\cos t)^2$

6. $x = \sec t, \quad y = \tan t$

7. $x = 3 \sin t, \quad y = 4 \cos t$

8. $x = 3 \sec t, \quad y = 4 \tan t$

9. $x = e^t, \quad y = e^{-t}$

10. $x = \ln t, \quad y = \ln \sqrt{t}$

11. $x = \sin t, \quad y = \sec t$

12. $x = \sin t, \quad y = \cot t$

13. $x = 4t^3, \quad y = 2t^2 - 1$

14. $x = e^{-t} - 2, \quad y = e^{2t} + 3$

In Exercises 15–18, sketch the graph of the conic described by the parametric equations.

15. a. $x = t, \quad y = t^2$ **b.** $x = t^2, \quad y = t$

   c. $x = t^2, \quad y = t^4$ **d.** $x = t^4, \quad y = t^2$

16. a. $x = 2t + 1, \quad y = 3t$ **b.** $x = 3t + 1, \quad y = 2t$

   c. $x = 2t, \quad y = 3t + 1$ **d.** $x = 3t, \quad y = 2t + 1$

17. a. $x = \sin t, \quad y = \cos t$

   b. $x = \sin t, \quad y = \cos t + 1$

   c. $x = \cos t + 1, \quad y = \sin t$

   d. $x = \cos t, \quad y = \sin t + 1$

18. a. $x = 1 - \sin t, \quad y = 1 - \cos t$

   b. $x = \sin t - 1, \quad y = \cos t - 1$

   c. $x = 1 - \cos t, \quad y = 1 - \sin t$

   d. $x = \cos t - 1, \quad y = \sin t - 1$

In Exercises 19–22 the graphs of the parametric equations in each part represents a portion of the same curve. Sketch the graphs of these equations, and label representative values of the parameter.

19. a. $x = \cos t, \quad y = \sin t$

   b. $x = \sin t, \quad y = \cos t$

   c. $x = t, \quad y = \sqrt{1 - t^2}$

   d. $x = -t, \quad y = \sqrt{1 - t^2}$

20. a. $x = t + 1, \quad y = 2t + 3$

   b. $x = t^2 - 1/2, \quad y = 2t^2$

   c. $x = \ln t, \quad y = 1 + \ln t^2$

   d. $x = \sin t, \quad y = 1 + 2\sin t$

21. a. $x = t, \quad y = \ln t$ **b.** $x = e^t, \quad y = t$

   c. $x = t^2, \quad y = 2\ln t$ **d.** $x = 1/t, \quad y = -\ln t$

22. a. $x = t, \quad y = 1/t$ **b.** $x = e^t, \quad y = e^{-t}$

   c. $x = \sin t, \quad y = \csc t$ **d.** $x = \tan t, \quad y = \cot t$

In Exercises 23 and 24, find parametric equations for the line satisfying the given conditions.

23. Slope $\frac{1}{3}$, passing through $(2, -1)$

24. Passing through $(5, 3)$ and $(-2, 7)$

25. Describe the differences in the curves defined by

$$x = \cos t, \qquad y = \sin t, \qquad \text{for } -\pi \le t \le \pi$$

and

$$x = \sin t, \qquad y = \cos t, \qquad \text{for } -\pi \le t \le \pi.$$

26. Find parametric equations in the parameter t, where $0 \le t \le 2\pi$, for the circle $x^2 + y^2 = r^2$ so that it is traced as described.

   a. Once around, counterclockwise, starting at $(r, 0)$

   b. Twice around, counterclockwise, starting at $(r, 0)$

   c. Three times around, counterclockwise, starting at $(-r, 0)$

   d. Twice around, clockwise, starting at $(0, r)$

   e. Three times around, clockwise, starting at $(0, r)$

27. The curve described by the rectangular equation $x^3 + y^3 = xy$ is called the *folium of Descartes*.

   a. Show that this curve is described by the parametric equations

$$x = \frac{t^2}{1 + t^3}, \qquad y = \frac{t}{1 + t^3}, \qquad \text{for } t \ne -1.$$

   b. Sketch the graph of the curve, and use arrows to indicate the direction of increasing t.

28. Use a graphing device to investigate the family of curves given by the parametric equations

$$x = (a - \sin t)\cos t \qquad \text{and} \qquad y = (a - \sin t)\sin t$$

where a is a real number and $0 \le t \le 2\pi$.

29. A *cycloid* can be described as the path traced out by a point on a circle as the circle rolls along a line. If the radius of the circle is a, the parametric equations of the cycloid are given by

$$x = a(t - \sin t) \qquad \text{and} \qquad y = a(1 - \cos t).$$

Use a graphing device to sketch the cycloid for different values of a. What effect does the parameter a have on the curve? For what values of t does the curve touch the x-axis?

30. Use a graphing device to investigate the family of curves given by the parametric equations

$$x = a(\cos t)^3 \qquad \text{and} \qquad y = a(\sin t)^3,$$

where a is a real number. What effect does the parameter a have on the curve? These curves are called *four-cusp hypocycloids*.

5.8   ROTATION OF AXES

At the beginning of the chapter we stated that all equations of the form

$$Ax^2 + Bxy + Cy^2 + Dx + Ey + F = 0$$

represent a conic section, which might possibly be degenerate. We saw in Section 5.2 that the graph of the quadratic equation

$$Ax^2 + Cy^2 + Dx + Ey + F = 0$$

is a parabola when $A = 0$ or $C = 0$; that is, when $AC = 0$. In Section 5.3 we found that the graph is an ellipse if $AC > 0$, and in Section 5.4 we saw that the graph is a hyperbola when $AC < 0$. So we have classified the situation when $B = 0$. Degenerate situations can occur; for example, the quadratic equation $x^2 + y^2 + 1 = 0$ has no solutions, and the graph of $x^2 - y^2 = 0$ is not a hyperbola, it is the pair of lines with equations $y = \pm x$. But excepting this type of situation, we have categorized the graphs of all the quadratic equations with $B = 0$.

When $B \neq 0$, a rotation can be performed to bring the conic into a form that will allow us to compare it with one that is in standard position. To set up the discussion, let us first suppose that a set of xy-coordinate axes has been rotated about the origin by an angle θ, where $0 < \theta < \pi/2$, to form a new set of $\hat{x}\hat{y}$-axes, as shown in Figure 5.60(a). We would like to determine the coordinates for a point P in the plane relative to the two coordinate systems.

Rotation techniques are used extensively in computer graphics.

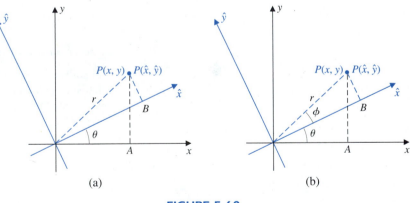

FIGURE 5.60

First introduce a new pair of variables r and ϕ to represent, respectively, the distance from P to the origin and the angle formed by the $\hat{x}$-axis and the line connecting the origin to P, as shown in Figure 5.60(b).

From the right triangle OBP shown in Figure 5.61(a), we see that

$$\hat{x} = r \cos \phi \quad \text{and} \quad \hat{y} = r \sin \phi,$$

and from the right triangle OAP shown in Figure 5.61(b), we see that

$$x = r \cos(\phi + \theta) \quad \text{and} \quad y = r \sin(\phi + \theta).$$

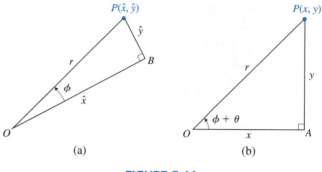

FIGURE 5.61

Using the fundamental trigonometric identities for the sum of the sine and the cosine, we have

$$x = r\cos(\phi + \theta) = r\cos\phi\cos\theta - r\sin\phi\sin\theta$$

and

$$y = r\sin(\phi + \theta) = r\cos\phi\sin\theta + r\sin\phi\cos\theta.$$

Since $\hat{x} = r\cos\phi$ and $\hat{y} = r\sin\phi$, we have the following result. The second pair of equations is derived by solving for $\hat{x}$ and $\hat{y}$ in the first pair.

Coordinate Rotation Formulas

If a rectangular xy-coordinate system is rotated through an angle θ to form an $\hat{x}\hat{y}$-coordinate system, then a point $P(x, y)$ will have coordinates $P(\hat{x}, \hat{y})$ in the new system, where (x, y) and $(\hat{x}, \hat{y})$ are related by

$$x = \hat{x}\cos\theta - \hat{y}\sin\theta \qquad \text{and} \qquad y = \hat{x}\sin\theta + \hat{y}\cos\theta$$

and

$$\hat{x} = x\cos\theta + y\sin\theta \qquad \text{and} \qquad \hat{y} = -x\sin\theta + y\cos\theta.$$

EXAMPLE 1   Show that the graph of the equation $xy = 1$ is a hyperbola by rotating the xy-axes through an angle of $\pi/4$.

Solution   Denoting a point in the rotated system by $(\hat{x}, \hat{y})$, we have

$$x = \hat{x}\cos\frac{\pi}{4} - \hat{y}\sin\frac{\pi}{4} = \frac{\sqrt{2}}{2}(\hat{x} - \hat{y})$$

and

$$y = \hat{x}\sin\frac{\pi}{4} + \hat{y}\cos\frac{\pi}{4} = \frac{\sqrt{2}}{2}(\hat{x} + \hat{y}).$$

Substituting these expressions into the original equation $xy = 1$ produces the equation

$$\frac{\sqrt{2}}{2}(\hat{x} - \hat{y})\frac{\sqrt{2}}{2}(\hat{x} + \hat{y}) = 1, \qquad \text{or} \qquad \hat{x}^2 - \hat{y}^2 = 2.$$

In the $\hat{x}\hat{y}$-coordinate system, then, we have a standard-position hyperbola whose asymptotes are $\hat{y} = \pm\hat{x}$. These are the same lines as the x- and y-axes, as seen in Figure 5.62. ∎

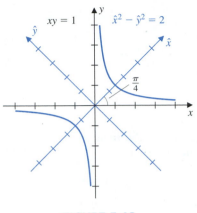

FIGURE 5.62

In Example 1 the appropriate angle of rotation was provided to eliminate the $\hat{x}\hat{y}$-term from the equation. Such an angle θ can always be found so that when the coordinate axes are rotated through this angle, the equation in the new coordinate system will not involve the product $\hat{x}\hat{y}$.

To determine the angle, suppose we have the general second-degree equation

$$Ax^2 + Bxy + Cy^2 + Dx + Ey + F = 0, \qquad \text{where} \qquad B \neq 0.$$

Introduce the variables

$$x = \hat{x}\cos\theta - \hat{y}\sin\theta \qquad \text{and} \qquad y = \hat{x}\sin\theta + \hat{y}\cos\theta,$$

and substitute for x and y in the original equation. This gives us the new equation in $\hat{x}$ and $\hat{y}$:

$$A(\hat{x}\cos\theta - \hat{y}\sin\theta)^2 + B(\hat{x}\cos\theta - \hat{y}\sin\theta)(\hat{x}\sin\theta + \hat{y}\cos\theta)$$
$$+ C(\hat{x}\sin\theta + \hat{y}\cos\theta)^2 + D(\hat{x}\cos\theta - \hat{y}\sin\theta) + E(\hat{x}\sin\theta + \hat{y}\cos\theta) + F = 0.$$

Performing the multiplication and collecting similar terms gives

$$\hat{x}^2 \left(A(\cos\theta)^2 + B(\cos\theta\sin\theta) + C(\sin\theta)^2 \right)$$
$$+ \hat{x}\hat{y} \left[-2A\cos\theta\sin\theta + B\left((\cos\theta)^2 - (\sin\theta)^2\right) + 2C\sin\theta\cos\theta \right]$$
$$+ \hat{y}^2 \left(A(\sin\theta)^2 - B(\sin\theta\cos\theta) + C(\cos\theta)^2 \right)$$
$$+ \hat{x}(D\cos\theta + E\sin\theta) + \hat{y}(-D\sin\theta + E\cos\theta) + F = 0.$$

To eliminate the $\hat{x}\hat{y}$-term from this equation, choose θ so that the coefficient of this term is zero, that is, so that

$$-2A\cos\theta\sin\theta + B\left((\cos\theta)^2 - (\sin\theta)^2\right) + 2C\sin\theta\cos\theta = 0.$$

Simplifying this equation, we have

$$B\left((\cos\theta)^2 - (\sin\theta)^2\right) = 2(A - C)\cos\theta\sin\theta$$

and

$$\frac{(\cos\theta)^2 - (\sin\theta)^2}{2\cos\theta\sin\theta} = \frac{A - C}{B}.$$

Using the double-angle formulas for sine and cosine gives

$$\frac{\cos 2\theta}{\sin 2\theta} = \frac{A - C}{B}, \qquad \text{or} \qquad \cot 2\theta = \frac{A - C}{B}.$$

Rotating Quadratic Equations

To eliminate the xy-term in the general quadratic equation

$$Ax^2 + Bxy + Cy^2 + Dx + Ey + F = 0, \qquad \text{where} \qquad B \neq 0,$$

rotate the coordinate axes through an angle θ that satisfies

$$\cot 2\theta = \frac{A - C}{B}.$$

EXAMPLE 2 Sketch the graph of $4x^2 - 4xy + 7y^2 - 24 = 0$.

Solution To eliminate the xy-term we first rotate the coordinate axes through the angle θ, where

$$\cot 2\theta = \frac{A - C}{B} = \frac{4 - 7}{-4} = \frac{3}{4}.$$

From the triangle in Figure 5.63, we see that

$$\cos 2\theta = \frac{3}{5}.$$

Using the half-angle formulas we have

$$\cos\theta = \sqrt{\frac{1 + \frac{3}{5}}{2}} = \frac{2\sqrt{5}}{5} \qquad \text{and} \qquad \sin\theta = \sqrt{\frac{1 - \frac{3}{5}}{2}} = \frac{\sqrt{5}}{5}.$$

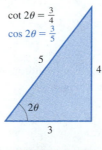

$\cot 2\theta = \frac{3}{4}$
$\cos 2\theta = \frac{3}{5}$

5

4

2θ

3

FIGURE 5.63

The rotation-of-axes formulas then give

$$x = \frac{2\sqrt{5}}{5}\hat{x} - \frac{\sqrt{5}}{5}\hat{y} \quad \text{and} \quad y = \frac{\sqrt{5}}{5}\hat{x} + \frac{2\sqrt{5}}{5}\hat{y}.$$

Substituting these into the original equation gives

$$4\left(\frac{2\sqrt{5}}{5}\hat{x} - \frac{\sqrt{5}}{5}\hat{y}\right)^2 - 4\left(\frac{2\sqrt{5}}{5}\hat{x} - \frac{\sqrt{5}}{5}\hat{y}\right)\left(\frac{\sqrt{5}}{5}\hat{x} + \frac{2\sqrt{5}}{5}\hat{y}\right)$$

$$+ 7\left(\frac{\sqrt{5}}{5}\hat{x} + \frac{2\sqrt{5}}{5}\hat{y}\right)^2 - 24 = 0.$$

After simplifying we have

$$\frac{\hat{x}^2}{5} + 8\hat{y}^2 = 24 \quad \text{or} \quad \frac{\hat{x}^2}{120} + \frac{\hat{y}^2}{3} = 1.$$

This is the graph of an ellipse in standard position with respect to the $\hat{x}\hat{y}$-coordinate system. The graph of the equation

$$4x^2 - 4xy + 7y^2 = 24$$

is consequently the graph of an ellipse that has been rotated through an angle $\theta = \arcsin\frac{\sqrt{5}}{5} \approx 27°$, as shown in Figure 5.64.   ■

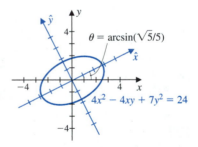

FIGURE 5.64

There is an easily applied formula that can be used to determine which conic will be produced once the rotation has been performed. Suppose that a rotation θ of the coordinate axes changes the equation

$$Ax^2 + Bxy + Cy^2 + Dx + Ey + F = 0$$

into the equation

$$\hat{A}\hat{x}^2 + \hat{B}\hat{x}\hat{y} + \hat{C}\hat{y}^2 + \hat{D}\hat{x} + \hat{E}\hat{y} + \hat{F} = 0.$$

It is not hard to show, although it is algebraically tedious, that the coefficients of these two equations satisfy

$$B^2 - 4AC = \hat{B}^2 - 4\hat{A}\hat{C}.$$

When the particular angle chosen is such that $\hat{B} = 0$, we have

$$B^2 - 4AC = -4\hat{A}\hat{C}.$$

However, except for the degenerate cases we know that the new equation

$$\hat{A}\hat{x}^2 + \hat{C}\hat{y}^2 + \hat{D}\hat{x} + \hat{E}\hat{y} + \hat{F} = 0$$

will be

1. An ellipse if $\hat{A}\hat{C} > 0$;
2. A hyperbola if $\hat{A}\hat{C} < 0$;
3. A parabola if $\hat{A}\hat{C} = 0$.

Applying this result to the original equation gives the following.

Classification of Conic Curves

Except for degenerate cases, the equation

$$Ax^2 + Bxy + Cy^2 + Dx + Ey + F = 0$$

will be

(i) An ellipse if $B^2 - 4AC = -4\hat{A}\hat{C} < 0$;
(ii) A hyperbola if $B^2 - 4AC = -4\hat{A}\hat{C} > 0$;
(iii) A parabola if $B^2 - 4AC = -4\hat{A}\hat{C} = 0$.

The beauty of this result is that we do not need to know the angle that produces the elimination of the $\hat{x}\hat{y}$-terms. We need to know only that such an angle exists, and we have already established this fact.

If we apply the result to the examples in this section we see that:

Example 1: The equation $xy = 1$ has

$$B^2 - 4AC = 1^2 - 4(0)(0) = 1 > 0$$

and is, therefore, a hyperbola.

Example 2: The equation $4x^2 - 4xy + 7y^2 - 24 = 0$ has

$$B^2 - 4AC = (-4)^2 - 4(4)(7) = -96 < 0$$

and is, therefore, an ellipse.

We strongly recommend that you apply this simple test when rotation is required to graph a general quadratic equation. It provides a final check on a result produced from a considerable amount of algebraic and trigonometric computation. Although it will not ensure that you are correct, it can tell you when you are certainly wrong, and this is often more important knowledge.

EXERCISE SET 5.8

In Exercises 1–4, determine the $\hat{x}\hat{y}$-coordinates of the given point if the coordinate axes are rotated through the given angle θ.

1. $(0, 1)$, $\theta = 60°$

2. $(1, 1)$, $\theta = 45°$

3. $(-3, 1)$, $\theta = 30°$

4. $(-2, -2)$, $\theta = 90°$

In Exercises 5–12, (a) determine whether the conic section is an ellipse, hyperbola, or parabola, and (b) perform a rotation and, if necessary, a translation, and sketch the graph.

5. $x^2 - xy + y^2 = 2$

6. $x^2 + 4xy + y^2 = 3$

7. $17x^2 - 6xy + 9y^2 = 0$

8. $x^2 - \sqrt{3}xy = 1$

9. $4x^2 - 4xy + 7y^2 = 24$

10. $5x^2 - 3xy + y^2 = 5$

11. $6x^2 + 4\sqrt{3}xy + 2y^2 - 9x + 9\sqrt{3}y - 63 = 0$

12. $x^2 + 2\sqrt{3}xy + 3y^2 + (4 + 2\sqrt{3})x + (4\sqrt{3} - 2)y + 12 = 0$

13. Find an equation of the parabola with axis $y = x$ passing through the points $(1, 0)$, $(0, 1)$, and $(1, 1)$ (a) in $\hat{x}\hat{y}$-coordinates and (b) in xy-coordinates.

REVIEW EXERCISES FOR CHAPTER 5

In Exercises 1–4, sketch the graph and show the vertex, focus, and directrix of the parabola.

1. $4y - x^2 = 0$

2. $y^2 + 12x = 0$

3. $4x - y^2 + 6y - 17 = 0$

4. $x^2 + 4x + 8y - 4 = 0$

In Exercises 5–8, sketch the graph and show the vertices and focal points of the ellipse.

5. $x^2 + 4y^2 = 4$

6. $4x^2 + y^2 = 16$

7. $4(x - 1)^2 + 9(y + 2)^2 = 36$

8. $2(x + 1)^2 + (y - 1)^2 = 2$

In Exercises 9–12, sketch the graph and show the vertices, focal points, and asymptotes of the hyperbola.

9. $x^2 - 2y^2 = 4$

10. $4y^2 - x^2 = 16$

11. $2x^2 - 4x - 4y^2 + 1 = 0$

12. $9x^2 - 4y^2 - 18x + 16y - 43 = 0$

In Exercises 13–22, identify the type of curve and sketch the graph.

13. $x^2 = -2(y - 5)$

14. $9(x + 2)^2 - 4(y - 5)^2 = 36$

15. $16x^2 + 25y^2 = 400$

16. $y^2 = -16(x + 1)$

17. $x^2 - 2x - 4y - 11 = 0$

18. $y^2 - 2x + 2y + 7 = 0$

19. $9x^2 - 16y^2 = 144$

20. $9x^2 + 4y^2 - 90x - 16y + 205 = 0$

21. $16x^2 - 64x - 25y^2 + 150y = 561$

22. $4x^2 - 9y^2 - 16x - 90y - 173 = 0$

In Exercises 23–30, find an equation of the conic.

23. A parabola with focus at $(0, 0)$ and directrix $y = 2$

24. An ellipse with foci at $(0, \pm 1)$ and vertices at $(0, \pm 3)$

25. A hyperbola with foci at $(\pm 3, 0)$ and vertex at $(1, 0)$

26. A parabola with focus at $(0, 1)$ and vertex at $(0, -1)$

27. An ellipse with foci at $(0, \pm 5)$ and passing through the point $(4, 0)$

28. A conic with focus at the origin, eccentricity $3/4$, and directrix $x = 2$

29. A conic with focus at the origin, eccentricity 3, and directrix $y = -2$

30. A conic with focus at the origin, eccentricity 1, and directrix $x = -3$

In Exercises 31–36, sketch the curve whose polar equation is given.

31. $r = 4 + 4\cos\theta$

32. $r = 1 + 3\sin\theta$

33. $r = 2\sin\theta$

34. $r = 2$

35. $r = 2\cos 2\theta$

36. $\theta = \dfrac{1}{2}$

In Exercises 37–40, sketch the conic section whose polar equation is given.

37. $r = \dfrac{3}{1 + \cos\theta}$

38. $r = \dfrac{3}{2 + 4\cos\theta}$

39. $r = \dfrac{2}{1 - \sin\theta}$

40. $r = \dfrac{4}{2 - \cos\theta}$

In Exercises 41–46, sketch the parametric curve and eliminate the parameter to find a rectangular equation for the curve.

41. $x = t^2 - 1, \quad y = t + 1$

42. $x = 2\cos t, \quad y = 3\sin t$

43. $x = e^t$, $y = 1 + e^{-t}$

44. $x = t + 2$, $y = (t - 1)^2 + 1$

45. $x = (\sin t)^2 + 1$, $y = (\cos t)^2$

46. $x = \ln t^2$, $y = \ln t^3 + 1$

In Exercises 47 and 48, (a) determine whether the conic section is an ellipse, hyperbola, or parabola, and (b) perform a rotation and, if necessary, a translation, and sketch the graph.

47. $2xy - \sqrt{2}x + \sqrt{2}y = 5$

48. $x^2 + 4xy + y^2 = 16$

49. Use a graphing device to sketch several curves from the family of curves

$$r = 1 + \sin 2n\theta + (\cos n\theta)^2$$

for positive integers n. How does the parameter n effect the curve?

50. Use a graphing device to sketch several curves from the family of curves

$$r = \sin m\theta + (\cos n\theta)^2$$

for m and n positive integers. Discuss the effect of the parameters m and n on the curve.

51. Use a graphing device to investigate the family of curves given by the parametric equations

$$x = (1 - a\sin t)\cos t, \qquad y = (1 - a\sin t)\sin t,$$

where a is a real number and $0 \le t \le 2\pi$. What effect does the parameter a have on the curves?

52. Use a graphing device to investigate the family of curves given by the parametric equations

$$x = (a - b)\cos t + b\cos \frac{a - b}{b}t,$$

$$y = (a - b)\sin t - b\sin \frac{a - b}{b}t$$

for $0 \le t \le 2\pi$ and a and b real numbers with $a > b$. These curves are called *hypocycloids*.

53. Use a graphing device to investigate the family of curves given by the parametric equations

$$x = (a + b)\cos t - b\cos \frac{a + b}{b}t,$$

$$y = (a + b)\sin t - b\sin \frac{a + b}{b}t$$

for $0 \le t \le 2\pi$ and a and b real numbers with $a > b$. These curves are called *epicycloids*.

CHAPTER 5: CALCULUS PREVIEW EXERCISES

1. a. Find an equation of the parabola with axis the y-axis and vertex $(0, 0)$ that passes through the point (x_1, y_1), where $x_1 \ne 0$ and $y_1 \ne 0$.

b. Find an equation of the parabola with axis parallel to the y-axis and vertex (h, k) that passes through (x_1, y_1), where $x_1 \ne h$ and $y_1 \ne k$.

2. a. Find an equation of the parabola with axis the x-axis and vertex $(0, 0)$ that passes through the point (x_1, y_1), where $x_1 \ne 0$ and $y_1 \ne 0$.

b. Use the result in part (a) to find an equation of the parabola with axis parallel to the x-axis and vertex (h, k) that passes through (x_1, y_1), where $x_1 \ne h$ and $y_1 \ne k$.

3. Find an equation of the parabola with vertex at (h, k) and passing through $(h + 1, k + 1)$ with (a) a vertical axis, and (b) a horizontal axis.

4. Find an equation of the parabola passing through the three points $(1, 0)$, $(0, 1)$ and $(2, 2)$ (a) with axis parallel to the y-axis, and (b) with axis parallel to the x-axis.

In Exercises 5 and 6, use the fact that the tangent line shown on the following figure to an ellipse of the form $x^2/a^2 + y^2/b^2 = 1$ at the point (x_0, y_0) has equation

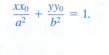

$$\frac{xx_0}{a^2} + \frac{yy_0}{b^2} = 1.$$

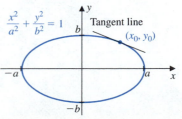

5. Find the equation of the tangent line to the ellipse $9x^2 + 4y^2 = 36$ at $(1, \frac{3\sqrt{3}}{2})$.

6. Find equations of the tangent lines to the ellipse $4x^2 + y^2 = 4$ that pass through the point $(3, 0)$.

In Exercises 7 and 8, use the fact that the tangent line shown in the following figure to a hyperbola of the form $\frac{x^2}{a^2} - \frac{y^2}{b^2} = 1$ at the point (x_0, y_0) has equation

$$\frac{xx_0}{a^2} - \frac{yy_0}{b^2} = 1.$$

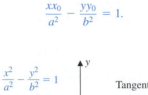

7. Find an equation of a tangent line to the hyperbola with equation $3x^2 - 4y^2 = 12$ at the point $(2\sqrt{2}, \sqrt{3})$.

8. Find equations of the tangent lines to the hyperbola $9x^2 - 4y^2 = 36$ that passes through $(1, 0)$. Sketch the hyperbola and show the tangent lines.

9. Describe the curve traced by a point 2 ft from the top of an 8-ft ladder as the bottom of the ladder moves away from a vertical wall.

10. A DSS satellite dish receiver has its amplifier in line with the edge of the dish, as shown in the figure. The diameter of the dish at the edge is 18 in. How deep is the dish?

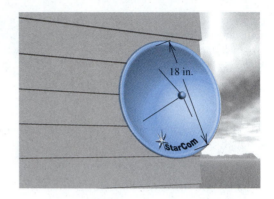

11. A reflecting dish is made by revolving the point of the parabola $4y = x^2$ below $y = 2$ about its axis of symmetry. Where should a light source be placed to produce a focused beam of light with parallel rays?

12. It can be shown that the shape of a flexible and inelastic cable supporting a load that is uniformly distributed horizontally, like the cables of a suspension bridge, is a parabola. Place the origin of an xy-coordinate system at the lowest point of a parabolic cable of a suspension bridge whose span is L and sag is h, as shown in the figure. Determine the equation for the cable. The George Washington bridge across the Hudson River in New York has a span of 3500 feet and a sag of 316 feet. What is the equation of the cable?

GEORGE WASHINGTON BRIDGE PLANS

13. The Hale telescope, named to honor the American astronomer George Ellery Hale (1868–1939), is one of the world's largest compound reflecting telescopes. It is located at the Palomar Mountain Observatory 45 mi northeast of San Diego, California. The main parabolic mirror of this telescope is 200 in. in diameter, with a depth from rim to vertex of 3.75 in. A small cylindrical platform is located within the tube of the telescope along the axis of the parabola for an observer to view and record the reflection from the telescope. How far from the center of the mirror is the observer's viewing area located?

14. A satellite is placed in a position to make a parabolic flight past the moon, with the center of the moon at the focal point of the parabola. When the satellite is 5783 km from the surface of the moon, it makes an angle of 60° with the axis of the parabola. The closest the satellite gets to the surface is 143 km. What is the diameter of the moon? (Assume that the gravitational center of the moon is at its center: the focus of the parabola. This assumption is not quite correct since the gravitational center is offset approximately 2 km from the center toward the earth.)

15. Show that for constants a and b, the polar curve with equation $r = a \cos \theta + b \sin \theta$ is the circle through the origin with Cartesian equation

$$\left(x - \frac{a}{2}\right)^2 + \left(y - \frac{b}{2}\right)^2 = \frac{a^2 + b^2}{4}.$$

16. The chambered nautilus (*Nautilus pompilus*) is a mollusk found in the Pacific and Indian oceans. The outside of the shell grows in the form of an exponential spiral. A typical equation of such a spiral is $r = 2e^{-0.2\theta}$. Use a graphing device to sketch the graph of this spiral, and compare the resulting curve to the curve in the photograph.

17. The planets travel in elliptical orbits about the sun, with the sun at a focus. Let a denote the aphelion, the greatest distance from the planet to the sun, and p the perihelion, the minimum distance from the planet to the sun.

   a. Verify that the eccentricity e of the orbit is given by

   $$e = \frac{a - p}{a + p}.$$

   b. Verify that $a = R(1 + e)$ and $p = R(1 - e)$, where $2R$ is the length of the major axis of the ellipse.

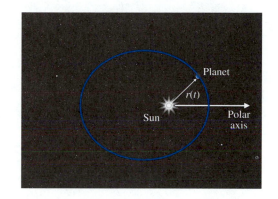

18. The comet Kahoutek has an elliptical orbit about the sun with eccentricity $e = 0.99993$ and a perihelion of 1.95×10^7 mi. Find a polar equation for the orbit, and determine the maximum distance of Kahoutek from the sun.

19. Show that the graph of

   $$\sqrt{x^2 + (y - 1)^2} + 1 = \sqrt{x^2 + (y + 1)^2}$$

   is a conic section.

20. Show that if the graph of the equation

   $$Ax^2 + Bxy + Cy^2 + Dx + Ey + F = 0$$

   is rotated through an angle θ, the coefficients in the new equation

   $$\hat{A}\hat{x}^2 + \hat{B}\hat{x}\hat{y} + \hat{C}\hat{y}^2 + \hat{D}\hat{x} + \hat{E}\hat{y} + \hat{F} = 0$$

   satisfy $\hat{B}^2 - 4\hat{A}\hat{C} = B^2 - 4AC$.

ANSWERS TO SELECTED EXERCISES

Exercise Set 1.2 (Page 11)

1. $-2 \le x \le 4$

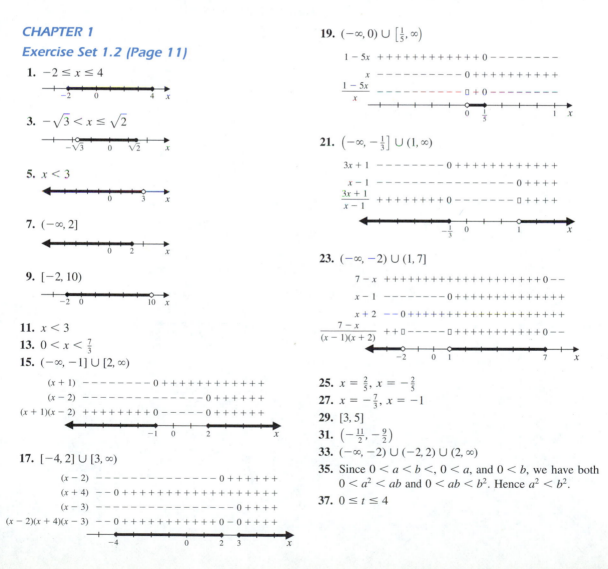

3. $-\sqrt{3} < x \le \sqrt{2}$

5. $x < 3$

7. $(-\infty, 2]$

9. $[-2, 10)$

11. $x < 3$

13. $0 < x < \frac{7}{3}$

15. $(-\infty, -1] \cup [2, \infty)$

$$
\begin{array}{l}
(x+1) \quad --------\;0+++++++++++ \\
(x-2) \quad --------------\;0++++++ \\
(x+1)(x-2) \quad +++++++++\;0-----\;0++++++
\end{array}
$$

17. $[-4, 2] \cup [3, \infty)$

$$
\begin{array}{l}
(x-2) \quad --------------\;0++++++ \\
(x+4) \quad --\;0++++++++++++++++++++ \\
(x-3) \quad ----------------\;0++++ \\
(x-2)(x+4)(x-3) \quad --\;0++++++++++++\;0-\;0++++
\end{array}
$$

19. $(-\infty, 0) \cup \left[\frac{1}{5}, \infty\right)$

$$
\begin{array}{l}
1 - 5x \quad ++++++++++++\;0-------- \\
x \quad ---------\;0++++++++++ \\
\dfrac{1-5x}{x} \quad ---------\;\square+0--------
\end{array}
$$

21. $\left(-\infty, -\frac{1}{3}\right] \cup (1, \infty)$

$$
\begin{array}{l}
3x+1 \quad --------\;0+++++++++++++ \\
x-1 \quad ----------------\;0++++ \\
\dfrac{3x+1}{x-1} \quad ++++++++\;0-------\;\square++++
\end{array}
$$

23. $(-\infty, -2) \cup (1, 7]$

$$
\begin{array}{l}
7-x \quad ++++++++++++++++++++++\;0-- \\
x-1 \quad --------\;0+++++++++++++++ \\
x+2 \quad --\;0+++++++++++++++++++++++ \\
\dfrac{7-x}{(x-1)(x+2)} \quad ++\;\square-----\;\square+++++++++++\;0--
\end{array}
$$

25. $x = \frac{2}{5}, \; x = -\frac{2}{5}$

27. $x = -\frac{7}{3}, \; x = -1$

29. $[3, 5]$

31. $\left(-\frac{11}{2}, -\frac{9}{2}\right)$

33. $(-\infty, -2) \cup (-2, 2) \cup (2, \infty)$

35. Since $0 < a < b <, \; 0 < a$, and $0 < b$, we have both $0 < a^2 < ab$ and $0 < ab < b^2$. Hence $a^2 < b^2$.

37. $0 \le t \le 4$

Exercise Set 1.3 (Page 16)

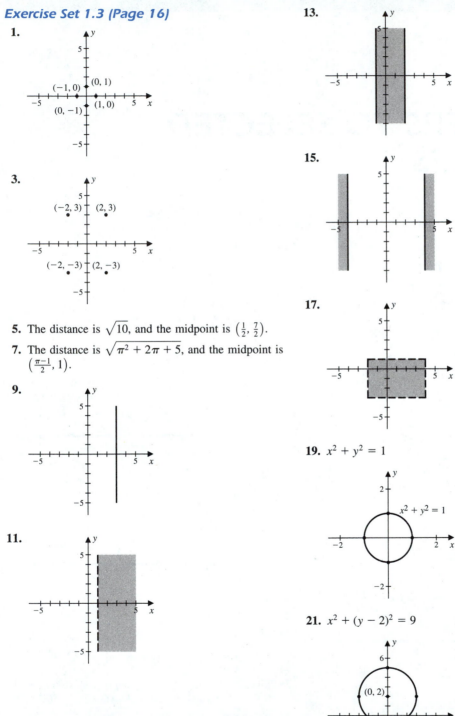

1.

3.

5. The distance is $\sqrt{10}$, and the midpoint is $\left(\frac{1}{2}, \frac{7}{2}\right)$.

7. The distance is $\sqrt{\pi^2 + 2\pi + 5}$, and the midpoint is $\left(\frac{\pi-1}{2}, 1\right)$.

9.

11.

13.

15.

17.

19. $x^2 + y^2 = 1$

21. $x^2 + (y - 2)^2 = 9$

23. Center is $(0, 0)$,   radius is 3.

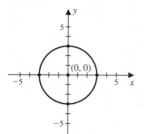

25. Center is $(2, 1)$,   radius is 3.

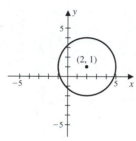

27.

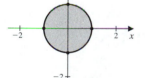

29.

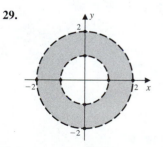

31. Since $d((-1, 4), (-3, -4)) = \sqrt{68}$,
$d((-3, -4), (2, -1)) = \sqrt{34}$, and
$d((2, -1), (-1, 4)) = \sqrt{34}$, the triangle is right with
the right angle at $(2, -1)$.

33. The unique point is $(-6, 1)$.

35. $x^2 + y^2 = 13$

37. $(x - 3)^2 + (y - 7)^2 = 9$

39. $(6, 3)$ is closer.

Exercise Set 1.4 (Page 22)

1. y-axis

3. x-axis, y-axis, origin

5. No symmetry

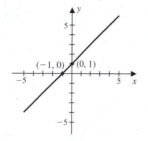

7. No symmetry

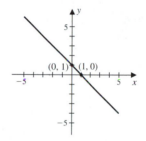

9. y-axis symmetry

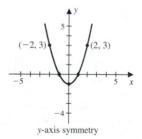

11. No symmetry

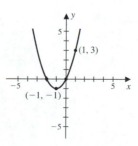

13. *x*-axis symmetry

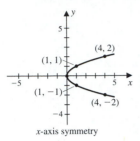

x-axis symmetry

15. No symmetry

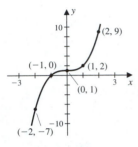

17. No symmetry

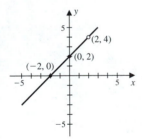

19. No symmetry

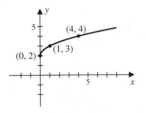

21. *x*-axis, *y*-axis, and origin symmetry

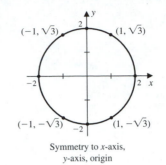

Symmetry to *x*-axis,
y-axis, origin

23. *y*-axis symmetry

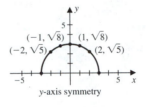

y-axis symmetry

25. *y*-axis symmetry

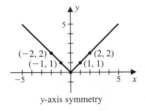

y-axis symmetry

27. *y*-axis symmetry

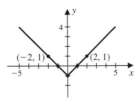

y-axis symmetry

Exercise Set 1.5 (Page 30)

1. a.

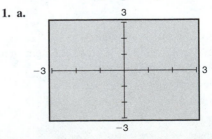

b.

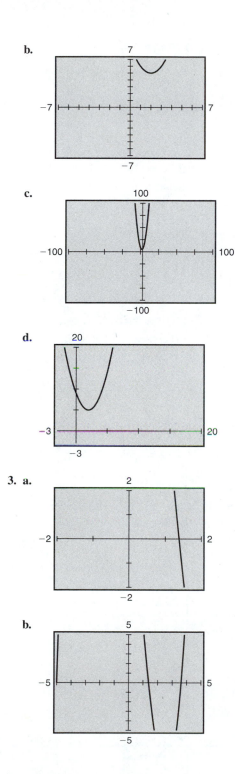

c.

d.

3. a.

b.

c.

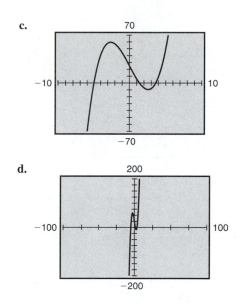

d.

5. a. $x \approx 0.7, -2.7, 2.5$ **b.** $x \approx -1.5, 2.1$

7. If $c > 0$, the graph has an appearance similar to that of $y = x^2$. If $c < 0$, then the graph crosses the x-axis three times, just touching $(0, 0)$. As c increases in magnitude the other two points where the graph crosses the x-axis move further from the origin, symmetric on either side of the origin.

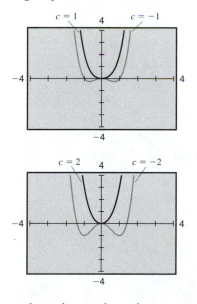

9. a. $\frac{1}{x^n} < \frac{1}{x^{n+2}}$ **b.** $\frac{1}{x^n} > \frac{1}{x^{n+2}}$
   c. n-even: $\frac{1}{x^n} < \frac{1}{x^{n+2}}$; n-odd: $\frac{1}{x^n} > \frac{1}{x^{n+2}}$
   d. n-even: $\frac{1}{x^n} > \frac{1}{x^{n+2}}$; n-odd: $\frac{1}{x^n} < \frac{1}{x^{n+2}}$

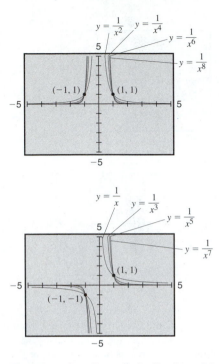

11. The value of a affects the inclination of the line. If $a > 0$, the line is increasing and if $a < 0$ the line is decreasing. The larger the magnitude of a, the steeper the inclination. The constant b determines where the line crosses the y-axis.

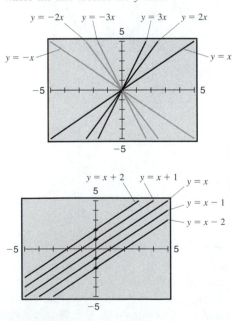

Exercise Set 1.6 (Page 43)

1. a. 21   **b.** 17   **c.** $33 + 16\sqrt{3}$ **d.** 38   **e.** $16x^2 + 5$
  f. $4x^2 - 8x + 9$

3. Yes

5. Yes

7. No

9. Domain: $(-\infty, 0) \cup (0, \infty)$; range: $(-\infty, 0) \cup (0, \infty)$

11. Domain: $(-\infty, \infty)$; range: $\{-1\} \cup (1, \infty)$

13. Domain; $(-\infty, \infty)$; range: $[0, \infty)$

15. Domain: $(-\infty, -1) \cup (-1, 1) \cup (1, \infty)$; range:
  $[0, 1) \cup (1, \infty)$

17. Domain: $(-\infty, 0] \cup [2, \infty)$; range: $[0, \infty)$

19. Domain: $(-\infty, \infty)$; range: $\{-1, 1\}$

21. Yes

23. No

25. $f(-x) = x^2 + 2$; $-f(x) = -x^2 - 2$; $f\left(\frac{1}{x}\right) = \frac{1}{x^2} + 2$;
  $\frac{1}{f(x)} = \frac{1}{x^2+2}$; $f(\sqrt{x}) = x + 2$; $\sqrt{f(x)} = \sqrt{x^2 + 2}$

27. $f(-x) = -\frac{1}{x}$; $-f(x) = -\frac{1}{x}$; $f\left(\frac{1}{x}\right) = x$; $\frac{1}{f(x)} = x$;
  $f(\sqrt{x}) = \frac{1}{\sqrt{x}} = \frac{\sqrt{x}}{x}$; $\sqrt{f(x)} = \sqrt{\frac{1}{x}} = \frac{\sqrt{x}}{x}$

29. $a = \pm 1$

31. $a = \frac{5}{4}$

33. $f(x + h) = 2x + 2h - 4$; $\frac{f(x+h)-f(x)}{h} = 2$

35. $f(x + h) = x^2 + 2hx + h^2$; $\frac{f(x+h)-f(x)}{h} = 2x + h$

37. $f(x + h) = \sqrt{x + h}$; $\frac{f(x+h)-f(x)}{h} = \frac{1}{\sqrt{x+h}+\sqrt{x}}$

39. Even

41. Neither

43. a. Even   **b.** Neither   **c.** Odd   **d.** Neither
  e. Neither   **f.** Neither   **g.** Even   **h.** Odd

45. a.

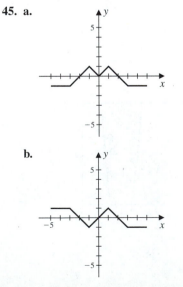

b.

47. a.

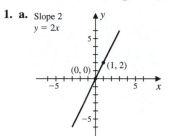

b.

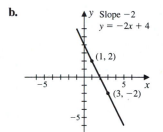

c.

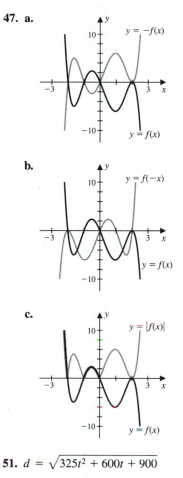

51. $d = \sqrt{325t^2 + 600t + 900}$

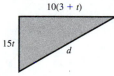

53. a. The perimeter is $2w + \frac{864}{w}$; domain: $(0, \infty)$.
  b. $w = l \approx 20.8$ ft

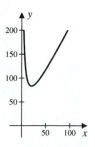

Exercise Set 1.7 (Page 54)

1. a. Slope 2
  $y = 2x$

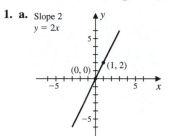

b.

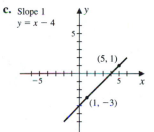

c. Slope 1
  $y = x - 4$

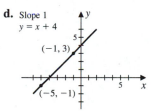

d. Slope 1
  $y = x + 4$

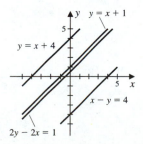

3. (a) and (f); (c) and (g); (b), (d), (i) and (j);
  (e) and (h)

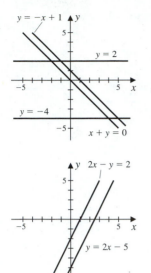

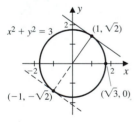

5. a. $y = 2x$;   **b.** $y = -\frac{1}{2}x$

7. a. $y = -2x$;   **b.** $y = \frac{1}{2}x + \frac{5}{2}$

9. $y = 3x - 5$

11. $y = -x + 2$

13. $y = -1$

15. $y = x$

17. $y = -\frac{\sqrt{2}}{2}x + \frac{3\sqrt{2}}{2}; (-1, -\sqrt{2})$

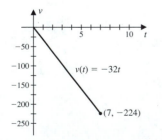

19. Velocity is -224 ft/s.

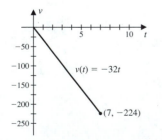

21. b. The total weight is $n(500 - 0.5n)$.
   c. There are no fish when $n \geq 1000$.

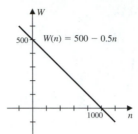

Exercise Set 1.8 (Page 65)

1. Range: $[1, \infty)$

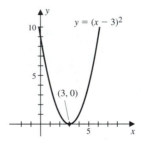

3. Range: $[0, \infty)$

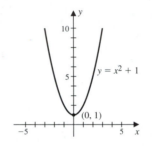

5. Range: $[0, \infty)$

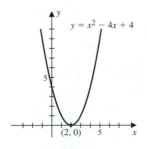

7. Range: $[1, \infty)$

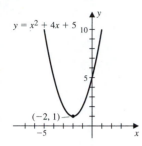

$y = x^2 + 4x + 5$

$(-2, 1)$

-5

9. Range: $(-\infty, 4]$

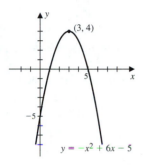

$(3, 4)$

5

-5

$y = -x^2 + 6x - 5$

11. Range: $[-3, \infty)$

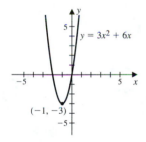

$y = 3x^2 + 6x$

-5

5

$(-1, -3)$

-5

13. Range $= [2, \infty)$

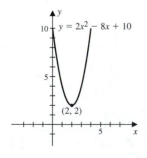

10

$y = 2x^2 - 8x + 10$

5

$(2, 2)$

5

15. Range: $[0, \infty)$

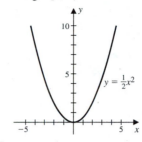

10

5

$y = \frac{1}{2}x^2$

-5

5

17. Range: $[0, \infty)$

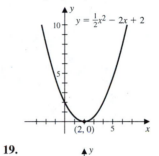

$y = \frac{1}{2}x^2 - 2x + 2$

10

5

$(2, 0)$

5

19.

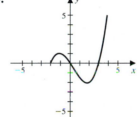

5

-5

5

-5

a.

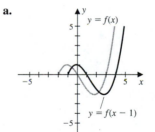

$y = f(x)$

5

-5

5

$y = f(x - 1)$

-5

b.

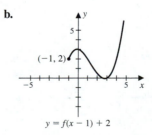

5

$(-1, 2)$

-5

5

$y = f(x - 1) + 2$

c.

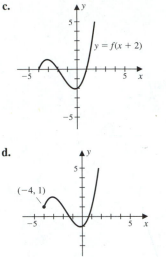

$y = f(x + 2)$

d.

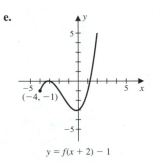

$(-4, 1)$

$y = f(x + 2) + 1$

e.

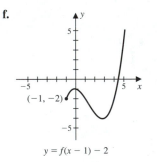

$(-4, -1)$

$y = f(x + 2) - 1$

f.

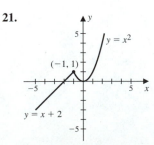

$(-1, -2)$

$y = f(x - 1) - 2$

21.

$y = x^2$

$(-1, 1)$

$y = x + 2$

23. a. $a + b + c = 1$ **b.** $c = 6$
   c. $-\frac{b}{2a} = 1$, and $\frac{4ac - b^2}{4a} = 1$
   d. $a = 5$, $b = -10$, $c = 6$

25. a.

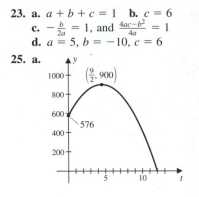

$\left(\frac{9}{2}, 900\right)$

576

b. $t = 12$ s

c. $t = \frac{9}{2}$ s, maximum height $= 900$ ft

d. The physical conditions imply that the domain is
   $[0, 12]$ and the range is $[0, 900]$.

27. Produce 60 items to minimize the cost at $400.00.

29. a. Using the data at 1965, 1980 and 1990 gives
   $y = \frac{88}{75}(x - 1965)^2 - \frac{44}{15}(x - 1965) + 30$ **b.** About
   1365 billion dollars

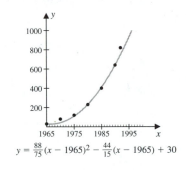

$y = \frac{88}{75}(x - 1965)^2 - \frac{44}{15}(x - 1965) + 30$

31. a.

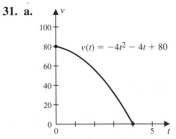

$v(t) = -4t^2 - 4t + 80$

b. 4 s

c.

t	0	0.5	1	1.5	2	2.5	3	3.5	4
$v(t)$	80	77	72	65	56	45	32	17	0

d. 77 ft/s; 80 ft/s   **e.** 72 ft/s; 77 ft/s; 65 ft/s;
   72 ft/s

f. 38.5 ft; 36 ft; 40 ft; 38.5 ft

g. 182 ft; 222 ft

Exercise Set 1.9 (Page 73)

1.

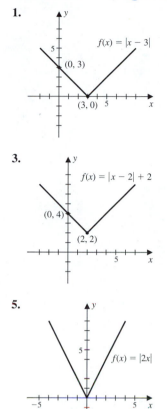

$f(x) = |x - 3|$

(0, 3)

(3, 0)   5

3.

$f(x) = |x - 2| + 2$

(0, 4)

(2, 2)

5

5.

$f(x) = |2x|$

−5   5   x

7. a.

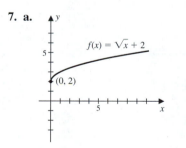

$f(x) = \sqrt{x} + 2$

(0, 2)

5

b. Domain: $[0, \infty)$; range: $[2, \infty)$

9. a.

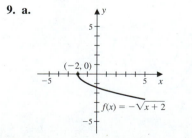

(−2, 0)

−5   5   x

$f(x) = -\sqrt{x + 2}$

−5

b. Domain: $[-2, \infty)$; range: $(-\infty, 0]$

11. a.–f.

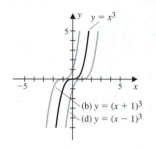

$y = x^3$

−5   5   x

(b) $y = (x + 1)^3$
(d) $y = (x - 1)^3$

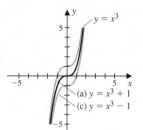

$y = x^3$

−5   5   x

(a) $y = x^3 + 1$
(c) $y = x^3 - 1$

−5

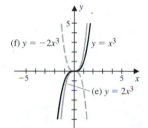

(f) $y = -2x^3$ $y = x^3$

−5   5   x

(e) $y = 2x^3$

13.

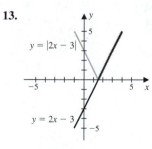

$y = |2x - 3|$

−5   5   x

$y = 2x - 3$

−5

15.

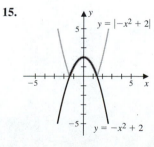

$y = |-x^2 + 2|$

−5   5   x

−5 $y = -x^2 + 2$

17.

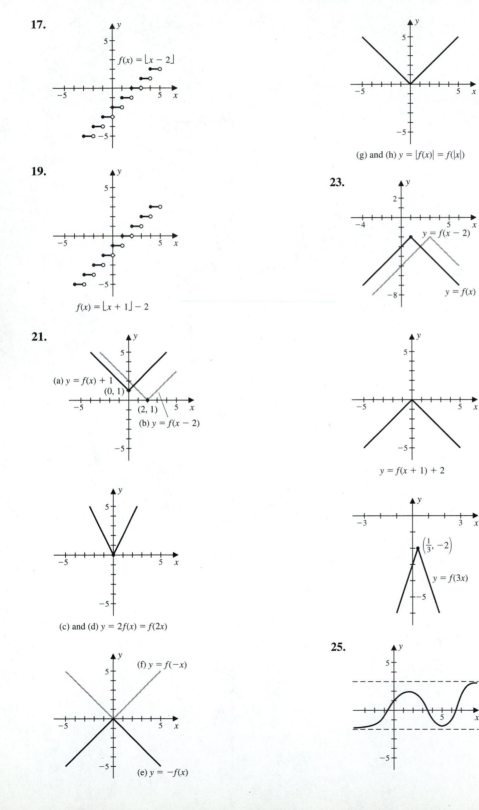

$f(x) = \lfloor x - 2 \rfloor$

(g) and (h) $y = |f(x)| = f(|x|)$

19.

$f(x) = \lfloor x + 1 \rfloor - 2$

23.

$y = f(x - 2)$

$y = f(x)$

21.

(a) $y = f(x) + 1$

(0, 1)

(2, 1)

(b) $y = f(x - 2)$

$y = f(x + 1) + 2$

(c) and (d) $y = 2f(x) = f(2x)$

$\left(\frac{1}{3}, -2\right)$

$y = f(3x)$

(f) $y = f(-x)$

(e) $y = -f(x)$

25.

27. $f(x) = \begin{cases} -x + 75, & \text{if } 0 \le x \le 75 \\ x - 75, & \text{if } 75 \le x \le 117.5 \\ -x + 160, & \text{if } 117.5 \le x \le 160 \\ x - 160, & \text{if } 160 \le x \le 241 \end{cases}$

$= \begin{cases} |x - 75|, & \text{if } 0 \le x \le 117.5 \\ |x - 160|, & \text{if } 117.5 \le x \le 241 \end{cases}$

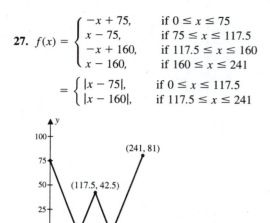

Exercise Set 1.10 (Page 81)

1. $(f + g)(x) = 4x - 2$; $(f - g)(x) = -2x + 2$;
$(f \cdot g)(x) = 3x^2 - 2x$; $(f/g)(x) = \frac{x}{3x-2}$;
domain of $f + g$, $f - g$, and $f \cdot g$: $(-\infty, \infty)$,
domain of f/g: $(-\infty, 2/3) \cup (2/3, \infty)$

3. $(f + g)(x) = \frac{1}{x} + \sqrt{x - 1}$; $(f - g)(x) = \frac{1}{x} - \sqrt{x - 1}$;
$(f \cdot g)(x) = \frac{\sqrt{x-1}}{x}$; $(f/g)(x) = \frac{1}{x\sqrt{x-1}}$; domain of
$f + g$, $f - g$, $f \cdot g$: $[1, \infty)$, domain of f/g: $(1, \infty)$

5. $(f + g)(x) = \sqrt{x + 2} + \sqrt{2 - x}$;
$(f - g)(x) = \sqrt{x + 2} - \sqrt{2 - x}$;
$(f \cdot g)(x) = \sqrt{4 - x^2}$; $(f/g)(x) = \frac{\sqrt{4-x^2}}{2-x}$;
domain of $f + g$, $f - g$, and $f \cdot g$: $[-2, 2]$;
domain of f/g: $[-2, 2)$

7. $(f + g)(x) = \begin{cases} 0, & \text{if } x < 0 \\ 1, & \text{if } x \ge 0 \end{cases}$

$(f - g)(x) = \begin{cases} -2, & \text{if } x < 0 \\ 1, & \text{if } x \ge 0 \end{cases}$

$(f \cdot g)(x) = \begin{cases} -1, & \text{if } x < 0 \\ 0, & \text{if } x \ge 0 \end{cases}$

$\left(\frac{f}{g}\right)(x) = -1$, if $x < 0$

domain $f + g$, $f - g$, $f \cdot g$: $(-\infty, \infty)$, domain f/g:
$(-\infty, 0)$

9.

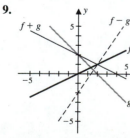

11. $f(x) = \frac{x^2 - 4}{x - 2} = \frac{(x - 2)(x + 2)}{x - 2} = x + 2$
for $x \ne 2$.
The point $(2, 4)$ is not on the graph of $y = f(x)$, but
is on the graph of $y = g(x) = x - 2$.

13.

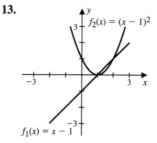

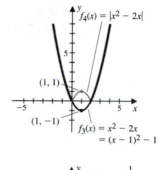

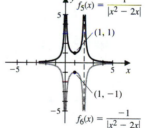

Exercise Set 1.11 (Page 89)

1. 9

3. 19

5. −17

7. $(f \circ g)(x) = 6x - 1$, $(g \circ f)(x) = 6x + 2$,
$(f \circ f)(x) = 4x + 3$, $(g \circ g)(x) = 9x - 4$
The domain of each is $(-\infty, \infty)$.

9. $(f \circ g)(x) = \frac{1}{x^2+2x}$, $(g \circ f)(x) = \frac{1+2x}{x^2}$, $(f \circ f)(x) = x$,
$(g \circ g)(x) = x^4 + 4x^3 + 6x^2 + 4x$. The domain $f \circ g$
is $(-\infty, -2) \cup (-2, 0) \cup (0, \infty)$, the domain $g \circ f$ and
$f \circ f$ is $(-\infty, 0) \cup (0, \infty)$, and the domain $g \circ g$ is
$(-\infty, \infty)$.

11. $f(x) = x^4$; $g(x) = 2 - 3x^2$

13. $f(x) = \frac{1}{x}$; $g(x) = x + 2$

15. a. $(f \circ g)(1) = 0$ **b.** $(g \circ f)(-1) = -1$
   c. $(g \circ f)(0) = 1$ **d.** $(f \circ g)(-2) = -2$

17. For example, if $f(x) = x$ and $g(x) = x + 1$, then
   $f(g(x)) = x + 1 = g(f(x))$.

19. Let f be an odd and g an even function, so that
   $f(-x) = -f(x)$ and $g(-x) = g(x)$. Then
   $(f \circ g)(-x) = f(g(-x)) = f(g(x)) = (f \circ g)(x)$ and
   $(g \circ f)(-x) = g(f(-x)) = g(-f(x)) = (g \circ f)(x)$.

21. a. $ad + b = bc + d$ **b.** $c = 1$ and $d = 0$
   c. $a = 1$ and $b = 0$

23. Volume is $\frac{4}{3}\pi(3 + 0.01t)^3$.

25. Volume is $\frac{4}{3}\pi(3\sqrt{t} + 5)^3$ cm³; surface area is
   $4\pi(3\sqrt{t} + 5)^2$ cm².

27. $F = \frac{Gm_1m_2}{r^2}$, where G is the constant of proportionality. The physical situation requires the domain
   to be $(0, \infty)$.

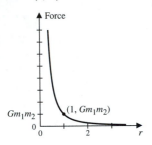

Exercise Set 1.12 (Page 100)

1. One-to-one

3. Not one-to-one

5. Not one-to-one

7. One-to-one

9. One-to-one

11. Not one-to-one

13.

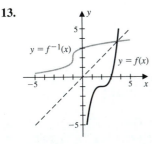

15.

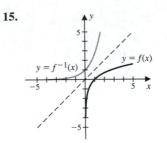

17. $f^{-1}(x) = \frac{x+1}{2}$

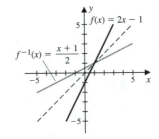

19. $f^{-1}(x) = x^2 + 3$, for $x \geq 0$

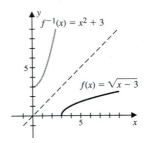

21. $f^{-1}(x) = \frac{1}{x^2}$, for $x > 0$

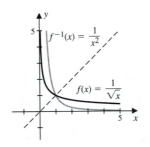

23. If the restricted domain of f is $[2/3, \infty)$, then
   $f^{-1}(x) = \frac{2-x}{3}$ with domain $[0, \infty)$.

25. If the restricted domain of f is $[0, \infty)$, then
$f^{-1}(x) = \sqrt{\frac{1-4x}{x}}$ with domain $(0, 1/4]$.

27. For $m \neq 0$ we have $f^{-1}(x) = \frac{1}{m}(x - b)$.

Review Exercises for Chapter 1 (Page 101)

1. $-1 \leq x \leq 7$

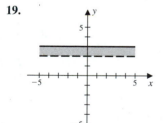

3. $-\infty < x < 7$

5. $(-\infty, -5)$

7. $[2, 10)$

9. $x \geq \frac{1}{2}$

11. $x \leq -2$ or $x \geq 0$

13. $-1 < x \leq -\frac{1}{4}$

15. $(-1, 4)$

17. $(-\infty, 0] \cup [4, \infty)$

19.

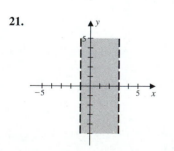

21.

23.

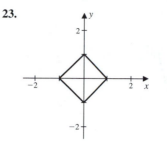

25. Domain: $(-\infty, \infty)$; range: $[-3, \infty)$

27. Domain: $(-\infty, 2) \cup (4, \infty)$; range: $(0, \infty)$

29. $f(x + h) = 7x + 7h + 4$; $\frac{f(x+h)-f(x)}{h} = 7$

31. $f(x + h) = x^2 + 2hx + h^2 - 1$; $\frac{f(x+h)-f(x)}{h} = 2x + h$

33. b. The distance is $\sqrt{13}$ **d.** $y = -\frac{3}{2}x + \frac{7}{2}$

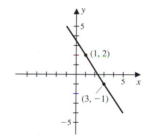

35. b. The distance is $\sqrt{10}$ **d.** $y = -\frac{1}{3}x - \frac{7}{3}$

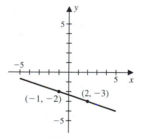

37. a. $y = 4x$ **b.** $y = -\frac{1}{4}x$

39. a. $y = -\frac{7}{5}x - \frac{22}{5}$ **b.** $y = \frac{5}{7}x - \frac{16}{7}$

41. Range: $[-4, \infty)$; minimum is -4 at $x = 2$.

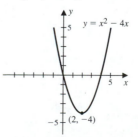

43. Range: $[0, \infty)$; minimum is 0 at $x = 3$.

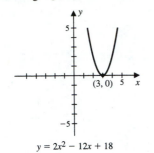

$y = 2x^2 - 12x + 18$

45. Range: $(-\infty, \frac{3}{2}]$; maximum is $\frac{3}{2}$ at $x = 3$.

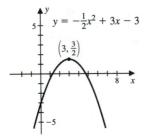

47.

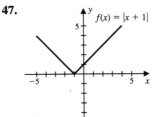

49.

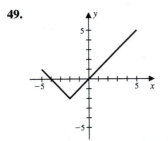

$f(x) = |x + 2| - 2$

51.

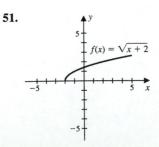

53.

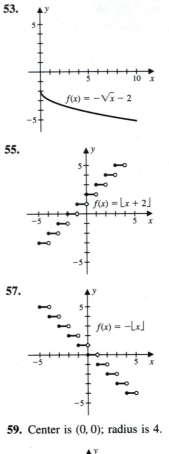

55.

57.

59. Center is $(0, 0)$; radius is 4.

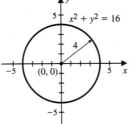

61. Center is $(-2, 1)$; radius is 3.

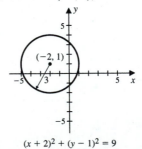

$(x + 2)^2 + (y - 1)^2 = 9$

63.

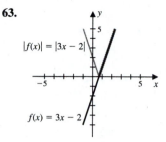

$|f(x)| = |3x - 2|$

$f(x) = 3x - 2$

65.

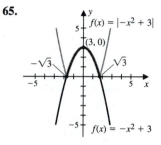

$f(x) = |-x^2 + 3|$

$(3, 0)$

$-\sqrt{3}$ $\sqrt{3}$

$f(x) = -x^2 + 3$

67. a, b.

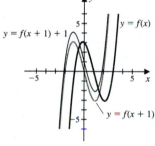

$y = f(x)$

$y = f(x + 1) + 1$

$y = f(x + 1)$

c, d.

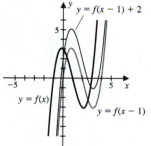

$y = f(x - 1) + 2$

$y = f(x)$

$y = f(x - 1)$

e, f.

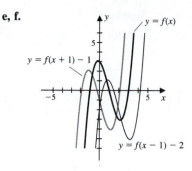

$y = f(x)$

$y = f(x + 1) - 1$

$y = f(x - 1) - 2$

69. The fourth vertex is $(1, 6)$.

71. $(x - 4)^2 + (y - 2)^2 = 4$

73.

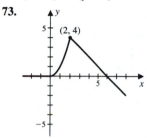

$(2, 4)$

75. a. One-to-one   **b.** Not one-to-one
c. Not one-to-one   **d.** One-to-one

77. a. Restricted domain: $[2, \infty)$

b.

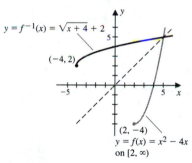

$y = f^{-1}(x) = \sqrt{x + 4} + 2$

$(-4, 2)$

$(2, -4)$

$y = f(x) = x^2 - 4x$
on $[2, \infty)$

c. $f^{-1}(x) = \sqrt{x + 4} + 2$; domain: $[-4, \infty)$

79. $43.00 per dinner

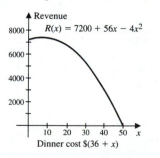

Revenue

8000

$R(x) = 7200 + 56x - 4x^2$

6000

4000

2000

$10 \quad 20 \quad 30 \quad 40 \quad 50 \quad x$

Dinner cost $(36 + x)$

81. $F(\ell) = 3\ell + \frac{864}{\ell}$; domain: $[0, \infty)$

83. a. $C(x) = 200{,}000x + 400{,}000\sqrt{(40 - x)^2 + 400}$;
  b. 28.5 miles

85. Part (d) appears to give the best representation.

a.

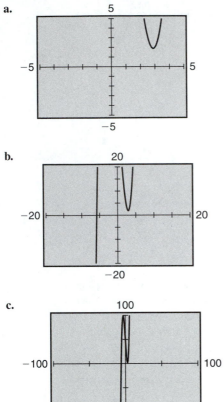

b.

c.

d.

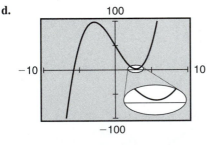

87. a. A and B and to the right of E **b.** C and D
  c. B and C **d.** D and E

89.

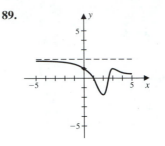

Chapter 1: Calculus Preview Exercises (Page 105)

1. $f(x) = \frac{1}{2}x^2$; $g(x) = 3x + 8$

3. $14 \le n \le 16$

5. n even: The graphs have a shape similar to $y = x^2$. The graphs all pass through $(0, 0)$, are above the x-axis for $x \ne 0$, are symmetric with respect to the y-axis, and are decreasing for $x < 0$ and increasing for $x > 0$. As n increases, the graph becomes flatter near the origin and increases more rapidly as x becomes larger in absolute value; all have an absolute minimum of 0 at $x = 0$ and no maximum. For $n > 0$, $f_{n+1}(x) > f_n(x)$, when $x > 1$, and $f_{n+1}(x) < f_n(x)$, when $0 < x < 1$. The same behavior holds for $x < -1$ or $-1 < x < 0$, by symmetry.

n odd: The graphs have a shape similar to $y = x^3$. The graphs all pass through $(0, 0)$, and are symmetric with respect to the origin, are always increasing. As n increases, the graph becomes flatter near the origin and increases more rapidly as x goes to ∞ and becomes negative more rapidly as x goes to $-\infty$; the graphs have no minimum nor maximum values. For $n > 0$, $f_{n+1}(x) > f_n(x)$, when $x > 1$, and $f_{n+1}(x) < f_n(x)$, when $0 < x < 1$.

7. a.

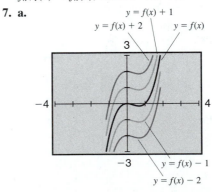

b.

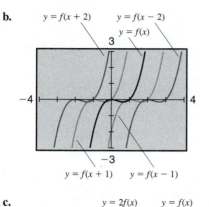

$y = f(x + 2)$ $y = f(x - 2)$
$y = f(x)$
$y = f(x + 1)$ $y = f(x - 1)$

c.

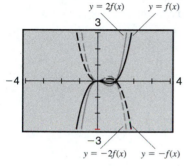

$y = 2f(x)$ $y = f(x)$
$y = -2f(x)$ $y = -f(x)$

d.

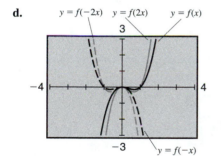

$y = f(-2x)$ $y = f(2x)$ $y = f(x)$
$y = f(-x)$

9. a.

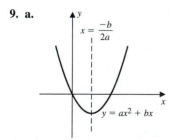

$x = \dfrac{-b}{2a}$

$y = ax^2 + bx$

b. *Hint:* Complete the square and then show that $x - b/(2a)$ always gives the same value of y as does $-x - b/(2a)$.

c. Adding c is just a vertical shift, which does not affect horizontal symmetry.

11.

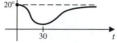

Temp
$20°$
30
t

13. a. Surface area is $2\pi r^2 + \dfrac{1800}{r}$.

b. $r \approx 5.3$ cm $; h \approx 10.2$ cm

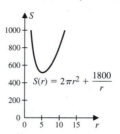

$S(r) = 2\pi r^2 + \dfrac{1800}{r}$

15. $L(t) = 2 + 22 \times 10^{-6}t; \; L(1000) = 2.022$ m

17. $V(t) = 15,000 - 1300t$

19. a. $g(x) = x^2, \; h(x) = x$

b. $g(x) = 1, \; h(x) = 1/x$

c. $g(x) = |x|, \; h(x) = x$

d. $g(x) = \frac{1}{2}[|x + 1| + |x - 1|],$
$h(x) = \frac{1}{2}[|x + 1| - |x - 1|]$

21. $f^{-1}(x) = \dfrac{a - bx}{x - 1}$

CHAPTER 2

Exercise Set 2.2 (Page 121)

1. Degree 3

3. Degree 4

5.

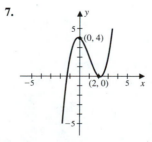

$(1, 2)$
$(3, -2)$

7.

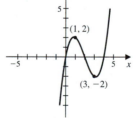

$(0, 4)$
$(2, 0)$

9.

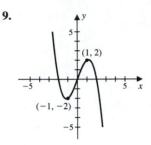

11.

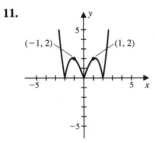

13. Odd degree at least 3

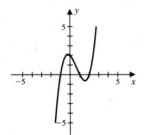

15. $P(x) = \frac{1}{2}x^3 - \frac{1}{2}x^2 - 2x + 2$

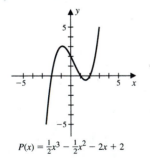

$P(x) = \frac{1}{2}x^3 - \frac{1}{2}x^2 - 2x + 2$

17. $P(x) = -\frac{2}{3}x^4 + \frac{8}{3}x^2$

19. a. Increasing: $-1.45 < x < 0$ or $x > 0.7$

  b. Local minimum: $(-1.45, -2.8), (0.7, -0.4)$;
       local maximum: $(0, 0)$

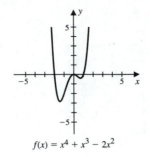

$f(x) = x^4 + x^3 - 2x^2$

21. a. Increasing: $x < -1$ or $1 < x < 2$ or $x > 3$

  b. Local minimum: $(1, -1.9), (3, 2.2)$;
       local maximum: $(-1, 6.4), (2, -1.3)$

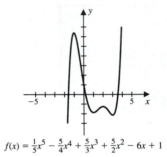

$f(x) = \frac{1}{5}x^5 - \frac{5}{4}x^4 + \frac{5}{3}x^3 + \frac{5}{2}x^2 - 6x + 1$

23.

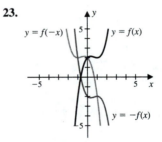

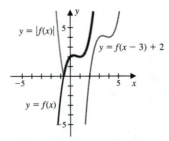

25.

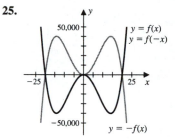

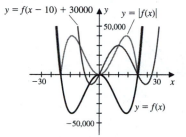

27. a. $V(x) = x(20 - 2x)^2$ **b.** $x = 5$ or $x \approx 1.91$

Exercise Set 2.3 (Page 132)

1. $Q(x) = 3x + 1; R(x) = 3$

3. $Q(x) = x^2 + 2x + 2; R(x) = 0$

5. $Q(x) = 2x^2 + 6; R(x) = 7x + 4$

7. $P(x) = (x - 1)(x - 2)^2$

9. $P(x) = (x + 1)(x + 3)(2x - 1)(x - 2)$

11. $\pm 1, \pm 2, \pm 3, \pm 6$

13. First divide all the terms by 2. Then the possibilities are $\pm 1, \pm 2, \pm \frac{1}{5}, \pm \frac{2}{5}$

15. The equivalent polynomial $2x^3 - 19x^2 + 54x - 45$ has possible rational roots at $\pm 1, \pm 3, \pm 5, \pm 9, \pm 15,$ $\pm 45, \pm \frac{1}{2}, \pm \frac{3}{2}, \pm \frac{5}{2}, \pm \frac{9}{2}, \pm \frac{15}{2}, \pm \frac{45}{2}.$

17. Maximum number of positive roots is 2; maximum number of negative roots is 2.

19. Maximum number of positive roots is 1; maximum number of negative roots is 2.

21. Zeros: $x = -1, x = 4, x = 2;$ $P(x) = (x + 1)(x - 4)(x - 2)$

23. Zeros: $x = 3; P(x) = (x - 3)(x^2 + 1)$

25. Zeros: $x = 2$ (of multiplicity 2), $x = -1, x = 1;$ $P(x) = (x - 2)^2(x + 1)(x - 1)$

27. $P(x) = \frac{1}{10}x^4 - \frac{1}{10}x^3 - \frac{3}{10}x^2 + \frac{1}{10}x + \frac{6}{5}$

29. $P(x) = x^5 - 5x^4 + 4x^3 - x^2 + 5x - 4$

31.

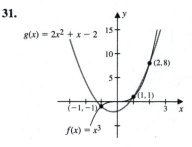

Exercise Set 2.4 (Page 146)

1. Domain: $(-\infty, 1) \cup (1, \infty)$; vertical asymptote: $x = 1$; horizontal asymptote: $y = 0$

3. Domain: $(-\infty, -1) \cup (-1, 1) \cup (1, \infty)$; vertical asymptotes: $x = -1, x = 1$; horizontal asymptote: $y = 0$

5. Domain: $(-\infty, 1)$; vertical asymptote: $x = 1$; horizontal asymptote: none

7. Domain: $(-\infty, -1) \cup (-1, \infty)$; x-intercept: $(3, 0)$; y-intercept: $(0, -3)$

9. Domain: $(-\infty, -2) \cup (-2, 2) \cup (2, \infty)$; x-intercepts: $(3, 0), (-2, 0)$; y-intercept: $(0, \frac{3}{2})$

11. Domain: $(-\infty, 0) \cup (0, \infty)$; x-intercepts: $(1, 0), (2, 0), (-1, 0)$; no y-intercept

13.

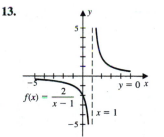

15.

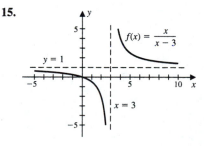

17.

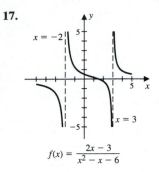

$$f(x) = \frac{2x - 3}{x^2 - x - 6}$$

19.

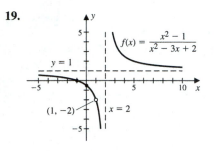

21.

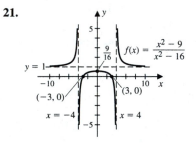

23.

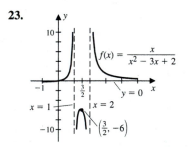

25.

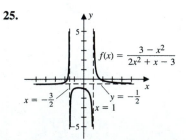

27.

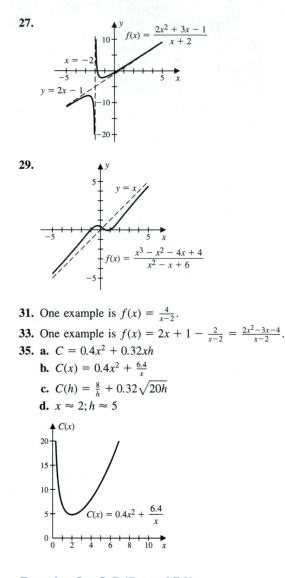

29.

31. One example is $f(x) = \frac{4}{x-2}$.

33. One example is $f(x) = 2x + 1 - \frac{2}{x-2} = \frac{2x^2 - 3x - 4}{x-2}$.

35. a. $C = 0.4x^2 + 0.32xh$

    b. $C(x) = 0.4x^2 + \frac{6.4}{x}$

    c. $C(h) = \frac{8}{h} + 0.32\sqrt{20h}$

    d. $x \approx 2; h \approx 5$

Exercise Set 2.5 *(Page 154)*

1. Domain: $(-\infty, -6] \cup [2, \infty)$

3. Domain: $(-\infty, -3) \cup [1, \infty)$

5. Domain: $(-3, -1] \cup (1, \infty)$

7.

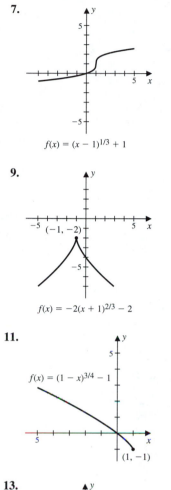

$$f(x) = (x - 1)^{1/3} + 1$$

9.

$$f(x) = -2(x + 1)^{2/3} - 2$$

11.

$$f(x) = (1 - x)^{3/4} - 1$$

$(1, -1)$

13.

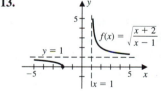

$$f(x) = \sqrt{\dfrac{x + 2}{x - 1}}$$

$y = 1$

$x = 1$

15.

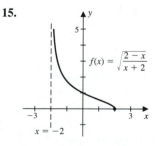

$$f(x) = \sqrt{\dfrac{2 - x}{x + 2}}$$

$x = -2$

17.

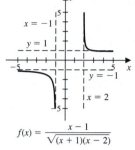

$x = -1$

$y = 1$

$y = -1$

$x = 2$

$$f(x) = \dfrac{x - 1}{\sqrt{(x + 1)(x - 2)}}$$

19. n even: Each graph is symmetric with respect to the y-axis, lies above the x-axis, and has vertical asymptote $x = 0$ and horizontal asymptote $y = 0$. As $x \longrightarrow 0^+$, $g_n(x) \longrightarrow \infty$; as $x \longrightarrow 0^-$, $g_n(x) \longrightarrow \infty$; as $x \longrightarrow \infty$, $g_n(x) \longrightarrow 0$; and as $x \longrightarrow -\infty$, $g_n(x) \longrightarrow 0$.

All the graphs pass through $(1, 1)$ and $(-1, 1)$. For $-1 < x < 0$ or $0 < x < 1$, $g_{n+1}(x) > g_n(x)$. For $x < -1$ or $x > 1$, $g_{n+1}(x) < g_n(x)$.

n odd: Each graph is symmetric with respect to the origin, and has vertical asymptote $x = 0$ and horizontal asymptote $y = 0$. The graphs do not cross the x- or y-axes. As $x \longrightarrow 0^+$, $g_n(x) \longrightarrow \infty$; as $x \longrightarrow 0^-$, $g_n(x) \longrightarrow -\infty$; as $x \longrightarrow \infty$, $g_n(x) \longrightarrow 0$; and as $x \longrightarrow -\infty$, $g_n(x) \longrightarrow 0$.

All the graphs pass through $(1, 1)$ and $(-1, -1)$. For $-1 < x < 0$ or $x > 1$, $g_{n+1}(x) < g_n(x)$. For $0 < x < 1$ or $x < -1$, $g_{n+1}(x) > g_n(x)$.

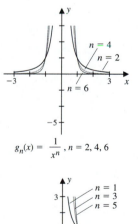

$n = 4$

$n = 2$

$n = 6$

$$g_n(x) = \dfrac{1}{x^n}, n = 2, 4, 6$$

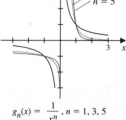

$n = 1$

$n = 3$

$n = 5$

$$g_n(x) = \dfrac{1}{x^n}, n = 1, 3, 5$$

Exercise Set 2.6 (Page 161)

1. $-1 - i$

3. $-5 + 8i$

5. $-2 + 6i$

7. $7 - i$

9. $-8 - 27i$

11. $-2 - 19i$

13. i

15. 1

17. Zeros: $\pm 2i$; $f(x) = (x - 2i)(x + 2i)$

19. Zeros: $1 \pm i$; $f(x) = (x - (1 + i))(x - (1 - i))$

21. Zeros: $\frac{1 \pm \sqrt{15}i}{4}$;
$f(x) = \left(x - \left(\frac{1+\sqrt{15}i}{4}\right)\right)\left(x - \left(\frac{1-\sqrt{15}i}{4}\right)\right)$

23. Solutions: $3i$, $-3i$, 2

25. Solutions: i, $-i$, 3, -1

27. Zeros: 3, $3i$, $-3i$; $f(x) = (x - 3)(x - 3i)(x + 3i)$

29. Zeros: 1, $-1 + i$, $-1 - i$;
$f(x) = (x - 1)^2(x - (-1 + i))(x - (-1 - i))$

31. $x^3 - 2x^2 + 4x - 8$

33. $2x^4 + 24x^2 + 54$

Review Exercises for Chapter 2 (Page 162)

1.

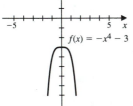

$f(x) = -2(x - 1)^2 + 2$

3.

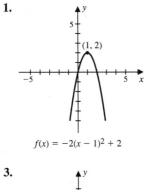

$f(x) = -x^4 - 3$

5.

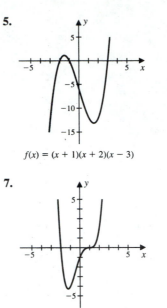

$f(x) = (x + 1)(x + 2)(x - 3)$

7.

$f(x) = \frac{1}{2}(x - 1)^3(x + 2)$

9.

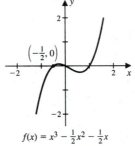

$f(x) = x^3 - \frac{1}{2}x^2 - \frac{1}{2}x$

11. Degree 3; positive leading coefficient

13. Degree 4; negative leading coefficient

15. $Q(x) = 4x + 10; R(x) = 19$

17. $Q(x) = 3x^2 - 4x + 6; R(x) = -11$

19. a. ± 1, ± 3, $\pm \frac{1}{3}$

  b. Maximum number of positive roots is 2; maximum number of negative roots is 2.

  d. $P(x) = (x - 3)(x - 1)(3x^2 + 3x + 1)$

21. a. ± 1, ± 2, ± 3, ± 4, ± 6, ± 12

  b. Maximum number of positive roots is 4; maximum number of negative roots is 1.

  d. $P(x) = (x + 3)(x - 1)^2(x - 2)^2$

23. Domain: $(-\infty, 1) \cup (1, \infty)$; x-intercept: $(4, 0)$; y-intercept: $(0, 4)$; vertical asymptote: $x = 1$; horizontal asymptote: $y = 1$

25. Domain: $(-\infty, -3) \cup (-3, 3) \cup (3, \infty)$; x-intercept: $(1, 0)$; y-intercept: $(0, -\frac{1}{18})$; vertical asymptote: $x = \pm 3$; horizontal asymptote: $y = \frac{1}{2}$

27. Domain: $(-\infty, 0) \cup (0, \infty)$; x-intercepts: $(1, 0), (-1, 0), (-2, 0)$; y-intercept: none; vertical asymptote: $x = 0$; horizontal asymptote: $y = 1$

29. Horizontal asymptote: $y = 0$; vertical asymptote: $x = 2$; x-intercept: none; y-intercept: $(0, -\frac{3}{2})$

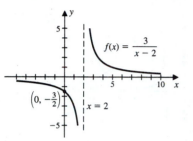

31. Horizontal asymptote: $y = 0$; vertical asymptotes: $x = 1, x = 2$; x-intercept: none; y-intercept: $(0, 1)$

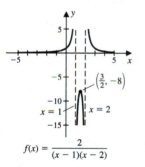

33. Horizontal asymptote: $y = 0$; vertical asymptotes: $x = -2, x = 2$; x-intercept: none; y-intercept: $(0, -1)$

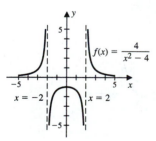

35. Horizontal asymptote: $y = 1$; vertical asymptote: $x = -2, x = 0$; x-intercepts: $(1, 0), (-1, 0)$; y-intercept: none;

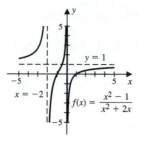

37. Vertical asymptote: $x = -1$; slant asymptote: $y = x - 3$; x-intercept: $(1, 0)$; y-intercept: $(0, 1)$

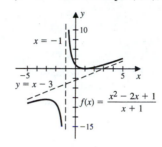

39.

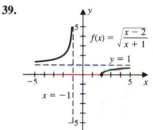

41.

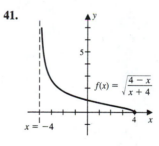

43.

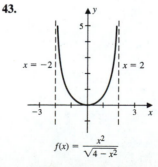

45. a. Increasing: $(-\infty, -0.2), (1.6, \infty)$; decreasing: $(-0.2, 1.6)$

  b. Local maximum: $(-0.2, 2.1)$; local minimum: $(1.6, 0.6)$

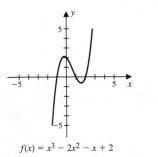

$f(x) = x^3 - 2x^2 - x + 2$

47. a. Increasing: $(-\infty, -4.8), (1.2, \infty)$; decreasing: $(-4.8, 1.2)$

  b. Local maximum: $(-4.8, 64.6)$; local minimum: $(1.2, -8.2)$

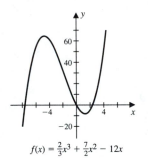

$f(x) = \frac{2}{3}x^3 + \frac{7}{2}x^2 - 12x$

49. $\frac{1}{2} + \frac{\sqrt{2}}{4}i$

51. $1 + 2i$

53. $3 - 4i$

55. 1

57. $-\frac{1}{5} + \frac{2}{5}i$

59.

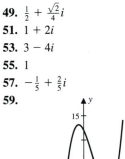

61. $P(x) = \frac{1}{3}x^3 - x^2 - \frac{1}{3}x + 1$

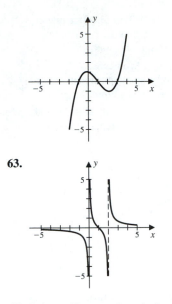

63.

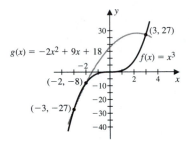

65. Points of intersection: $(-3, -27), (-2, -8), (3, 27)$

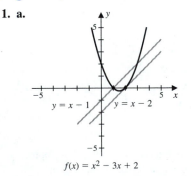

67. $P(x) = x^3 - x^2 + x - 1$

69. $P(x) = x^4 + 6x^2 + 8$

Chapter 2: Calculus Preview Exercises (Page 164)

1. a.

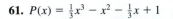

$f(x) = x^2 - 3x + 2$

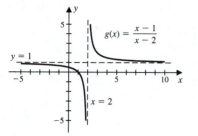

$g(x) = \dfrac{x - 1}{x - 2}$

$y = 1$

$x = 2$

3. $P(x) = x^2(2x - 3)$; $C < 0$: one real zero; $C = 0$: two real zeros; $0 < C < 1$: three real zeros; $C = 1$: two real zeros; $1 < C$: one real zero

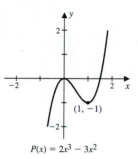

$(1, -1)$

$P(x) = 2x^3 - 3x^2$

5. Let $P(x) = mx + b$. Then
$Q(x) = m(mx + b) + b = m^2x + mb + b$, and $Q(x)$ is a linear polynomial with positive slope m^2.

7. One example is $f(x) = -\dfrac{4}{x-2}$.

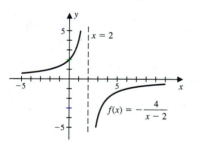

$x = 2$

$f(x) = -\dfrac{4}{x - 2}$

9. One example is $f(x) = \dfrac{2x^2}{(x+2)(x-1)}$.

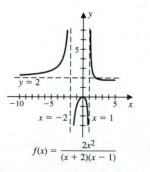

$y = 2$

$x = -2$ $x = 1$

$f(x) = \dfrac{2x^2}{(x + 2)(x - 1)}$

11. a. With (iii) **b.** With (iv) **c.** With (i)
d. With (ii)

13. There are either four, two, or no real zeros, and there are no rational zeros.

15. b. $P(x) = \dfrac{x-1}{-2}(6) + \dfrac{x+1}{2}(-2) = -4x + 2$

17.

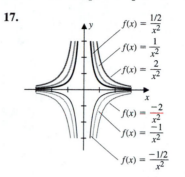

$f(x) = \dfrac{1/2}{x^2}$

$f(x) = \dfrac{1}{x^2}$

$f(x) = \dfrac{2}{x^2}$

$f(x) = \dfrac{-2}{x^2}$

$f(x) = \dfrac{-1}{x^2}$

$f(x) = \dfrac{-1/2}{x^2}$

19.

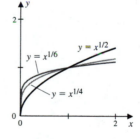

$y = x$

$y = x^{1/5}$

$y = x^{1/3}$

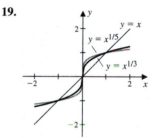

$y = x^{1/2}$

$y = x^{1/6}$

$y = x^{1/4}$

21.

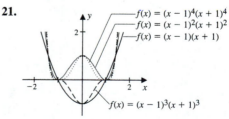

$f(x) = (x - 1)^4(x + 1)^4$

$f(x) = (x - 1)^2(x + 1)^2$

$f(x) = (x - 1)(x + 1)$

$f(x) = (x - 1)^3(x + 1)^3$

23. $V(x) = x(10 - x)(50 - 3x)$

CHAPTER 3
Exercise Set 3.2 (Page 182)

1.

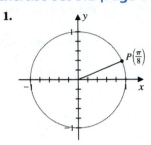

3.

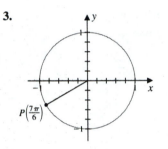

5.

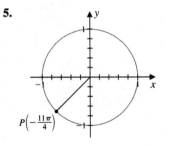

7.

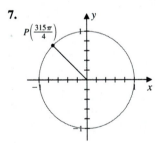

9.

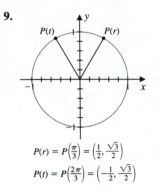

$$P(r) = P\left(\frac{\pi}{3}\right) = \left(\frac{1}{2}, \frac{\sqrt{3}}{2}\right)$$

$$P(t) = P\left(\frac{2\pi}{3}\right) = \left(-\frac{1}{2}, \frac{\sqrt{3}}{2}\right)$$

11.

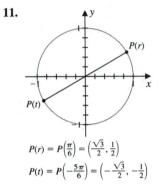

$$P(r) = P\left(\frac{\pi}{6}\right) = \left(\frac{\sqrt{3}}{2}, \frac{1}{2}\right)$$

$$P(t) = P\left(-\frac{5\pi}{6}\right) = \left(-\frac{\sqrt{3}}{2}, -\frac{1}{2}\right)$$

13.

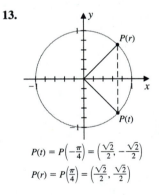

$$P(t) = P\left(-\frac{\pi}{4}\right) = \left(\frac{\sqrt{2}}{2}, -\frac{\sqrt{2}}{2}\right)$$

$$P(r) = P\left(\frac{\pi}{4}\right) = \left(\frac{\sqrt{2}}{2}, \frac{\sqrt{2}}{2}\right)$$

15. $\sin \frac{\pi}{6} = \frac{1}{2}$; $\cos \frac{\pi}{6} = \frac{\sqrt{3}}{2}$

17. $\sin \frac{13\pi}{4} = -\frac{\sqrt{2}}{2}$; $\cos \frac{13\pi}{4} = -\frac{\sqrt{2}}{2}$

19. $\sin \left(-\frac{\pi}{3}\right) = -\frac{\sqrt{3}}{2}$; $\cos \left(-\frac{\pi}{3}\right) = \frac{1}{2}$

21. $\sin \left(-\frac{5\pi}{6}\right) = -\frac{1}{2}$; $\cos \left(-\frac{5\pi}{6}\right) = -\frac{\sqrt{3}}{2}$

23. $t = 0, 2\pi$

25. $t = \frac{2\pi}{3}$

27. Even

29. Odd

31. a. $\left(-\frac{3}{5}, -\frac{4}{5}\right)$ **b.** $\left(\frac{3}{5}, -\frac{4}{5}\right)$ **c.** $\left(-\frac{3}{5}, -\frac{4}{5}\right)$
   d. $\left(-\frac{3}{5}, \frac{4}{5}\right)$

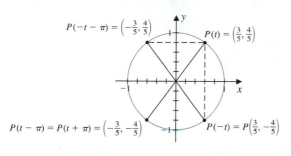

33. $-\frac{4}{5}$

35. $-\frac{\sqrt{5}}{3}$

37. $t = 0, 2\pi$

39. $t = 0, \frac{\pi}{2}, 2\pi$

Exercise Set 3.3 (Page 192)

1.

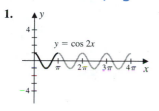

3.

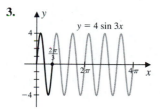

5.

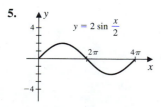

7.

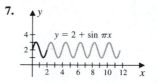

9.

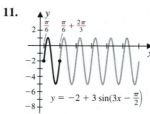

11.

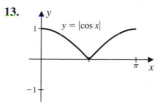

13.

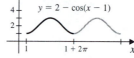

15. $y = 3 \cos 2x$

17. $y = \sin \frac{\pi}{3}(x - 1) + 2$

19. $[-\pi/50, \pi/50] \times [-1, 1]$

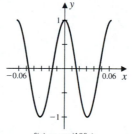

$f(x) = \cos(100x)$

21. $[-\pi/10, \pi/10] \times [-5, 5]$

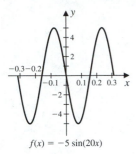

$f(x) = -5 \sin(20x)$

23. $x \approx 0.74$

25. $x \approx 1.26$

27. a. $t = \frac{\pi}{6}$; **b.** $t = \frac{5\pi}{12}$; **c.** $t = \frac{11\pi}{12}$

29. $y = 1 + 2 \sin \frac{\pi x}{2}$

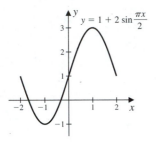

$y = 1 + 2 \sin \frac{\pi x}{2}$

31. a.

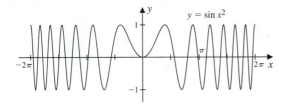

$y = \sin x^2$

b.

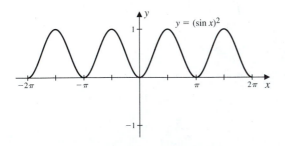

$y = (\sin x)^2$

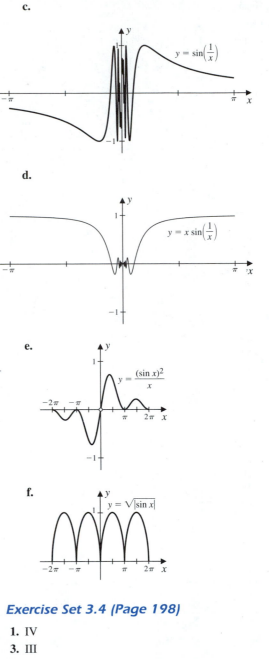

c.

$y = \sin\left(\frac{1}{x}\right)$

d.

$y = x \sin\left(\frac{1}{x}\right)$

e.

$y = \frac{(\sin x)^2}{x}$

f.

$y = \sqrt{|\sin x|}$

Exercise Set 3.4 (Page 198)

1. IV

3. III

5. I

7. $\sin t = \frac{4}{5}$; $\cos t = \frac{3}{5}$; $\tan t = \frac{4}{3}$; $\cot t = \frac{3}{4}$; $\sec t = \frac{5}{3}$; $\csc t = \frac{5}{4}$

9. $\sin t = \frac{1}{3}$; $\cos t = -\frac{2\sqrt{2}}{3}$; $\tan t = -\frac{\sqrt{2}}{4}$; $\cot t = -2\sqrt{2}$; $\sec t = -\frac{3\sqrt{2}}{4}$; $\csc t = 3$

11. $\sin t = -\frac{1}{2}$; $\cos t = \frac{\sqrt{3}}{2}$; $\tan t = -\frac{\sqrt{3}}{3}$; $\cot t = -\sqrt{3}$; $\sec t = \frac{2\sqrt{3}}{3}$; $\csc t = -2$

13.

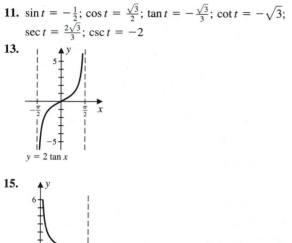

$y = 2 \tan x$

15.

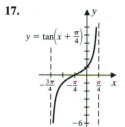

$y = \frac{1}{2} \cot x$

17.

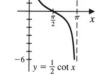

$y = \tan\left(x + \frac{\pi}{4}\right)$

19.

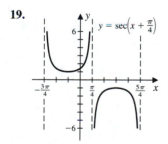

$y = \sec\left(x + \frac{\pi}{4}\right)$

21.

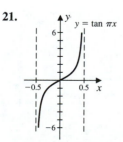

$y = \tan \pi x$

23.

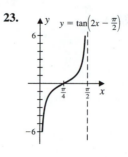

$y = \tan\left(2x - \frac{\pi}{2}\right)$

25. $t = \frac{\pi}{4}, \frac{3\pi}{4}, \frac{5\pi}{4}, \frac{7\pi}{4}$

27. $t = 0, \frac{\pi}{8}, \frac{\pi}{2}, \frac{7\pi}{8}, \pi, \frac{9\pi}{8}, \frac{3\pi}{2}, \frac{15\pi}{8}, 2\pi$

29. $\left[-\frac{\pi}{5}, \frac{\pi}{5}\right] \times [-5, 5]$

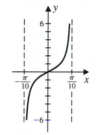

31. $\left[-\frac{\pi}{50}, \frac{\pi}{50}\right] \times [-5, 5]$

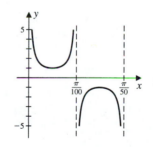

33. $[-100\pi, 100\pi] \times [-5, 5]$

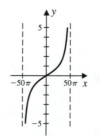

35. $\sin t = -\frac{2\sqrt{5}}{5}$; $\cos t = \frac{\sqrt{5}}{5}$; $\tan t = -2$; $\cot t = -\frac{1}{2}$; $\sec t = \sqrt{5}$; $\csc t = -\frac{\sqrt{5}}{2}$

Exercise Set 3.5 (Page 208)

1. $\frac{\sqrt{2}}{4}(1 - \sqrt{3})$

3. $\frac{\sqrt{2}}{4}(\sqrt{3} + 1)$

5. $\frac{\sqrt{3}-1}{\sqrt{3}+1}$

7. $\frac{\sqrt{2-\sqrt{3}}}{2}$

9. $\frac{\sqrt{2-\sqrt{2}}}{2}$

11. **a.** $\cos 2t = -\frac{7}{25}$ **b.** $\sin 2t = \frac{24}{25}$
   c. $\cos \frac{t}{2} = \frac{2\sqrt{5}}{5}$ **d.** $\sin \frac{t}{2} = \frac{\sqrt{5}}{5}$

19. $(\cos 2x)^2 = \frac{1+\cos 4x}{2}$

21. $(\cos x)^4 = \frac{3}{8} + \frac{\cos 4x}{8} + \frac{\cos 2x}{2}$

23. $\frac{1}{2}(\sin t + \sin 11t)$

25. $\frac{1}{2}(\cos t + \cos 5t)$

27. $2 \sin \frac{5t}{2} \cos \frac{t}{2}$

29. $2 \cos \frac{7t}{2} \cos \frac{3t}{2}$

31. $x = 0, \frac{\pi}{3}, \pi, \frac{5\pi}{3}, 2\pi$

33. $x = \frac{\pi}{3}, \frac{2\pi}{3}, \frac{4\pi}{3}, \frac{5\pi}{3}$

43. An identity

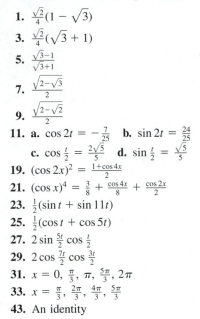

$(\sin x - \cos x)^2 = 1 - \sin 2x$

45. An identity

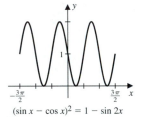

$\frac{\sin 2x}{1 + \cos 2x} = \tan x$

47. Not an identity

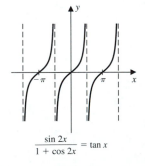

$(\sin x - \cos x)^2 \neq 1$

Exercise Set 3.6 (Page 216)

1. $\frac{\pi}{3}$

3. $\frac{5\pi}{4}$

5. $-\frac{2\pi}{5}$

7. $135°$

9. $120°$

11. $\sin \theta = \frac{3}{5}$; $\cos \theta = \frac{4}{5}$; $\tan \theta = \frac{3}{4}$; $\cot \theta = \frac{4}{3}$;
   $\sec \theta = \frac{5}{4}$; $\csc \theta = \frac{5}{3}$

13. $\sin \theta = \frac{1}{2}$; $\cos \theta = \frac{\sqrt{3}}{2}$; $\tan \theta = \frac{\sqrt{3}}{3}$; $\cot \theta = \sqrt{3}$;
   $\sec \theta = \frac{2\sqrt{3}}{3}$; $\csc \theta = 2$

15. $x = 8\sqrt{3}$

17. $x = \frac{4\sqrt{3}}{3}$

19. $\beta = 60°$; $\overline{AB} = 4\sqrt{3}$; $\overline{AC} = 8$

21. $\alpha = 45°$; $\overline{AB} = \overline{BC} = \frac{5\sqrt{2}}{2}$

23. $\beta = 62.4°$; $\overline{AB} \approx 29.3$; $\overline{AC} \approx 33.0$

25. $40 \tan 70° \approx 109.9$ ft

27. $80 \cot 10.5° \approx 431.64$ ft

Exercise Set 3.7 (Page 227)

1. $\frac{\pi}{6}$

3. $\frac{\pi}{2}$

5. $\frac{\pi}{3}$

7. π

9. $\frac{3\pi}{4}$

11. No solution

13. $\frac{1}{2}$

15. $\frac{\sqrt{2}}{2}$

17. $\frac{\pi}{2}$

19. $\frac{2\pi}{3}$

21. $-\frac{\pi}{4}$

23. $\frac{4}{5}$

25. $\frac{12}{13}$

27. $\frac{7}{25}$

29. $x = \arctan(-1) = -\frac{\pi}{4} \approx -0.785$;
   $x = \arctan 2 \approx 1.107$

Exercise Set 3.8 (Page 240)

1. $a = 15.4$; $\beta = 34.1°$; $\gamma = 110.9°$

3. $\alpha = 50.4°$; $\gamma = 99.6°$; $b = 16.2$

5. $\alpha = 31.1°$; $\beta = 108.6°$; $\gamma = 40.3°$

7. $\gamma = 54°$; $a = 94.7$; $b = 119.9$

9. $\beta = 65°$; $a = 10$; $c = 8.5$

11. $\alpha = 50°$; $a = 30.6$; $b = 39.4$

13. $\beta = 90°$; $\gamma = 30°$; $b = 6.9$

17. No such triangle exists.

19. a. Avoid the swamp.   **b.** $\approx \$115,881.00$

21. 83.8 mi

23. 9.7 mi

Review Exercises for Chapter 3 (Page 241)

1. a. $\left(\frac{1}{2}, \frac{\sqrt{3}}{2}\right)$ **b.** $\frac{\pi}{3}$

  c. $\sin\frac{\pi}{3} = \frac{\sqrt{3}}{2}$; $\cos\frac{\pi}{3} = \frac{1}{2}$; $\tan\frac{\pi}{3} = \sqrt{3}$;
  $\cot\frac{\pi}{3} = \frac{\sqrt{3}}{3}$; $\sec\frac{\pi}{3} = 2$; $\csc\frac{\pi}{3} = \frac{2\sqrt{3}}{3}$

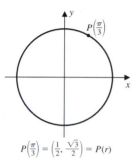

$$P\left(\frac{\pi}{3}\right) = \left(\frac{1}{2}, \frac{\sqrt{3}}{2}\right) = P(r)$$

3. a. $\left(-\frac{\sqrt{2}}{2}, -\frac{\sqrt{2}}{2}\right)$ **b.** $\frac{\pi}{4}$

  c. $\sin\frac{5\pi}{4} = -\frac{\sqrt{2}}{2}$; $\cos\frac{5\pi}{4} = -\frac{\sqrt{2}}{2}$; $\tan\frac{5\pi}{4} = 1$;
  $\cot\frac{5\pi}{4} = 1$; $\sec\frac{5\pi}{4} = -\sqrt{2}$; $\csc\frac{5\pi}{4} = -\sqrt{2}$

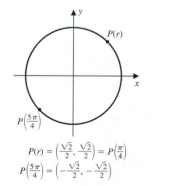

$$P(r) = \left(\frac{\sqrt{2}}{2}, \frac{\sqrt{2}}{2}\right) = P\left(\frac{\pi}{4}\right)$$
$$P\left(\frac{5\pi}{4}\right) = \left(-\frac{\sqrt{2}}{2}, -\frac{\sqrt{2}}{2}\right)$$

5. a. $\left(-\frac{\sqrt{2}}{2}, \frac{\sqrt{2}}{2}\right)$ **b.** $\frac{\pi}{4}$

  c. $\sin\frac{-21\pi}{4} = \frac{\sqrt{2}}{2}$; $\cos\frac{-21\pi}{4} = -\frac{\sqrt{2}}{2}$; $\tan\frac{-21\pi}{4} = -1$;
  $\cot\frac{-21\pi}{4} = -1$; $\sec\frac{-21\pi}{4} = -\sqrt{2}$;
  $\csc\frac{-21\pi}{4} = \sqrt{2}$

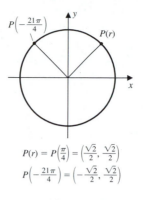

$$P(r) = P\left(\frac{\pi}{4}\right) = \left(\frac{\sqrt{2}}{2}, \frac{\sqrt{2}}{2}\right)$$
$$P\left(-\frac{21\pi}{4}\right) = \left(-\frac{\sqrt{2}}{2}, \frac{\sqrt{2}}{2}\right)$$

7. $\sin t = -\frac{4}{5}$; $\cos t = \frac{3}{5}$; $\tan t = -\frac{4}{3}$; $\cot t = -\frac{3}{4}$;
  $\sec t = \frac{5}{3}$; $\csc t = -\frac{5}{4}$

9. $\sin t = \frac{\sqrt{17}}{17}$; $\cos t = \frac{4\sqrt{17}}{17}$; $\tan t = \frac{1}{4}$; $\cot t = 4$;
  $\sec t = \frac{\sqrt{17}}{4}$; $\csc t = \sqrt{17}$

11. $x = \pi$

13. $x = 0$, π, $\arctan 2 \approx 1.11$, $\pi - \arctan 2 \approx 2.03$

15. No solutions

17. Odd

19. Even

21.

$y = 5\sin\frac{1}{2}x$

23.

$y = -3\cos 2x$

25.

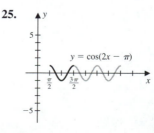

$y = \cos(2x - \pi)$

27.

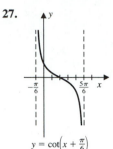

$$y = \cot\left(x + \frac{\pi}{6}\right)$$

29.

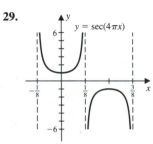

$y = \sec(4\pi x)$

31. $y = \sin 4x = \cos 4\left(x - \frac{\pi}{8}\right)$

33. $y = 4\sin \pi(x + \frac{1}{2}) = 4\cos \pi x$

35. $x = 0,\ 0.88$

37. $\left[-\frac{2\pi}{125}, \frac{2\pi}{125}\right] \times [-4, 4]$

43. $\frac{1}{2} + \frac{1}{2}\cos 6x$

45. $\frac{3}{8} - \frac{1}{2}\cos 2x + \frac{1}{8}\cos 4x$

47. $\frac{1}{2}\sin 9t - \frac{1}{2}\sin t$

49. $\frac{1}{2}\cos 2t + \frac{1}{2}\cos 6t$

51. $2\sin 4t \cos 2t$

53. $2\cos 3t \cos t$

59. $\frac{\sqrt{3}}{2}$

61. $\frac{16}{65}$

63. $\beta = 29.1°;\ \gamma = 125.9°; a = 10.4$

65. $\alpha = 36.9°;\ \beta = 53.1°;\ \gamma = 90°$

67. $\alpha = 54°;\ a = 8.8;\ c = 8.3$

69. $A = b^2(1 + \cos \theta)\sin \theta$

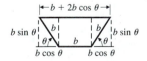

71. Miles east: 475; miles south: about 822.7

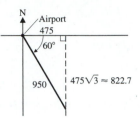

$475\sqrt{3} \approx 822.7$

Chapter 3: Calculus Preview Exercises (Page 243)

1. The curves coincide.

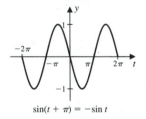

$\sin(t + \pi) = -\sin t$

3. The curves coincide.

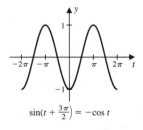

$\sin\left(t + \frac{3\pi}{2}\right) = -\cos t$

5. Absolute maximum: about 2.9 at about $x = 3$; absolute minimum: about -1.1 at about $x = -2$

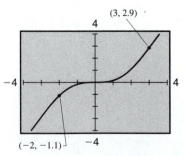

11. $t = \frac{11\pi}{6}$. Extraneous solution at $t = \frac{7\pi}{6}$.

13. In each case the curve $y = h(x)$ is bounded between the curves $y = f(x)$ and $y = g(x)$. The curve $y = h(x)$ touches but never crosses the curves $y = f(x)$ and $y = g(x)$.

a.

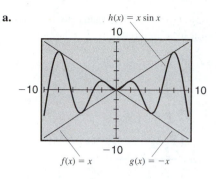

$h(x) = x \sin x$

$f(x) = x$ $g(x) = -x$

b.

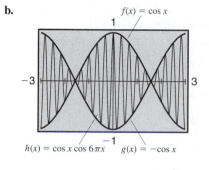

$f(x) = \cos x$

$h(x) = \cos x \cos 6\pi x$ $g(x) = -\cos x$

c.

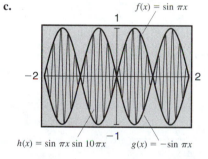

$f(x) = \sin \pi x$

$h(x) = \sin \pi x \sin 10 \pi x$ $g(x) = -\sin \pi x$

15. points of intersection: $x = \pm \frac{2}{(4k+1)\pi}$, or $x = \pm \frac{2}{(4k+3)\pi}$, for $k = 0, 1, 2, \ldots$

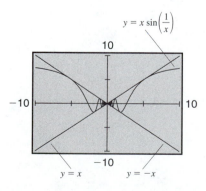

$y = x \sin\left(\frac{1}{x}\right)$

$y = x$ $y = -x$

17. The factor a causes an elongation of the graph in the y-direction if $a > 1$ and a compression if $0 < a < 1$.

If $a < 0$, the graph is reflected through the x-axis. The factor b determines the period $\frac{\pi}{b}$ of the tangent curve. The factor c causes a horizontal shift c units to the left, x-intercepts: $x = \pm \frac{k\pi}{b} - c$, vertical asymptotes: $x = \pm \frac{(2k+1)\pi}{b} - c$, for $k = 0, 1, 2, \ldots$.

21. $\tan(t_1 + t_2) = \frac{11}{2}$ and $\tan(t_1 - t_2) = \frac{1}{2}$

25. $A = 2r^2 \cos \theta \sin \theta = r^2 \sin 2\theta$

27. 22,000 ft

31. 392 ft

CHAPTER 4

Exercise Set 4.2 (Page 258)

1.

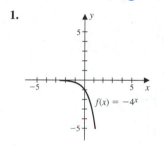

$f(x) = -4^x$

3.

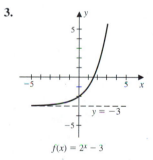

$y = -3$

$f(x) = 2^x - 3$

5.

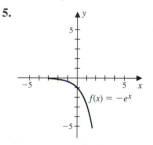

$f(x) = -e^x$

7.

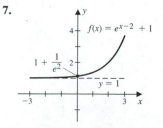

$f(x) = e^{x-2} + 1$

$1 + \frac{1}{e^2}$

$y = 1$

9.

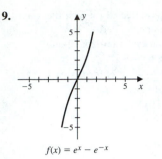

$$f(x) = e^x - e^{-x}$$

11.

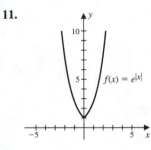

13. $x \approx 0.1586$, $x \approx 3.146$

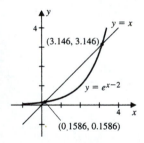

15. $x \approx 1.536$, $x \approx 2.315$

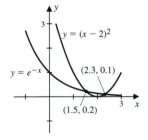

17. Decreasing: $(-\infty, -1)$; increasing: $(-1, \infty)$; no local maximum; local minimum: $(-1, -0.37)$

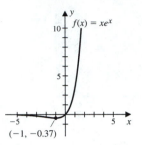

19. Decreasing: $(-0.5, \infty)$; increasing: $(-\infty, -0.5)$; local maximum: $(-0.5, 1.28)$; local minimum: none

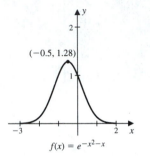

21. $k \approx 1.098$

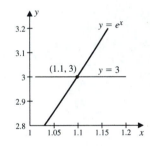

23. $x \approx 1.177$, $x \approx 22.44$

a.

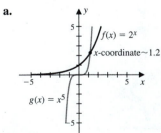

b.

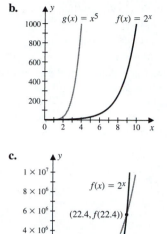

c.

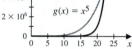

25. $0 < x < 2.7$

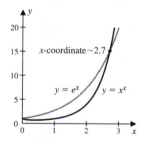

27. a. \$6850.43   **b.** \$6914.09   **c.** \$6919.95
d. \$6920.15

Exercise Set 4.3 (Page 266)

1. 5

3. 3

5. -3

7. $\frac{1}{2}$

9. 5

11. π

13. $\ln x + \ln(x + 2)$

15. $5 \log_2(2x - 1)$

17. $\ln 3 + 2 \ln x - 4 \ln(x + 3)$

19. $\frac{3}{2} \log_3(3x + 2) + 3 \log_3(x - 1) - \log_3 x - \frac{1}{2} \log_3(x + 1)$

21. $x = 81$

23. $x = 4$

25. $x = 2$

27. $x = \log_4 3 = \frac{\ln 3}{\ln 4}$

29. $x = 2 - e^4$

31. $x = \frac{9}{2}$

33. $x = 2$

35. $x = \frac{4}{3}$

37. $x = \frac{1}{2}, x = -9$

39. $x = \frac{4 \ln 3}{\ln 3 - 2}$

41.

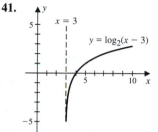

43.

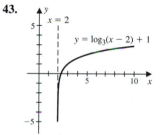

45.

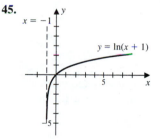

47.

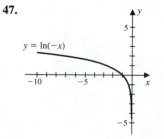

49.

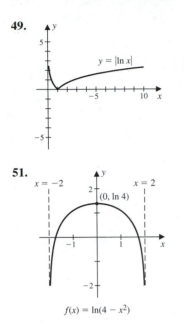

51.

$f(x) = \ln(4 - x^2)$

53. Vertical asymptotes: $x = 0$; horizontal asymptote: $y = 0$

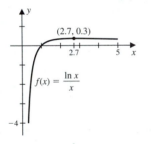

$f(x) = \dfrac{\ln x}{x}$

55. The function $g(x) = \sqrt[n]{x}$ will grow more rapidly than $f(x) = a + \ln x$ for all $n > 0$ and real numbers a as $x \longrightarrow \infty$.

57. $t = \dfrac{100 \ln 2}{9} \approx 7.7$ years

Exercise Set 4.4 (Page 271)

1. a. $Q(t) = 1000e^{[(\ln 2)/4]t}$

b. $Q(7) = 1000e^{(7 \ln 2)/4} \approx 3364$

c. $t = \dfrac{4 \ln 20}{\ln 2} \approx 17.3$ h

3. a. $Q(t) = 200e^{[(\ln 2)/22]t}$

b. $Q(600) = 200e^{[(300 \ln 2)/11]t} \approx 3.24 \times 10^{10}$

c. $t = \dfrac{22 \ln 50}{\ln 2} \approx 124$ min

5. $t = \dfrac{2 \ln 2}{\ln 10 - \ln 9} \approx 13.16$ years

7. The half life is $-\dfrac{5 \ln 2}{\ln 0.97} \approx 113.8$ years.

9. a. Population in 2000 $\approx$ 6372 million; population in 2050 $\approx$ 16,158 million

b. Population in 2000 $\approx$ 6323 million; population in 2050 $\approx$ 14,978 million

c. To plot the data points along with the exponential models on the same set of axes, shift the exponential functions so that time $t = 0$ corresponds to the initial year used to determine the model.

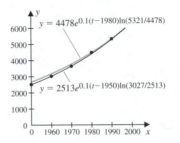

11. $t = \dfrac{5(\ln 36.6 - \ln 5)}{6(\ln 4 - \ln 5)} \approx -7.4$ h. Death occurred about 7.4 h before 11:00 A.M., or about 3:36 A.M.

Review Exercises for Chapter 4 (Page 272)

1.

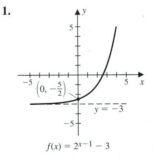

$f(x) = 2^{x-1} - 3$

3.

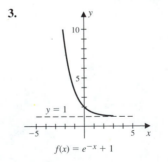

$f(x) = e^{-x} + 1$

5.

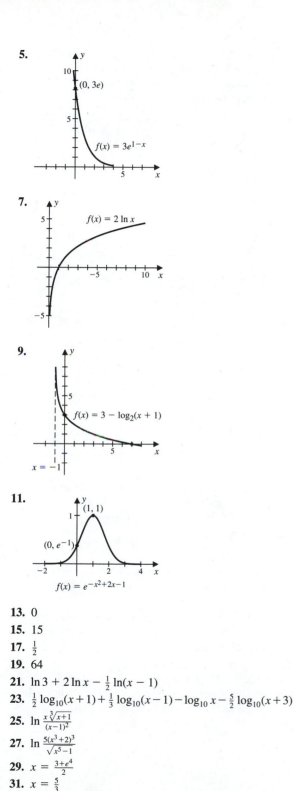

7.

9.

11.

13. 0

15. 15

17. $\frac{1}{2}$

19. 64

21. $\ln 3 + 2 \ln x - \frac{1}{2} \ln(x - 1)$

23. $\frac{1}{2} \log_{10}(x+1) + \frac{1}{3} \log_{10}(x-1) - \log_{10} x - \frac{5}{2} \log_{10}(x+3)$

25. $\ln \frac{x\sqrt[3]{x+1}}{(x-1)^2}$

27. $\ln \frac{5(x^3+2)^3}{\sqrt{x^5-1}}$

29. $x = \frac{3+e^4}{2}$

31. $x = \frac{5}{3}$

33. $\frac{2\ln 5}{\ln 5 - 3\ln 3} \approx -1.91$

35. $x = -\frac{1}{2}, \; x = 1$

37. $-0.81 < x < 1.43$ or $x > 8.61$

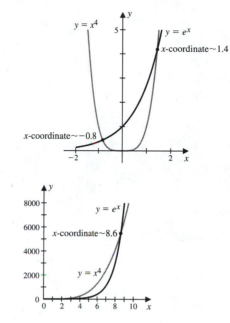

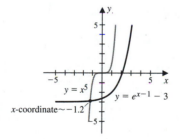

39. $-1.24 < x < 14.3$

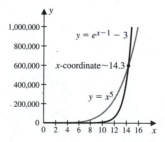

41. Increasing: $(-\infty, -1)$ and $(0, 1)$; decreasing: $(-1, 0)$ and $(1, \infty)$; local maxima: $(1, 1)$ and $(-1, 1)$; local minimum: $(0, 0)$

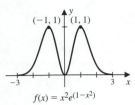

$$f(x) = x^2 e^{(1-x^2)}$$

43. a. 7.9×10^{-7} g   **b.** 2.9×10^9 years

45.

Interest Compounded	Value of CD
a. Annually	$21,435.89
b. Monthly	$22,181.18
c. Daily	$22,253.00
d. Continuously	$22,255.41

Chapter 4: Calculus Preview Exercises (Page 273)

1.

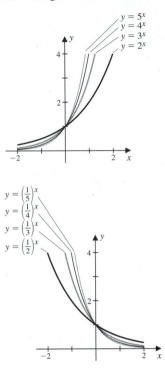

3.

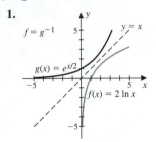

5.

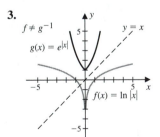
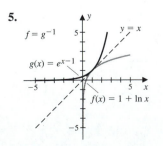

7. For $a, b > 1$, if $b > a$, then b^x grows faster than a^x as $x \longrightarrow \infty$ and both go to ∞. For $0 < a, b < 1$, if $b > a$, then b^x decreases slower then a^x as $x \longrightarrow \infty$ and both go to 0.

9. When $a > 1$, $f(x) \longrightarrow \infty$ and $g(x) \longrightarrow \infty$ as $x \longrightarrow \infty$, with $g(x)$ growing faster than $f(x)$. When $0 < a < 1$, $f(x) \longrightarrow \infty$ and $g(x) \longrightarrow 0$ as $x \longrightarrow \infty$.

11. As $x \longrightarrow \infty$,
$$\frac{x^{10}}{e^x} < \frac{1}{x^4} < \ln x < x^{\frac{1}{20}} < x^{20} < e^{3x} < \frac{e^{6x}}{x^8} < x^x.$$

13. $y = 3e^{x-2} = \frac{3}{e^2}e^x$. So, the graph of $y = 3e^{x-2}$ is a vertical scaling by a factor $\frac{3}{e^2}$ of the graph of $y = e^x$.

15. $y = 3 + \ln 2x = (3 + \ln 2) + \ln x$. So, the graph of $y = 3 + \ln 2x$ is a vertical translation, by $3 + \ln 2$ units, of the graph of $y = \ln x$.

17. The parameter a scales the graph of $y = \ln x$ vertically. If, in addition, $a < 0$, the graph is reflected about the x-axis. Since $\ln(bx) = \ln b + \ln x$, the parameter b vertically shifts the graph $-\ln b$

units to the right if $\ln b < 0$ and $\ln b$ units to the left if $\ln b > 0$.

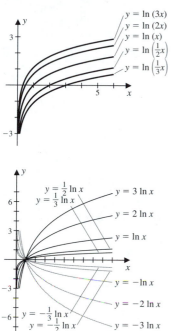

19. The age is $-\frac{5730\ln(0.78)}{\ln 2} \approx 2054$ years.

CHAPTER 5

Exercise Set 5.2 (Page 283)

1.

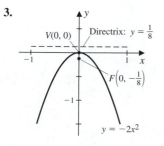

3.

5.

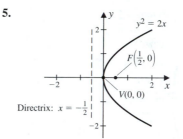

7.

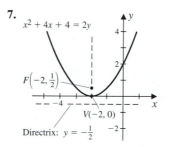

9.

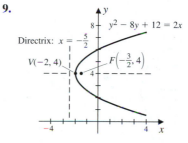

11.

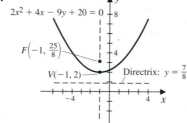

13. $y = \frac{1}{4}x^2$

15. $x = \frac{1}{4}y^2$

17. $y = \frac{1}{8}(x+2)^2$

19. $x + 2 = -\frac{1}{24}(y-2)^2$

21. $y = \frac{3}{8}x^2$

23. $y - 2 = \frac{3}{8}x^2$

25. $x = \frac{1}{9}y^2$

27. $x = \frac{1}{9}(y-2)^2$

29. $y - b = (1-b)x^2$

31. a. 200 ft   **b.** $125\sqrt{58} \approx 952$ ft

Exercise Set 5.3 (Page 292)

1.

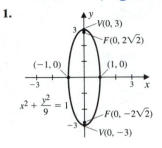

3.

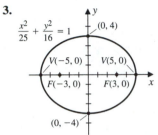

5.

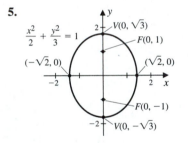

7.

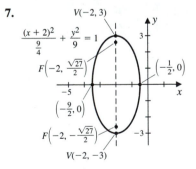

9.

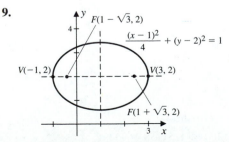

11. $\frac{x^2}{16} + \frac{y^2}{9} = 1$

13. $\frac{x^2}{9} + \frac{y^2}{5} = 1$

15. $\frac{x^2}{3} + \frac{y^2}{4} = 1$

17. $\frac{(x-4)^2}{4} + \frac{(y-2)^2}{3} = 1$

19. $\frac{4x^2}{25} + \frac{4y^2}{9} = 1$

21. $\frac{2b^2}{a}$

23. $\frac{x^2}{57600} + \frac{y^2}{19600} = 1$

Exercise Set 5.4 (Page 300)

1.

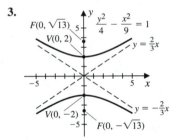

3.

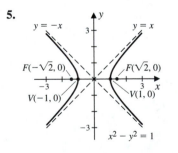

5.

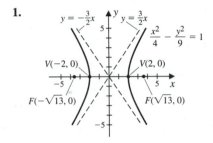

7.

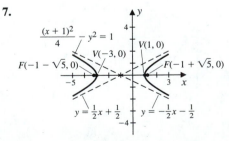

9. $y = -\sqrt{3}x + \sqrt{3}$ $y = \sqrt{3}x - \sqrt{3}$

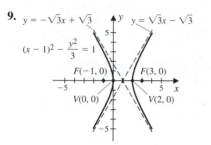

$(x-1)^2 - \frac{y^2}{3} = 1$

$F(-1, 0)$ $F(3, 0)$

$V(0, 0)$ $V(2, 0)$

11.

$y = -\frac{3}{2}x + \frac{1}{2}$ $y = \frac{3}{2}x - \frac{5}{2}$

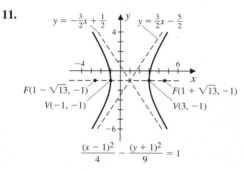

$F(1 - \sqrt{13}, -1)$ $F(1 + \sqrt{13}, -1)$

$V(-1, -1)$ $V(3, -1)$

$\frac{(x-1)^2}{4} - \frac{(y+1)^2}{9} = 1$

13. $\frac{x^2}{9} - \frac{y^2}{16} = 1$

15. $\frac{y^2}{16} - \frac{x^2}{9} = 1$

17. $\frac{(x-2)^2}{4} - \frac{(y-4)^2}{5} = 1$

19. $\frac{y^2}{4} - \frac{x^2}{3} = 1$

21. $\frac{x^2}{9} - \frac{16y^2}{81} = 1$

23. $\frac{2b^2}{a}$

25. The source is the point of intersection of the two hyperbolas described by the equations, $d(S, C) - d(S, A) = 660$ and $d(S, B) - d(S, A) = 330$. There are two possible locations for the point S, one to the north of the detection points and one to the south.

27. The conjugate has equation $4x^2 - 9y^2 = 36$. The hyperbolas have the same asymptotes, the foci are the same distance from the origin, and the axes are perpendicular.

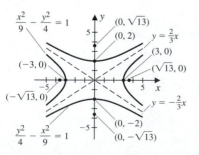

$\frac{x^2}{9} - \frac{y^2}{4} = 1$ $(0, \sqrt{13})$

$(0, 2)$ $y = \frac{2}{3}x$

$(3, 0)$

$(-3, 0)$ $(\sqrt{13}, 0)$

$(-\sqrt{13}, 0)$

$(0, -2)$ $y = -\frac{2}{3}x$

$\frac{y^2}{4} - \frac{x^2}{9} = 1$ $(0, -\sqrt{13})$

Exercise Set 5.5 (Page 310)

1.

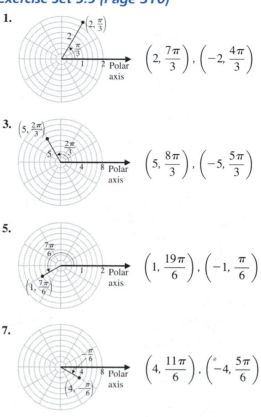

$\left(2, \frac{7\pi}{3}\right), \left(-2, \frac{4\pi}{3}\right)$

3. $\left(5, \frac{2\pi}{3}\right)$

$\left(5, \frac{8\pi}{3}\right), \left(-5, \frac{5\pi}{3}\right)$

5.

$\left(1, \frac{19\pi}{6}\right), \left(-1, \frac{\pi}{6}\right)$

7.

$\left(4, \frac{11\pi}{6}\right), \left(-4, \frac{5\pi}{6}\right)$

9. $(\frac{1}{2}, \frac{\sqrt{3}}{2})$

11. $(0, 0)$

13. $(\sqrt{2}, \sqrt{2})$

15. $(2, 0)$

17. $(4, \frac{\pi}{2})$

19. $(2\sqrt{2}, \frac{3\pi}{4})$

21. $x^2 + y^2 = 16$

23. $(x - 1)^2 + y^2 = 1$

25. $\theta = \pi/4$

27. $r = 4$

29. $r = \tan \theta \sec \theta$

31.

$r = 3$ Polar axis

33.

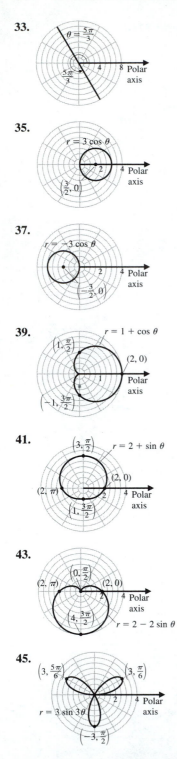

35.

37.

39.

41.

43.

45.

47.

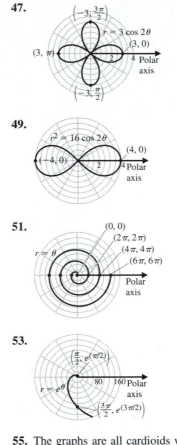

49.

51.

53.

55. The graphs are all cardioids with axis along the
x-axis. The graphs in (a) and (c) coincide, as do the
graphs in (b) and (d). The graph in (b) is the
reflection of graph (a) about the y-axis.

a.

b. $r = 1 - \cos\theta$

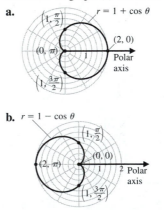

c.

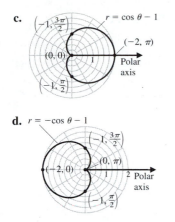

$r = \cos\theta - 1$

$\left(-1, \frac{3\pi}{2}\right)$ $(-2, \pi)$ $(0, 0)$ Polar axis $\left(-1, \frac{\pi}{2}\right)$

d. $r = -\cos\theta - 1$

$\left(-1, \frac{3\pi}{2}\right)$ $(0, \pi)$ $(-2, 0)$ Polar axis $\left(-1, \frac{\pi}{2}\right)$

57. Graph (a) is a circle with center $(\frac{1}{2}, 0)$ and radius $\frac{1}{2}$ and graphs (b)–(d) are cardioids with axis along the x-axis that approach a circle.

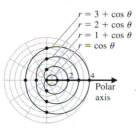

$r = 3 + \cos\theta$
$r = 2 + \cos\theta$
$r = 1 + \cos\theta$
$r = \cos\theta$

Polar axis

59. Graph (a) is a circle with center $(0, \frac{1}{2})$ and radius $\frac{1}{2}$ and graphs (b)–(d) are cardioids with axis along the y-axis which approach a circle.

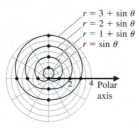

$r = 3 + \sin\theta$
$r = 2 + \sin\theta$
$r = 1 + \sin\theta$
$r = \sin\theta$

Polar axis

61. $r = \frac{a}{2}(1 + \sin\theta)$, where a is the height of the apple.

63. Points of intersection: $(0, 0)$, $(\frac{1}{2}, \frac{2\pi}{3})$, $(\frac{1}{2}, \frac{4\pi}{3})$

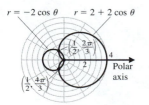

$r = -2\cos\theta$ $r = 2 + 2\cos\theta$

$\left(\frac{1}{2}, \frac{2\pi}{3}\right)$ Polar axis $\left(\frac{1}{2}, \frac{4\pi}{3}\right)$

65. Points of intersection: $(1, \frac{\pi}{4})$, $(1, \frac{3\pi}{4})$, $(1, \frac{5\pi}{4})$, $(1, \frac{7\pi}{4})$

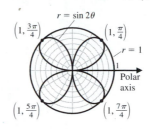

$r = \sin 2\theta$

$\left(1, \frac{3\pi}{4}\right)$ $\left(1, \frac{\pi}{4}\right)$ $r = 1$ Polar axis

$\left(1, \frac{5\pi}{4}\right)$ $\left(1, \frac{7\pi}{4}\right)$

67.

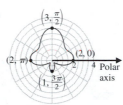

$\left(3, \frac{\pi}{2}\right)$ $(2, \pi)$ $(2, 0)$ Polar axis $\left(1, \frac{3\pi}{2}\right)$

$n = 1$:
$r = 1 + \sin(\theta) + (\cos(2\theta))^2$

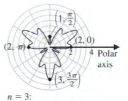

$\left(2, \frac{\pi}{2}\right)$ $(2, \pi)$ $(2, 0)$ Polar axis $\left(2, \frac{3\pi}{2}\right)$

$n = 2$:
$r = 1 + \sin 2\theta + (\cos 4\theta)^2$

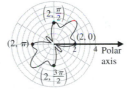

$\left(1, \frac{\pi}{2}\right)$ $(2, 0)$ $(2, \pi)$ Polar axis $\left(3, \frac{3\pi}{2}\right)$

$n = 3$:
$r = 1 + \sin 3\theta + (\cos 6\theta)^2$

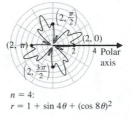

$\left(2, \frac{\pi}{2}\right)$ $(2, 0)$ $(2, \pi)$ Polar axis $\left(2, \frac{3\pi}{2}\right)$

$n = 4$:
$r = 1 + \sin 4\theta + (\cos 8\theta)^2$

69. The difference between the graphs of $r = \sin m\theta$ and $r = |\sin m\theta|$ is that the graph of $r = \sin m\theta$ is not symmetric with respect to the pole, but the absolute value results in symmetry with respect to the pole.

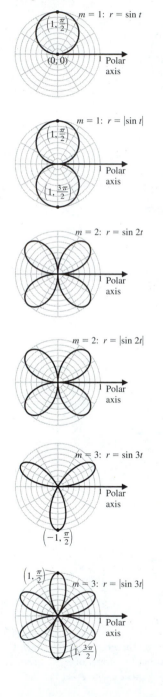

Exercise Set 5.6 (Page 316)

1.

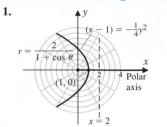

3.

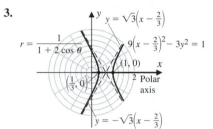

5.

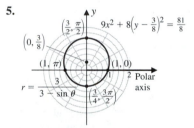

7. $r = \dfrac{8}{1 - 2\cos\theta}$

9. $r = \dfrac{1}{4 - 4\sin\theta}$

11. $r = \dfrac{3}{2 + \cos\theta}$

13. $r = \dfrac{4303}{1 - 0.048\cos\theta}$

Exercise Set 5.7 (Page 320)

1.

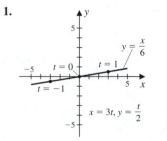

3.

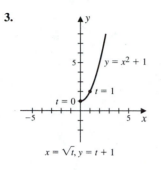

$x = \sqrt{t}, y = t + 1$

5.

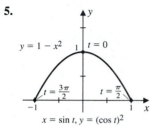

$x = \sin t, y = (\cos t)^2$

7.

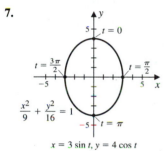

$x = 3 \sin t, y = 4 \cos t$

9.

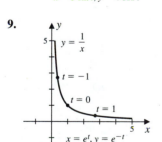

$x = e^t, y = e^{-t}$

11.

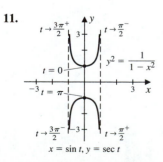

$x = \sin t, y = \sec t$

13.

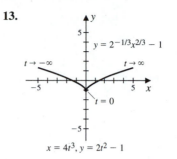

$x = 4t^3, y = 2t^2 - 1$

15. a.

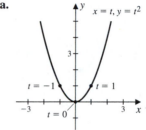

b.

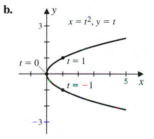

c.

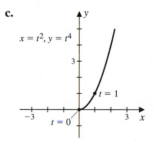

d.

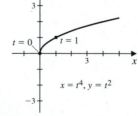

$x = t^4, y = t^2$

17. a.

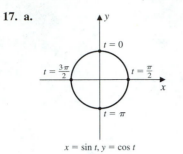

$x = \sin t, y = \cos t$

b.

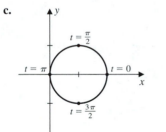

$x = \sin t, y = \cos t + 1$

c.

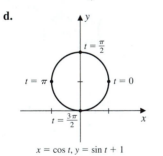

$x = \cos t + 1, y = \sin t$

d.

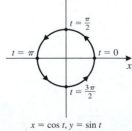

$x = \cos t, y = \sin t + 1$

19. a.

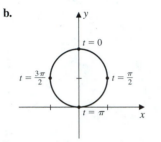

$x = \cos t, y = \sin t$

b.

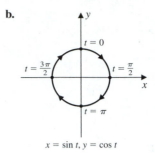

$x = \sin t, y = \cos t$

c.

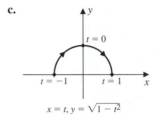

$x = t, y = \sqrt{1 - t^2}$

d.

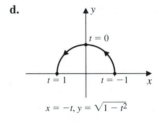

$x = -t, y = \sqrt{1 - t^2}$

21. a.

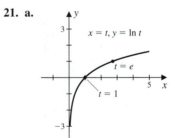

$x = t, y = \ln t$

b.

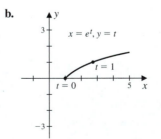

$x = e^t, y = t$

c.

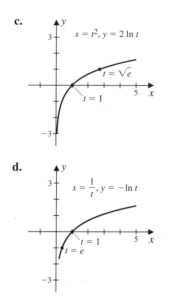

$x = t^2, y = 2 \ln t$

$t = \sqrt{e}$

$t = 1$

d.

$x = \dfrac{1}{t}, y = -\ln t$

$t = 1$

$t = e$

23. One choice is $x = t + 2$, $y = \frac{1}{3}t - 1$.

25. They both trace the circle $x^2 + y^2 = 1$. The first curve starts at $(-1, 0)$ and traces one revolution counterclockwise. The second curve starts at $(0, -1)$ and traces one revolution clockwise.

27.

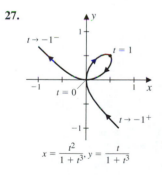

$t \to -1^-$

$t = 1$

$t = 0$

$t \to -1^+$

$x = \dfrac{t^2}{1 + t^3}, y = \dfrac{t}{1 + t^3}$

29. The parameter a determines the maximum height $2a$ of the curve. The curve touches the x-axis when $t = \pm 2n\pi$ for integers n.

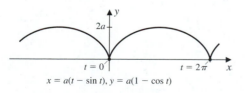

$2a$

$t = 0$ $t = 2\pi$

$x = a(t - \sin t), y = a(1 - \cos t)$

Exercise Set 5.8 (Page 327)

1. $\left(\dfrac{\sqrt{3}}{2}, \dfrac{1}{2} \right)$

3. $\left(\dfrac{1 - 3\sqrt{3}}{2}, \dfrac{3 + \sqrt{3}}{2} \right)$

5. a. Ellipse   **b.** The rotated equation is $\hat{x}^2 + 3\hat{y}^2 = 4$.

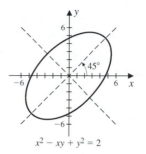

$45°$

$x^2 - xy + y^2 = 2$

7. a. Ellipse   **b.** The rotated equation is $8\hat{x}^2 + 18\hat{y}^2 = 0$, whose graph is the single point $(0, 0)$.

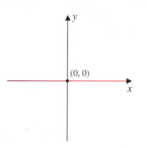

$(0, 0)$

9. a. Ellipse   **b.** The rotated equation is $3\hat{x}^2 + 8\hat{y}^2 = 24$.

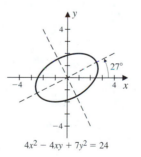

$27°$

$4x^2 - 4xy + 7y^2 = 24$

11. a. Parabola   **b.** The rotated equation is $8\hat{x}^2 + 18\hat{y}^2 - 63 = 0$.

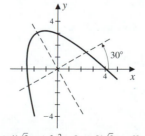

$30°$

$6x^2 + 4\sqrt{3}xy + 2y^2 - 9x + 9\sqrt{3}y - 63 = 0$

13. a. $\hat{x} - \sqrt{2} = -\sqrt{2}\hat{y}^2$

　　b. $x + y - 2 = -(x - y)^2$

Review Exercises for Chapter 5 (Page 328)

1.

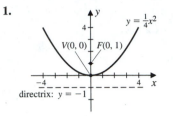

$y = \frac{1}{4}x^2$

$V(0, 0)$　$F(0, 1)$

directrix: $y = -1$

3.

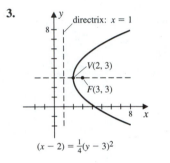

directrix: $x = 1$

$V(2, 3)$

$F(3, 3)$

$(x - 2) = \frac{1}{4}(y - 3)^2$

5.

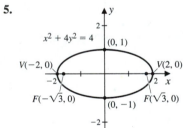

$x^2 + 4y^2 = 4$

$(0, 1)$

$V(-2, 0)$　$V(2, 0)$

$F(-\sqrt{3}, 0)$　$F(\sqrt{3}, 0)$

$(0, -1)$

7.

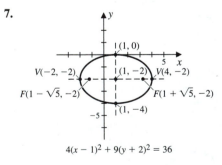

$(1, 0)$

$V(-2, -2)$　$(1, -2)$　$V(4, -2)$

$F(1 - \sqrt{5}, -2)$　$F(1 + \sqrt{5}, -2)$

$(1, -4)$

$4(x - 1)^2 + 9(y + 2)^2 = 36$

9.

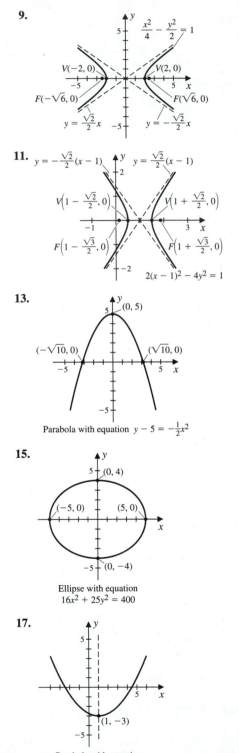

$\frac{x^2}{4} - \frac{y^2}{2} = 1$

$V(-2, 0)$　$V(2, 0)$

$F(-\sqrt{6}, 0)$　$F(\sqrt{6}, 0)$

$y = \frac{\sqrt{2}}{2}x$　$y = -\frac{\sqrt{2}}{2}x$

11. $y = -\frac{\sqrt{2}}{2}(x - 1)$　$y = \frac{\sqrt{2}}{2}(x - 1)$

$V\left(1 - \frac{\sqrt{2}}{2}, 0\right)$　$V\left(1 + \frac{\sqrt{2}}{2}, 0\right)$

$F\left(1 - \frac{\sqrt{3}}{2}, 0\right)$　$F\left(1 + \frac{\sqrt{3}}{2}, 0\right)$

$2(x - 1)^2 - 4y^2 = 1$

13.

$(0, 5)$

$(-\sqrt{10}, 0)$　$(\sqrt{10}, 0)$

Parabola with equation $y - 5 = -\frac{1}{2}x^2$

15.

$(0, 4)$

$(-5, 0)$　$(5, 0)$

$(0, -4)$

Ellipse with equation

$16x^2 + 25y^2 = 400$

17.

$(1, -3)$

Parabola with equation

$y + 3 = \frac{1}{4}(x - 1)^2$

19.

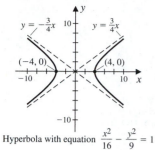

Hyperbola with equation $\dfrac{x^2}{16} - \dfrac{y^2}{9} = 1$

21.

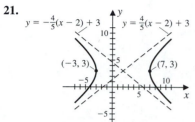

Hyperbola with equation

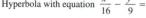

$$\dfrac{(x-2)^2}{25} - \dfrac{(y-3)^2}{16} = 1$$

23. $y - 1 = -\dfrac{1}{4}x^2$

25. $x^2 - \dfrac{y^2}{8} = 1$

27. $\dfrac{y^2}{41} + \dfrac{x^2}{16} = 1$

29. $r = \dfrac{6}{1 - 3\sin\theta}$

31.

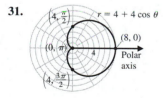

33.

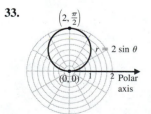

35.

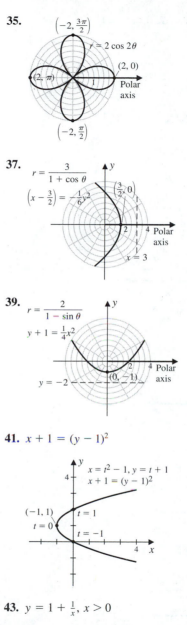

$r = 2\cos 2\theta$

37. $r = \dfrac{3}{1 + \cos\theta}$

$\left(x - \dfrac{3}{2}\right) = -\dfrac{1}{6}y^2$

39. $r = \dfrac{2}{1 - \sin\theta}$

$y + 1 = \dfrac{1}{4}x^2$

41. $x + 1 = (y - 1)^2$

$x = t^2 - 1,\ y = t + 1$
$x + 1 = (y - 1)^2$

43. $y = 1 + \dfrac{1}{x},\ x > 0$

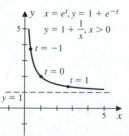

$x = e^t,\ y = 1 + e^{-t}$
$y = 1 + \dfrac{1}{x},\ x > 0$

45. $y = 2 - x,\ 1 \le x \le 2,\ 0 \le y \le 1$

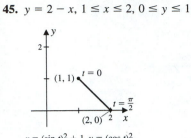

$x = (\sin t)^2 + 1,\ y = (\cos t)^2$
$y = 2 - x,\ 1 \le x \le 2,\ 0 \le y \le 1$

47. a. Hyperbola

b.

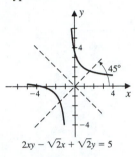

$2xy - \sqrt{2}x + \sqrt{2}y = 5$

49.

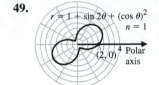

$r = 1 + \sin 2\theta + (\cos \theta)^2$
$n = 1$

$r = 1 + \sin 4\theta + (\cos 2\theta)^2$
$n = 2$

$r = 1 + \sin 6\theta + (\cos 3\theta)^2$
$n = 3$

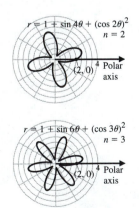

51.

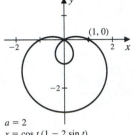

$a = 1$
$x = \cos t\,(1 - \sin t)$
$y = \sin t\,(1 - \sin t)$

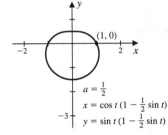

$a = 2$
$x = \cos t\,(1 - 2\sin t)$
$y = \sin t\,(1 - 2\sin t)$

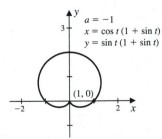

$a = \frac{1}{2}$
$x = \cos t\,(1 - \frac{1}{2}\sin t)$
$y = \sin t\,(1 - \frac{1}{2}\sin t)$

$a = -1$
$x = \cos t\,(1 + \sin t)$
$y = \sin t\,(1 + \sin t)$

$(1, 0)$

53.

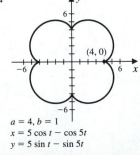

$a = 4,\ b = 1$
$x = 5\cos t - \cos 5t$
$y = 5\sin t - \sin 5t$

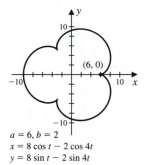

$a = 6, b = 2$
$x = 8 \cos t - 2 \cos 4t$
$y = 8 \sin t - 2 \sin 4t$

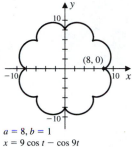

$a = 8, b = 1$
$x = 9 \cos t - \cos 9t$
$y = 9 \sin t - \sin 9t$

Chapter 5: Calculus Preview Exercises (Page 329)

1. a. $y = \frac{y_1}{x_1^2} x^2$ **b.** $y - k = \frac{y_1 - k}{(x_1 - h)^2}(x - h)^2$

3. a. $y - k = (x - h)^2$ **b.** $x - h = (y - k)^2$

5. $y = -\frac{\sqrt{3}}{2} x + 2\sqrt{3}$

7. $y = \frac{\sqrt{6}}{2} x - \sqrt{3}$

9. $y = 3\sqrt{4 - x^2}$ for $0 \le x \le 2$

11. One unit along the axis from the vertex.

13. $\frac{2000}{3}$ in.

INDEX

Abel, Niels, 123
Absolute value
    function (| |), 67
    properties, 10
    of a real number, 8
Absolute zero, 101
Acute angle, 210
Addition of complex numbers, 156
Adjacent side, 210
Algebraic function, 110
Ambiguous case, 234
Amplitude of trigonometric
        functions, 186
Angle
    acute, 210
    definition, 169
    degree measure, 212
    initial side, 169
    obtuse, 210
    radian measure, 171
    reference, 210
    right, 210
    standard position, 209
    terminal side, 169
    trigonometric function of, 173
    vertex of, 169
Apogee of an orbit, 315
Approximately equal to ($\approx$), 29
Arccosecant function, 227
Arccosine function, 221
Arccotangent function, 227
Arcsecant function, 226
Arcsine function, 219
Arctangent function, 223
Area formulas, Back end pages

Asymptote
    definition of, 135
    horizontal, 140
    of a hyperbola, 295
    slant, 143, 295
    vertical, 136
Axis (axes)
    coordinate, 12
    of an ellipse, 285
    of a hyperbola, 293
    of a parabola, 278
    rotation of, 322

Base
    of an exponential function, 248
    of a logarithm function, 260
Bell-shaped curve, 255
Binomial Theorem, Front end pages
Bounds
    cosine function, 174
    sine function, 174

C, the complex numbers, 155
Carbon dating, 246
Cardioid, 307
Cartesian coordinate system, 13
Ceiling function ($\lceil\ \rceil$), 71
Celsius to Fahrenheit conversion, 11
Celsius to Kelvin conversion, 101
Chambered nautilus, 331
Circle
    parametric equations of, 319
    standard equation of, 16
    unit, 16
Closed interval, 6

Common logarithm, 265
Completeness, 3
Completing the square, 57
Complex conjugate
    definition, 156
    results, 158
Complex number(s) (C)
    addition of, 156
    conjugates of, 156
    definition, 155
    imaginary part, 155
    product of, 156
    quotient of, 162
    real part, 155
    subtraction of, 156
Composition of functions, 83
Compound interest, 256
Compression of a graph, 79, 85
Computer algebra system, 23
Conditional equation, 17
Conic section(s)
    classification of, 327
    definition, 276
    directrix of, 312
    eccentricity of, 312
    focus of, 312
Conjugate, complex, 157
Conjugate hyperbola, 300
Constant function, 109
Constant of proportionality,
        267
Constant term, 110
Continuity, 116
Continuously compound interest,
        257

Conversions
  degrees to radians, 212
  exponential to natural exponential,
    265
  logarithm to natural logarithm,
    265
  radians to degrees, 212
Coordinate line, 4
Coordinate plane, 12
Coordinate rotation formulas, 323
Coordinate system, 13, 300
Cosecant function, 193
Cosine function
  of an angle, 173
  bounds on, 174
  combination formula, 237
  definition, 173
  double-angle formula, 203
  graph of, 184
  half-angle formula, 205
  period of, 176
  products of, 208
  sums and differences of, 200
  values of, 179
Cosines, Law of, 229
Cotangent function, 193
Cube root function, 74
Cubing function, 111
Cycloid, 321

Decreasing function, 55
Degenerate conic section, 276
Degree measure of an angle, 212
Degree to radian conversion, 212
Dependent variable, 31
Derive, 23
Descartes, René, 13
Descartes' rule of signs, 131
Difference of functions, 75
Difference quotient, 34, 251
Directrix
  of a conic section, 312
  of a parabola, 278
Discriminant, 64
Distance
  between points in the plane, 14
  between real numbers, 9
Division algorithm, 124
Division of polynomials, 124
Domain of a function
  definition, 32
  finding, 39

of rational functions, 133
of trigonometric functions,
  194
Double-angle formulas, 203

e, the real number, 252
Eccentricity
  of a conic section, 312
  of an ellipse, 291
  of a hyperbola, 299
Ellipse
  axis of, 285
  definition of, 285
  eccentricity of, 291
  focal points of, 285
  latus rectum of, 292
  polar equation of, 312
  reflection property of, 289
  standard-position equation, 287
  tangent line to, 329
  vertices of, 285
Elongation of a graph, 79, 85
End behavior
  of a polynomial, 114
  of a rational function, 138
Epicycloid, 329
Equation(s)
  algebraic, 110
  conditional, 17
  extraneous solution of, 205
  graph of, 18
  identity, 17
  linear, 54
  parametric, 317
  polar, 304
  root of, 115
  transcendental, 110
Euler, Leonhard, 159
Even function, 40
Exponential(s)
  converted to natural exponential,
    265
  decay, 267
  growth, 267
  properties of, 247
Exponential function,
  definition, 248
  general, 247
  graph, 249
  natural, 251
Extraneous solution, 205, 264
Extrema, local, 117

Factor Theorem, 125
Factoring formulas, Front end pages
Fahrenheit to Celsius conversion, 11
Ferrari, Ludovico, 123
Floor function ($\lfloor\ \rfloor$), 71
Focal point(s) (focus)
  of a conic section, 312
  of an ellipse, 285
  of a hyperbola, 293
  of a parabola, 278
Folium of Descartes, 321
Fontana, Nicolo, 123
Four-leafed rose, 310
Frequency, of harmonic motion, 235
Function(s)
  absolute value, 67
  algebraic, 110
  arccosecant, 227
  arccosine, 221
  arccotangent, 227
  arcsecant, 226
  arcsine, 219
  arctangent, 223
  ceiling, 71
  compositions of, 83
  constant, 109
  continuity, 116
  cosecant, 193
  cosine, 173
  cotangent, 193
  cube root, 74
  cubing, 11
  decreasing, 55
  definition of, 31
  domain of, 32
  even, 40
  exponential, 248
  floor, 71
  graph, 32
  greatest integer, 71
  horizontal line test, 93
  hyperbolic cosine, 259
  hyperbolic sine, 259
  increasing, 55
  inverse, 95
  inverse cosecant, 227
  inverse cosine, 221
  inverse cotangent, 227
  inverse secant, 226
  inverse sine, 219
  inverse tangent, 223
  linear, 47, 109

logarithm, 260
natural exponential, 247, 251
natural logarithm, 263
odd, 40
one-to-one, 92
periodic, 176
piecewise defined, 39
polynomial, 109
product of, 75
quadratic, 55, 109
quotient of, 75
range, 32
rational, 110, 133
rational power, 148
secant, 193
sine, 173
square root, 69
squaring, 55
sum of, 75
tangent, 193
transcendental, 110
vertical line test, 38
zero, 115
Fundamental Theorem of Algebra,
    159

Galois, Evariste, 123
Gauss, Carl Friedrich, 159
General to natural exponential
    conversion, 265
General to natural logarithm
    conversion, 265
George Washington bridge, 330
Girard, Albert, 159
Global positioning system, 274, 297
Graph(s)
    arccosecant function, 227
    arccosine function, 221
    arccotangent function, 227
    arcsecant function, 226
    arcsine function, 220
    arctangent function, 223
    axis intercepts, 19
    bell-shaped, 255
    compression of, 79, 85
    cosecant function, 196
    cosine function, 184
    cotangent function, 197
    elongation of, 79, 85
    end behavior, 114, 138
    of an equation, 18
    general exponential functions, 249

general logarithm functions, 261
hole in, 137
horizontal asymptote of, 141
horizontal shift, 56
of inequalities, 13
linear equation, 48
natural exponential function, 253
natural logarithm function, 263
one-to-one function, 92
quadratic equation, 59
reciprocal of a function, 80
secant function, 197
sign, 7
sine function, 183
slant asymptote of, 143
tangent function, 195
vertical asymptote of, 136
Graph reflection
    about the origin, 21, 40
    about the x-axis, 21, 40, 79
    about the y-axis, 21, 40, 85
    about $y = x$, 97
Graphing calculator, 23
Graphing device, 23
Greater than ($>$), 4
Greater than or equal to ($\geq$), 4
Greatest integer function ($\lfloor\ \rfloor$) or ($[\ ]$),
    71
Greek alphabet, Back end pages
Growth and decay, 267

Hale, George Ellery, 331
Half-angle formulas, 205
Half-life, 269
Half-open interval, 6
Halley, Edmund, 292
Harmonic motion, 235
Heron's formula, 238
Hole in a graph, 137
Horizontal
    asymptote, 140
    compression of a graph, 85
    elongation of a graph, 85
    line, 54
    line test, 93
    shifts of a graph, 56
Hyperbola(s)
    asymptotes of, 295
    axis of, 293
    conjugate, 300
    definition of, 293
    eccentricity of, 299

focal points of, 293
latus rectum of, 300
polar equation of, 312
reflection property of, 299
standard-position equation, 295
tangent line to, 330
vertices of, 293
Hyperbolic cosine function, 259
Hyperbolic sine function, 259
Hypocycloid, 321, 329
Hypotenuse, 210

i, the complex number, 155
Identity, 17
Image, 32
Imaginary part of a complex number,
    155
Increasing function, 55
Independent variable, 31
Inequalities
    definition, 4
    graph of, 7, 13
    properties, 5
Infinity symbol (∞), 6
Initial side of an angle, 169
Integers (Z), 3
Intercept form of a line, 54
Intercepts of a graph, 19
Interest compound, 256, 257
Interior of an interval, 6
Intermediate Value Theorem, 116
Intersection of sets ($\cap$), 8
Intervals, 5
Inverse cosecant function, 227
Inverse cosine function, 221
Inverse cotangent function, 227
Inverse function, 95
Inverse secant function, 226
Inverse sine function, 219
Inverse tangent function, 223
Inverse trigonometric functions, 218
Irrational numbers, 3

Kelvin to Celsius conversion, 101
Kepler, Johannes, 284

Latus rectum
    of an ellipse, 292
    of a hyperbola, 300
Law of cosines, 229
Law of sines, 232
Leading coefficient, 110

Length formulas, Back end pages
Less than ($<$), 4
Less than or equal to ($\leq$), 4
Libby, Willard, 246
Limaçon, 308
Line(s)
  general equation of, 54
  horizontal, 54
  intercept form of, 54
  normal, 52
  parallel, 52
  perpendicular, 52
  point-slope form of, 48
  slope of, 48
  slope-intercept form of, 49
  vertical, 54
Linear equation, 54
Linear function, 47, 109
Local
  extrema, 117
  maximum, 117
  minimum, 117
Logarithm(s)
  converted to natural logarithm, 265
  function, 260
  properties of, 263

Maple, 23
Mathematica, 23
Maximum, local, 117
Midpoint
  of two numbers, 10
  of two points, 15
Minimum, local, 117
Mormon Tabernacle, 289
Multiplicity of a zero, 118

N, the natural numbers, 2
Natural exponential function
  applications, 267
  definition, 251
  graph, 253
Natural logarithm function, 263
Natural numbers (N), 2
Newton, Sir Isaac, 159, 284
Newton's law of cooling, 272
Nonrepeating decimals, 3
Normal line, 52
Not equal to ($\neq$), 10
Number(s)
  complex (C), 155

integers (Z), 3
  irrational, 3
  natural (N), 2
  rational (Q), 3
  real ($\mathbb{R}$), 2

Oblique asymptote, 143, 295
Obtuse angle, 210
Odd function, 40
Olympic Stadium, Montreal, 292
One-to-one function, 92
Open interval, 6
Opposite side, 210
Ordered pair, 12
Origin, symmetry to, 21

π, the real number, 172
Parabola(s)
  axis of, 278
  definition of, 19, 278
  directrix of, 278
  focal point of, 278
  polar equation of, 312
  reflection property of, 282
  standard-position equation, 280
  vertex of, 57, 278
Parallel lines, 52
Parameter, 316
Parametric equations
  of a circle, 319
  of a cycloid, 321
  definition, 317
Perigee of an orbit, 315
Period
  definition, 176
  of trigonometric functions, 176, 197
Periodic function, 176
Perpendicular lines, 52
Phase shift of trigonometric
  functions, 187
Piecewise-defined function, 39
Planetary orbits, 291
Plus or minus ($\pm$), 17
Point-slope equation of a line, 48
Polar axis, 301
Polar coordinate system
  definition, 300
  related to rectangular coordinates, 303
Polar equation(s)
  of a cardioid, 307

of a circle, 331
  of conics, 312
  definition, 304
  of an ellipse, 312
  four-leafed rose, 310
  of an hyperbola, 312
  limaçon, 307
  of a parabola, 312
Pole, 301
Polynomial(s)
  definition, 109
  degree of, 64, 109
  Descartes' rule of signs, 131
  division of, 123
  factoring, 132
  finding zeros, 132
  leading coefficient of, 109
  rational zero test, 129
Population growth, 267
Principal square root ($\sqrt{}$), 69
Product of complex numbers, 156
Product of functions, 75
Pythagorean identities, 173, 199
Pythagorean Theorem, 14

Q, the rational numbers, 3
Quadratic equation(s)
  definition, 59, 276
  rotating, 325
Quadratic formula, 64
Quadratic function, 55, 109
Quotient
  of complex numbers, 162
  of functions, 75
  of polynomials, 124

$\mathbb{R}$, the real numbers, 3
Radian to degree conversion, 212
Radian measure of an angle, 172
Radioactive carbon dating, 246
Radioactive decay, 269
Range of a function
  definition, 32
  finding, 39
Rational function
  definition, 110, 133
  end behavior of, 138
Rational numbers (Q), 3
Rational power function, 148
Rational zero test, 129
Ray, 169, 301
Real line, 4

Real numbers (ℝ), 2
Real part of a complex number, 155
Reciprocal of a function, 80
Rectangular coordinates
    definition, 13
    related to polar coordinates, 303
Reference
    angle, 210
    number, 180
Reflection of a graph
    about the origin, 21, 40
    about the x-axis, 21, 40, 79
    about the y-axis, 21, 40, 85
    about $y = x$, 97
Reflection property
    of an ellipse, 289
    of a hyperbola, 299
    of a parabola, 282
Repeating decimals, 3
Right angle, 210
Right triangle, 210
Root of an equation, 115
Rotating quadratic equations, 325
Rotation of axes, 322

Satellite orbits, 315
Scatter plot, 47
Secant function, 193
Semiperimeter of a triangle, 238
Set(s)
    intersection of, 8
    interval notation, 7
    union of, 8
Shifts of graphs
    horizontal, 56
    vertical, 58
Sign graph, 7
Signs of trigonometric functions,
    194
Simple harmonic motion, 235
Sine function
    of an angle, 173
    bounds on, 174
    combination formula, 236
    definition, 173
    double-angle formula, 203
    graph of, 183
    half-angle formula, 205
    period of, 176
    products of, 208

sums and differences of, 200
    values of, 179
Sines, Law of, 232
Sketch of a graph, 19
Slant asymptote, 143, 295
Slope of a line, 48
Slope-intercept form, 49
Sputnik I, 316
Square, completing the, 57
Square root
    approximating, 75
    function, 69
    of -1, 155
Squaring function, 55
St. Paul's Cathedral, 289
Standard-position equation(s)
    of a conic section, 313
    ellipse, 287
    hyperbola, 295
    parabola, 280
Statuary Hall, 289
Subtraction of complex numbers,
    156
Sum of functions, 75
Surface area formulas, Back end
    pages
Symmetry
    to the origin, 21
    to the polar axis, 308
    of polar equations, 308
    to the pole, 308
    to the x-axis, 20
    to the y-axis, 20

Tangent function, 193
Tangent line
    to an ellipse, 329
    to a hyperbola, 330
Tartaglia (Nicolo Fontana), 123
Terminal side of an angle, 169
Transcendental function, 110
Translations of graphs
    horizontal, 56
    vertical, 58
Triangle
    area of, 238, Back end pages
    right, 210
    semiperimeter of, 238
Trigonometric function(s)
    addition of, 200

difference of, 200
    domains of, 194
    double-angle formulas, 203
    half-angle formulas, 205
    period of, 176, 197
    products of, 208
    ranges of, 194
    in a right triangle, 211
    signs of, 194
    sum of, 200
    on a unit circle, 173

Union of sets (∪), 8
Unit circle, 16

Variable
    dependent, 31
    independent, 31
Vertex (vertices)
    of an angle, 169
    of an ellipse, 285
    of a hyperbola, 293
    of a parabola, 57, 278
Vertical
    asymptote, 136
    compression of a graph, 79
    elongation of a graph, 79
    line, 54
    line test, 38
    shifts of a graph, 58
Viewing rectangle, 23
Volume formulas, Back end
    pages

Whispering galleries, 289
Wrapping function, 171
Wren, Christopher, 289

x-axis, symmetry to, 20
x-intercept, 19
xy-plane, 13

y-axis, symmetry to, 20
y-intercept, 19

Z, the integers, 3
Zero
    of conjugate pairs, 159
    of a function, 115
    multiplicity of, 118

Turn the graphing calculator into a powerful tool for your success!

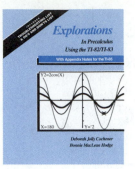

Explorations in Precalculus Using the TI-82/TI-83: With Appendix Notes for the TI-85
by Deborah J. Cochener and Bonnie M. Hodge, both of Austin Peay State University

$19.00 Single Copy Price. 432 pages. Spiralbound. 8 1/2 x 11.
ISBN: 0-534-34227-2. © 1997. Published by Brooks/Cole.

You can quickly learn to use the graphing calculator to develop problem-solving and critical-thinking skills that will help improve your performance in your precalculus course!

Designed to help you succeed in your precalculus course, this unique and student-friendly workbook improves both your understanding and retention of precalculus concepts—using the graphing calculator. By integrating technology into mathematics, the authors help you develop problem-solving and critical-thinking skills.

To guide you in your explorations, you'll find:
- hands-on applications with solutions
- correlation charts that relate course topics to the workbook units
- key charts (specific to the TI-82, TI-83, and TI-85) that show which units introduce keys on the calculator
- a "Troubleshooting Section" to help you avoid common errors

I think this is an excellent workbook. . . . It is definitely easier to follow than the book that comes with the calculator. The examples are clear and the keystrokes are easy to follow. . . . The correlation charts are extremely helpful.
Terry Teegarden, San Diego Mesa College

I particularly like the writing style. It is clear and easy to follow. The correlation charts and the introduction of the keys is really nice. I like the summary questions which provide the opportunity for students to write and collect their thoughts about a given topic.
Gladys Crates, Chattanooga State Technical Community College

Topics

This text contains 50 units divided into the following subsections:
Textbook Correlation Charts
    Basic Calculator Operations
    Graphically Solving Equations and Inequalities
    Graphing and Applications of Equations in Two Variables
    Trigonometric Functions
    Conic Sections
    Miscellaneous (Matrices, Combinatorics and Probability, and Sequences and Series)
    Stat Plots
    Programming
Trouble Shooting
Calculator Menus

Order your copy today!

To receive your copy of **Explorations in Precalculus Using the TI-82/TI-83: With Appendix Notes for the TI-85,** please call Brooks/Cole at: (800) 354-9706.

TRIGONOMETRY

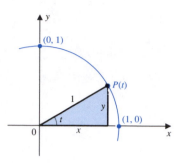

$$\sin t = y$$

$$\cos t = x$$

$$\tan t = \frac{\sin t}{\cos t} = \frac{y}{x}$$

$$\cot t = \frac{\cos t}{\sin t} = \frac{x}{y}$$

$$\sec t = \frac{1}{\cos t} = \frac{1}{x}$$

$$\csc t = \frac{1}{\sin t} = \frac{1}{y}$$

$$\sin^2 t + \cos^2 t = 1$$
$$1 + \tan^2 t = \sec^2 t$$
$$1 + \cot^2 t = \csc^2 t$$

$$\sin(-t) = -\sin t$$
$$\cos(-t) = \cos t$$
$$\tan(-t) = -\tan t$$

$$\cot(-t) = -\cot t$$
$$\sec(-t) = \sec t$$
$$\csc(-t) = -\csc t$$

$$\sin(t_1 \pm t_2) = \sin t_1 \cos t_2 \pm \cos t_1 \sin t_2$$
$$\cos(t_1 \pm t_2) = \cos t_1 \cos t_2 \mp \sin t_1 \sin t_2$$

$$\tan(t_1 + t_2) = \frac{\tan t_1 \pm \tan t_2}{1 \mp \tan t_1 \tan t_2}$$

$$\sin 2t = 2 \sin t \cos t$$
$$\cos 2t = \cos^2 t - \sin^2 t = 1 - 2\sin^2 t = 2\cos^2 t - 1$$

$$\cos^2 t = \frac{1 + \cos 2t}{2} \qquad \sin^2 t = \frac{1 - \cos 2t}{2}$$

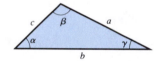

LAW OF COSINES

$$a^2 = b^2 + c^2 - 2bc \cos \alpha$$
$$b^2 = a^2 + c^2 - 2ac \cos \beta$$
$$c^2 = a^2 + b^2 - 2ab \cos \gamma$$

LAW OF SINES

$$\frac{\sin \alpha}{a} = \frac{\sin \beta}{b} = \frac{\sin \gamma}{c}$$

SPECIAL VALUES OF TRIGONOMETRIC FUNCTIONS

SPECIAL RIGHT TRIANGLES

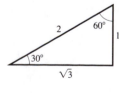

θ (radians)	θ (degrees)	$\sin \theta$	$\cos \theta$	$\tan \theta$	$\cot \theta$	$\sec \theta$	$\csc \theta$
0	$0°$	0	1	0	—	1	—
$\dfrac{\pi}{6}$	$30°$	$\dfrac{1}{2}$	$\dfrac{\sqrt{3}}{2}$	$\dfrac{\sqrt{3}}{3}$	$\sqrt{3}$	$\dfrac{2\sqrt{3}}{3}$	2
$\dfrac{\pi}{4}$	$45°$	$\dfrac{\sqrt{2}}{2}$	$\dfrac{\sqrt{2}}{2}$	1	1	$\sqrt{2}$	$\sqrt{2}$
$\dfrac{\pi}{3}$	$60°$	$\dfrac{\sqrt{3}}{2}$	$\dfrac{1}{2}$	$\sqrt{3}$	$\dfrac{\sqrt{3}}{3}$	2	$\dfrac{2\sqrt{3}}{3}$
$\dfrac{\pi}{2}$	$90°$	1	0	—	0	—	1